SMART MATERIALS

Integrated Design, Engineering Approaches,
and Potential Applications

SMART MATERIALS

Integrated Design, Engineering Approaches, and Potential Applications

Edited by
Anca Filimon, PhD

Apple Academic Press Inc.	Apple Academic Press Inc.
3333 Mistwell Crescent	9 Spinnaker Way
Oakville, ON L6L 0A2 Canada	Waretown, NJ 08758 USA

© 2019 by Apple Academic Press, Inc.

First issued in paperback 2021

Exclusive worldwide distribution by CRC Press, a member of Taylor & Francis Group

No claim to original U.S. Government works

ISBN 13: 978-1-77463-166-9 (pbk)
ISBN 13: 978-1-77188-687-1 (hbk)

Library and Archives Canada Cataloguing in Publication

Smart materials : integrated design, engineering approaches, and potential applications / edited by Anca Filimon, PhD.

Includes bibliographical references and index.
Issued in print and electronic formats.
ISBN 978-1-77188-687-1 (hardcover).--ISBN 978-1-351-16796-3 (PDF)

1. Smart materials. 2. Polymers. I. Filimon, Anca, editor

TA418.9.S62S63 2018	620.1'12	C2018-902771-1	C2018-902772-X

Library of Congress Cataloging-in-Publication Data

Names: Filimon, Anca, editor.
Title: Smart materials : integrated design, engineering approaches, and
   potential applications / editor, Anca Filimon, PhD.
Other titles: Smart materials (Apple Academic Press)
Description: Toronto ; New Jersey : Apple Academic Press, 2018. | Includes
   bibliographical references and index.
Identifiers: LCCN 2018023158 (print) | LCCN 2018025266 (ebook) | ISBN
   9781351167963 (ebook) | ISBN 9781771886871 (hardcover : alk. paper)
Subjects: LCSH: Smart materials.
Classification: LCC TA418.9.S62 (ebook) | LCC TA418.9.S62 S485 2018 (print) |
   DDC 620.1/12--dc23
LC record available at https://lccn.loc.gov/2018023158

Apple Academic Press also publishes its books in a variety of electronic formats. Some content that appears in print may not be available in electronic format. For information about Apple Academic Press products, visit our website at **www.appleacademicpress.com** and the CRC Press website at **www.crcpress.com**

ABOUT THE EDITOR

Anca Filimon, PhD, is a Scientific Researcher in the Department of Physical Chemistry of Polymers at the "Petru Poni" Institute of Macromolecular Chemistry, Iasi, Romania, of the Romanian Academy. She has authored over 30 papers for peer-reviewed journals and has written several books and book chapters (as author or co-author). She has participated at national and international conferences (as author and/or speaker), has led several research projects, and has acted as reviewer of several prestigious scientific journals. She is a member of the Romanian Society of Chemistry and of the Romanian Society of Rheology. Dr. Filimon's scientific contributions are situated at the interdisciplinary interface of polymer chemistry, physics, and biochemistry, and include theoretical and experimental substantiations of the processes and technologies at nano-, micro-, and macro-scales. Her research is focused on the development of new strategies to achieve the complex architectures with well-defined functionality and various applications. The traditional synthesis tools, new theoretical and experimental physicochemical approaches, surface science, and biopolymer engineering are combined to realize the design of targeted materials. She earned her PhD in chemistry from the Romanian Academy, "Petru Poni" Institute of Macromolecular Chemistry, Iasi, Romania.

CONTENTS

LIST OF CONTRIBUTORS

Iulian Vasile Antoniac
Materials Science and Engineering Faculty, Biomaterials Group, University Politehnica of Bucharest, Splaiul Independentei 313, 060032 Bucharest, Romania

Alexandra Bargan
Inorganic Polymers Department, "Petru Poni" Institute of Macromolecular Chemistry, 41A Grigore Ghica Voda Alley, 700487 Iasi, Romania

Maria Bercea
Department of Electroactive Polymers and Plasmochemistry, "Petru Poni" Institute of Macromolecular Chemistry, 41A Grigore Ghica Voda Alley, 700487 Iasi, Romania

Cristina-Eliza Brunchi
Department of Electroactive Polymers and Plasmochemistry, "Petru Poni" Institute of Macromolecular Chemistry, 41A Grigore Ghica Voda Alley, 700487 Iasi, Romania

Maria Cazacu
Inorganic Polymers Department, "Petru Poni" Institute of Macromolecular Chemistry, 41A Grigore Ghica Voda Alley, 700487 Iasi, Romania

Ionel Adrian Dinu
"Mihai Dima" Department of Functional Polymers, "Petru Poni" Institute of Macromolecular Chemistry, 41A Grigore Ghica Voda Alley, 700487 Iasi, Romania

Maria Valentina Dinu
"Mihai Dima" Department of Functional Polymers, "Petru Poni" Institute of Macromolecular Chemistry, 41A Grigore Ghica Voda Alley, 700487 Iasi, Romania

Adina Maria Dobos
Department of Physical Chemistry of Polymers, "Petru Poni" Institute of Macromolecular Chemistry, 41A Grigore Ghica Voda Alley, 700487 Iasi, Romania

Simona Dunca
Microbiology Department, Faculty of Biology, "Alexandru Ioan Cuza" University, 20A Carol I Bvd., 700505 Iasi, Romania

Anca Filimon
Physical Chemistry of Polymers Department, "Petru Poni" Institute of Macromolecular Chemistry, 41A Grigore Ghica Voda Alley, 700487 Iasi, Romania

Ionela Gugoasa
Faculty of Chemical Engineering and Environmental Protection, "Gheorghe Asachi" Technical University of Iasi, Prof. Dr. Docent Dimitrie Mangeron Street, No. 73, 700050 Iasi, Romania

Daniela Ivanov
Department of Bioactive and Biocompatible Materials, "Petru Poni" Institute of Macromolecular Chemistry, 41A Grigore Ghica Voda Alley, 700487 Iasi, Romania

Mihai Lomora
Department of Chemistry, University of Basel, Klingelbergstrasse 80, 4056 Basel, Switzerland

Wolfgang Meier
Department of Chemistry, University of Basel, Klingelbergstrasse 80, 4056 Basel, Switzerland

Georgeta Mocanu
Department of Natural Polymers, Bioactive and Biocompatible Materials, "Petru Poni" Institute of Macromolecular Chemistry, 41A Grigore Ghica Voda Alley, 700487 Iasi, Romania

Simona Morariu
Department of Electroactive Polymers and Plasmochemistry, "Petru Poni" Institute of Macromolecular Chemistry, 41A Grigore Ghica Voda Alley, 700487 Iasi, Romania

Marieta Nichifor
Department of Natural Polymers, Bioactive and Biocompatible Materials, "Petru Poni" Institute of Macromolecular Chemistry, 41A Grigore Ghica Voda Alley, 700487 Iasi, Romania

Mihaela-Dorina Onofrei
Department of Physical Chemistry of Polymers, "Petru Poni" Institute of Macromolecular Chemistry, 41A Grigore Ghica Voda Alley, 700487 Iasi, Romania

Cornelia G. Palivan
Department of Chemistry, University of Basel, Klingelbergstrasse 80, 4056 Basel, Switzerland

Marcel Popa
Faculty of Chemical Engineering and Environmental Protection, "Gheorghe Asachi" Technical University of Iasi, Prof. Dr. Docent Dimitrie Mangeron Street, No. 73, 700050 Iasi, Romania
Academy of Romanian Scientists, Spaiul Independentei Str. 54, 050094 Bucharest, Romania

Dumitru Popovici
Laboratory of Polycondensation and Thermostable Polymers, "Petru Poni" Institute of Macromolecular Chemistry, 41A Grigore Ghica Voda Alley, 700487 Iasi, Romania

Carmen Racles
Inorganic Polymers Department, "Petru Poni" Institute of Macromolecular Chemistry, 41A Grigore Ghica Voda Alley, 700487 Iasi, Romania

Stefania Racovita
"Mihai Dima" Functional Polymers Department, "Petru Poni" Institute of Macromolecular Chemistry, 41A Grigore Ghica Voda Alley, 700487 Iasi, Romania

Cristina Magdalena Stanciu
Department of Natural Polymers, Bioactive and Biocompatible Materials, "Petru Poni" Institute of Macromolecular Chemistry, 41A Grigore Ghica Voda Alley, 700487 Iasi, Romania

Mirela Teodorescu
Department of Electroactive Polymers and Plasmochemistry, "Petru Poni" Institute of Macromolecular Chemistry, 41A Grigore Ghica Voda Alley, 700487 Iasi, Romania

Silvia Vasiliu
"Mihai Dima" Functional Polymers Department, "Petru Poni" Institute of Macromolecular Chemistry, 41A Grigore Ghica Voda Alley, 700487 Iasi, Romania

Cristina Doina Vlad
"Mihai Dima" Functional Polymers Department, "Petru Poni" Institute of Macromolecular Chemistry, 41A Grigore Ghica Voda Alley, 700487 Iasi, Romania

Mirela-Fernanda Zaltariov
Inorganic Polymers Department, "Petru Poni" Institute of Macromolecular Chemistry, 41A Grigore Ghica Voda Alley, 700487 Iasi, Romania

LIST OF ABBREVIATIONS

AA	acrylic acid
ABQn	PDMS-*b*-PDMAEMA containing ONB pendant moieties
AFM	atomic force microscopy
AFP	alpha-fetoprotein
Alg	alginate
AM	acrylamide
AN	acrylonitrile
Anis	anisamide
APBA	3-aminophenylboronic acid
ASM	acid sphingomyelinase
ATP	adenosine-5′-triphosphate
ATRP	atom transfer radical polymerization
BMP	bone morphogenetic proteins
BSA	bovine serum albumin
Bz	benzyl
CA	contrast agent
CAB	cellulose acetate butyrate
CAC	critical aggregation concentration
CAD	computer-aided design
CAGR	compound annual growth rate
CAM	computer-aided manufacturing
CAP	cellulose acetate phthalate
CAT	cellulose acetate trimellitate
CD	cyclodextrin
CEA	carcinoembryonic antigen
CH	cholesterol
CHD	coronary heart disease
CMA	coumarin methacrylate
CMC	critical micelle concentration
CPT	camptothecin
CRP	controlled radical polymerization
CS	chitosan
CSS	carbon capture and sequestration

CT	X-ray computed tomography
CTAB	cetyltrimethylammonium bromide
CTE	coefficient of thermal expansion
Cys	cystamine
DCC	N,N-dicyclohexylcarbodiimide
DEAEMA	diethylaminoethyl methacrylate
DEG	diethylene glycol
DEGDMA	diethylene glycol dimethacrylate
DEX	dexamethasone
DLS	dynamic light scattering
DMAEMA	2-dimethylaminoethyl methacrylate
DMAM	N, N-dimethylacrylamide
DMIBM	4-(3,4-dimethylmaleimido) butyl methacrylate
DOCA	deoxycholic acid
DOX	doxorubicin
DS	degrees of substitution
dPG	dendritic polyglycerol
DTPA	diethylenetriaminepentaacetic acid
DTT	DL-dithiothreitol
DTX	docetaxel
DVB	divinylbenzene
DVS	divinyl sulfone
Dx	dextran
EA	ethyl acrylate
EC	ethyl cellulose
ECH	epichlorohydrin
EGDMA	ethylene glycol dimethacrylate
ELR	biomimetic elastin-like recombinamer
FA	folic acid
Fc	ferrocene
FDA	Food and Drug Administration
FPBA	2-formylphenylboronic acid
GEL	gelatin
GLL	gellan
GMA	glycidyl methacrylate
GSH	glutathione
HA	hyaluronic acid
HDL	high-density lipoprotein

HDPE	high-density polyethylene
Hep	heparin
His	histidine
HLB	hydrophilic–lipophilic balance
HMAM	N-hydroxymethylacrylamide
HPMC	hydroxypropyl methylcellulose
HPMCP	hydroxypropyl methylcellulose phthalate
HRP	horseradish peroxidase
IBD	including ion beam deposition
IBID	ion-beam-induced deposition
IBSD	ion beam sputtering deposition
IMD	implantable medical devices
LCST	lower critical solution temperature
LDH	layered double hydroxide
LDL	low-density lipoprotein
LDL-CH	low-density lipoprotein cholesterol
MAA	methacrylic acids
MAM	methacrylamide
MBA	N,N'-methylenebisacrylamide
MC	methyl cellulose
MEA	monoethanol amine
MIC	minimal inhibitory concentration
MOPs	microporous organic polymers
MRI	magnetic resonance imaging
MTX	methotrexate
NaCMC	sodium carboxymethyl cellulose
NASA	National Aeronautics and Space Administration
NIPAM	N-isopropylacrylamide
NIPMAM	N-isopropylmethacrylamide
NMAP	N,N,-dimethylaminopyridine
NMG	nano-/microgels
NMP	nitroxide-mediated radical polymerization
NPs	nanoparticles
OFET	organic field-effect transistors
ONB	2-nitrobenzyl
OVA	ovalbumin
P4VP	poly(4-vinylpyridine)
PAA	poly(acrylic acid)

PAChol	poly(cholesteryl acryloyoxy ethyl carbonate)
PAMAM	poly(amidoamine)
PAMPA	poly(N-(3-aminopropyl) methacrylamide hydrochloride)
PAPBA	poly(3-acrylamidophenylboronic acid)
PAsp	poly(aspartic acid)
PAsp(DET)	poly{N-[N-(2-aminoethyl)-2-aminoethyl] aspartamide}
PBC	poly(benzyl carbamate)
PBM	poly[butyl methacrylate-co-3-(trimethoxysilyl)propyl methacrylate]
PBMA	poly(butyl methacrylate)
PCL	poly(ε-caprolactone)
PDEAEMA	poly(2-diethylamino ethyl methacrylate)
PDI	polydispersity index
PDLLA	poly(D-lactic acid)
PDMAEMA	poly(2-(dimethylamino)ethyl methacrylate)
PDMAM	poly(dimethyl acrylamide)
PDMS	poly(dimethysiloxane)
PDPA	poly[2-(diisopropylamino)ethyl methacrylate]
PDSA	poly(distearin acrylate)
PE	pentaerythritol
PEG	poly(ethylene glycol)
PEI	polyethylenimine
PEO	poly(ethylene oxide)
PEtOXA	poly(2-ethyl-2-oxazoline)
PGMA	poly(glycerol monomethacrylate)
PHEMA	poly(2-hydroxyethyl methacrylate)
PHEMAmLac	poly[N-(2-hydroxyethyl)methacrylamide-oligolactate]
PHis	poly(histidine)
PHPMA	poly(2-hydroxypropylmethacrylate)
pHPMAmLac	poly[N-(2-hydroxypropyl) methacrylamide-oligolactate]
PiPrOx	poly(2-isopropyl-2-oxazoline)
PISA	polymerization-induced self-assembly
PLA	polylactic acid
PLGA	poly(L-glycolic acid)
PLLA	poly(L-lactide)
PLLys	poly(L-lysine)
PMCL	poly(methyl caprolactone)
PMMA	polymethyl methacrylate

PMOXA	poly(2-methyl-2-oxazoline)
PMPC	poly[2-(methacryloyloxy)ethyl phosphorylcholine]
PNBOC	poly(nitrobenzyloxy carbonyl aminoethylmethacrylate)
PNIPAM	poly(N-isopropylacrylamide)
PPDA	poly(2-(piperidino)ethyl methacrylate
PPDSM	poly(pyridyldisulfide ethylmethacrylate)
PpEGMA	poly((ethylene glycol)methacrylate))
PS	polystyrene
PSPA	poly(spiropyran)
PSpMA	poly(spiropyran-methacrylate)
PtBuA	poly(*t*-butyl acrylate)
PTTAMA	poly(2-(((((5-methyl-2-(2,4,6-trimethoxyphenyl)-1,3-di-oxan-5-yl)methoxy)carbonyl)amino)ethyl methacrylate)
PTTMA	poly(2,4,6-trimethoxybenzylidene-1,1,1-tris (hydroxy-methyl)ethane methacrylate)
PTX	paclitaxel
PVA	polyvinyl alcohol
PVCL	poly(N-vinylcaprolactam)
PVD	physical vapor deposition
PVPON	poly(N-vinylpyrrolidone)
PySS	pyridyldisulfide
RAFT	reversible addition-fragmentation chain transfer polymerization
ROP	ring-opening polymerization
SAIB	sucrose ethyl acetate
SANS	small-angle neutron scattering
SAXS	small-angle X-ray
SDS	sodium dodecyl sulfate
SPIONs	superparamagnetic iron oxide nanoparticles
SRB	sulforhodamine B
ssDNA	single-stranded DNA
SSP	seed swelling polymerization
TA	tannic acid
TEG	triethylene glycol
TEGDMA	triethylene glycol dimethacrylate
TEM	transmission electron microscopy
TEPA	tetraethylenepentamine
TMAEMA	trimethyl aminoethyl methacrylate

Vim	vinylimidazole
VLDL	very low-density lipoprotein
VP	vinylpyridine
WR	water amount retained
XAN	xanthan

PREFACE

The development of smart materials represents an innovative and promising approach that plays an essential role in solving many of today's global challenges, from environment to biomedicine and healthcare. With the advent of the industrial revolution, due to a multitude of existent information, it is difficult to find a resource that provides a complete overview of the different types of smart materials available.

It might be surprising for those who recognize that our modern lifestyle is dependent, to a large extent, on the use of polymer-based smart materials and to find that specific materials may also be designed to accomplish the multi-performance and multi-function objectives in a fully integrated system that will have a significant influence on many of the present-day technologies. Consequently, polymer-based smart materials are a fast-expanding area of research, part of the growing field of polymer technology.

This book, *Smart Materials: Integrated Design, Engineering Approaches, and Potential Applications*, aims to answer the questions: *How we distinguish "smart materials"?* or/and *How do they work?* in the next 13 chapters. Furthermore, it lays the groundwork for assimilation and exploitation of this technological advancement. The questions we should consider include determining the root need or the underlying problem. Four of the key aspects of the approach that we have developed throughout this book are highlighted, namely, the multidisciplinary exchange of knowledge, exploration of the relationships between multiple scales and their different behaviors, understanding that material's properties are dictated at the smallest scale, and therefore, the recognition that macroscale behavior can be controlled by nano-scale design. This last aspect might be not only the most provocative but also the most indicative of the design impact of smaller scale technologies, compared to our more normative and visible technologies.

Our foray into the world of smart materials and new technologies was not only a recounting of properties and products but also an attempt to use a transactional language that would allow us to answer new questions with direct impact on daily life. Additionally, by providing a clear framework of knowledge and overlaying it with applications, we have tried to present

an open-ended map that allows the researcher to make more than a selection of systems and provide directions for the implementation of these new
materials and technologies. Based on this knowledge, a researcher should
be able to use and develop an assembly/system of any component that has
a dynamic behavior.

Dr. Anca Filimon
"Petru Poni" Institute of Macromolecular Chemistry
Iasi, Romania

CHAPTER 1

IMPACT OF SMART STRUCTURES ON DAILY LIFE: AN ALTERNATIVE FOR PROGRESS TOWARD REVOLUTIONARY DISCOVERIES

ANCA FILIMON*

Physical Chemistry of Polymers Department, "Petru Poni" Institute of Macromolecular Chemistry, 41A Grigore Ghica Voda Alley, 700487 Iasi, Romania

E-mail: capataanca@yahoo.com

CONTENTS

ABSTRACT

Advances in the field of polymer-based smart materials have allowed the design and adaptation of various structures, by multiscale modeling, engineering, and thermodynamic aspects, for specific applications which satisfy the today's society's needs. Furthermore, the rapid expansion from problem-solving to "technology push" has led to achievement of multi-performance and multi-function objectives in a fully integrated system that will have a significant influence on many of the present-day technologies. In exploration of smart materials and new technologies must be taken into account the challenges from research field and knowledge of the fundamental roots of the barriers to implementation. Smart materials have shown promising characteristics and in the same time, with further research and development it will be superior to use smart materials, without fail, in various applications.

1.1 PREDICTIVE KNOWLEDGE, SMART DISCOVERY, AND INNOVATIVE TECHNOLOGIES

Over the last few decades, the technological developments have been related to changes in the use of material. Sustainable development is achieved when the present needs and challenges are met without endangering the ability of future generations to meet their own needs and challenges. Because most of our resources are limited, this means that we need to achieve more with less the development of materials, systems, and structures that are designed to accomplish multi-performance and multi-function objectives in a fully integrated system.

The concept of *"smart materials"* has been promoted by the scientific community, spanning from the ancient to the modern world, and the technological field has evolved over the past decades with increasing pace during the 1990s to become what it is today, at the transition to the next millennium (Shivaji et al., 2015). The effective development of multifunctional and smart materials, systems, and structures has the potential to contribute significantly to meeting abovementioned goals. As the scientific improvements of the modern era progressed, several attempts were made to synthesize and develop highly efficient smart polymeric from the practical point of view due to the increasing need in modern technologies and industrial applications (biomedical, environmental, communication, defense, space, and nanotechnological fields of research with tremendous progress). It includes more conventional material systems such as composite materials, hybrid materials, nanomaterials and smart materials, systems, and structures for sustainability. So, beginning in the 20th century with the advent of the industrial revolution, the role of materials changed dramatically, having a strategic and integrated approach to innovation that maximizes research and application potential. Materials transitioned from their premodern role of being subordinate to needs into a means to expand functional performance and open up new formal responses (Fig. 1.1).

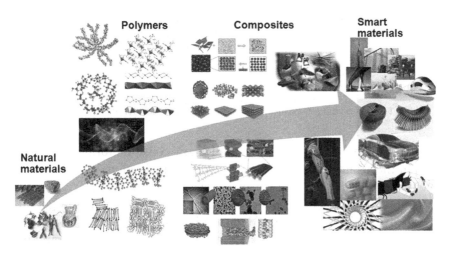

FIGURE 1.1   (See color insert.) Evolution of materials in time.

Therefore, smart materials defined as "highly engineered materials that respond intelligently to their environment," have become the *"targeted-towards"* answer for the 20th century's technological needs. The development of smart materials, systems, and structures represents a highly innovative and promising approach that satisfies the society's needs and, in the same time, plays an essential role in solving many of today's global challenges, from environment—energy and sustainability—to medicine and healthcare (Shailendra et al., 2015). Consequently, the smart materials, systems, and structures are designed to accomplish the multi-performance and multifunction objectives in a fully integrated system that will have a significant influence on many of present-day technologies (Fig. 1.2).

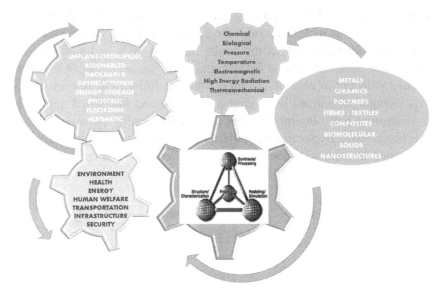

FIGURE 1.2   Everything is connected in the world of smart materials.

Fundamental to the development of a new construct for the exploitation of these materials in the design professions is an understanding of the origin and determinants of their behaviors. The behavior of smart materials results from a molecular adjustment to the changes caused by an external field applied. Historically, there have been two main approaches to modeling the smart material behavior, namely, the micromechanical and phenomenological modeling. The former approach starts from the material physics, and is generally based on energy, the obtained results from the material-specific models being represented by a low number of physically significant parameters. Phenomenological or black box model assumes a model structure and uses the system identification to fit the model parameters to experimental data. These models, which represent the material behavior without regard to its origin, typically have a high number of parameters that lack physical relevance (Gupta and Srivastava, 2010). Based on the general approach described above and fundamental characteristics that were defined as distinguishing a smart material from traditional materials (Addington and Schodek, 2005; Drossel et al., 2015), smart materials may be grouped into (Fig. 1.3):

- A first class to refer to materials that undergo changes in one or more of their properties (chemical, mechanical, electrical, magnetic, or thermal) in direct response to a change in the external stimuli associated with the environment surrounding the material. Changes are direct and reversible; there is no need for an external control system to cause these changes to occur. Among the materials in this category are thermochromic, magnetorheological, thermotropic, and shape memory.

- A second class is comprised of those materials that transform energy from one form to output energy in another form and again do so directly and reversibly (piezoelectrics, thermoelectrics, photovoltaics, pyroelectrics, photoluminescents, and others). Thus, an electrorestrictive material transforms electrical energy into mechanical energy which in turn results in a physical shape change.

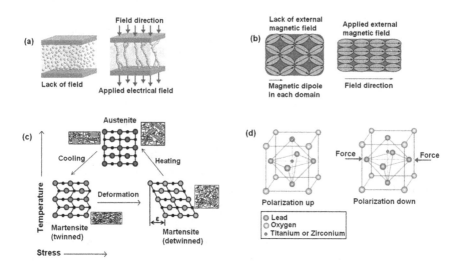

FIGURE 1.3 (See color insert.) Schematic representations of shape fixing and recovery mechanisms for some types of smart materials: (a) magneto- and electro-rheological, (b) magnetostriction (electrostriction), (c) shape–memory alloys, and (d) piezo materials.

Just as the responses of these materials are discrete and direct, then our interaction with these must ultimately function at the same scale, whether atomic, molecular, or microstructural. Knowledge of atomic and molecular structure is essential to understand the intrinsic properties of any material, and particularly so for smart materials. Thus, in order to understand how these different internal structures ultimately determine the resultant properties of materials, it is useful to analyze on the one hand, different kinds of bonding forces that exist between atoms that ultimately comprise the basic building blocks of any material, and on the other hand, the ways individual atoms aggregate into crystalline, amorphous, or polycrystalline structures (Preumont, 1997). As we have known, most smart materials actually work at the microscale and are thus not visible to the human eye. Nevertheless, the effects produced by these mechanisms are often at the meso- and macro-scale, whereas the physical mechanism, how the material works, is entirely dependent upon the material composition, the results produced by the action of the material (the phenomenological effects) are determined by many factors independent of the material composition, including quantity and environment. As a result, very similar effects can often be produced from seemingly dissimilar materials. In addition, developments taking place, as result of the diversity of molecular structures, complex architectures, and physical characteristics, as well as of the practical implications of the advanced systems from macro- to nano-scales, allow the designer to overcome some disadvantages of the materials and/ or to improve the performance of the structures realized from these materials, so-called *smart* or *active*. Desired behaviors, as defined by material properties, have no preference for specific materials or technologies. And as a result, it will be more suitable for more open to experimentation and novel solutions. The technological progress of scientific fields brings to light new challenging discoveries, which explains the constant motivation for an objective evaluation of new techniques. In conformity with the technology development and the recent concern about biomedical and environmental issues, more attention is given to using of polymer-based smart materials, each with its own advantages and disadvantages, characteristics, and potential (Aguilar et al., 2007; Roy and Gupta, 2003). The phenomenological perspective just presented emphasizes the immediately tangible results of the actions of smart materials. In order to accomplish these ends in real products or devices that are targeted for specific applications or uses, it is necessary that smart materials be made available in forms useful to the designer, for example, filaments, paints, strands, or films (Meléndez-Ortiz et al., 2015; Pucci et al., 2011; Ji et al., 2006). Improved processes for making polymeric products have particularly had a profound impact on making some smart materials ubiquitously available, notably processes for making polymer films and strands and processes for depositing thin layers of different materials on various substrates. Consequently, development of new material and technology allow integrated design at multiscale. The availability of the advanced polymer-based materials, composites, and active materials offers new design concepts for the industry. In composites materials, the design may address simultaneously some of several structural properties such as strength, stiffness, fracture toughness, and damping. In hybrid materials, in addition or exclusive to a combination of structural properties, the design may include other nonstructural properties and functions such as thermal and electrical conductivity, energy absorption, and electromagnetic properties. In nanomaterials and smart materials, in addition or exclusive to the above properties others such as optical, shape memory effects, sensing and actuation, electromagnetic interference shielding, radiation shielding, energy harvest and storage, self-healing and health monitoring capabilities, morphing, and recyclability may be system designed. In addition, the materials systems may combine the mimicking of nature biostructures and biomedical properties, such as biodegradability, and may include nanoparticles for dispensing drugs and diagnostics.

In forthcoming exploration of smart materials and new technologies must be ever mindful of the unique challenges presented by research field, and knowledge of the fundamental roots of the barriers to implementation. To achieve a specific objective for a particular function or application, a new material, composite, or alloy has to satisfy specific qualifications related to the various properties. Smart materials have all the possible potentials to fulfill maximum requirements of the changing trend which ultimately resulted in the use of smart materials in almost all the sectors of engineering and biomedical field. Smart materials have shown promising characteristics and with further research and development it will be superior to use smart materials in various applications without fail. Consequently, the potential benefits of smart materials, structures, and systems are amazing in their scope. This technology gives promise of optimum responses to highly complex problem areas by, for example, providing early warning of the problems or adapting the response to cope with unforeseen conditions, thus enhancing the survivability of the system and improving its life cycle. Moreover, enhancements to many products could provide better control by minimizing distortion and increasing precision (Shailendra et al., 2015). Another possible benefit is enhanced preventative maintenance of systems and thus better performance of their functions. By its nature, the technology of smart materials and structures is an inter-disciplinary field, encompassing the basic sciences—physics, chemistry, mechanics, computing, and electronics—as well as the applied sciences and engineering, such as aeronautics and mechanical engineering (Gandhi and Thompson, 1992).

This may explain the slow progress of the application of smart struc-tures in engineering systems, even if the science of smart materials is moving very fast and is the subject of future, even as it has already perco-lated into many aspects of our daily lives.

According to those mentioned, the book "*Smart Materials: Integrated Design, Engineering Approaches, and Potential Applications*" brings together the recent findings in this area and provides a critical analysis of the different materials available and how they can be applied as advanced systems in the industry. Also provides a wide range of information concerning the recent advances in polymer-based smart materials, based on an interdisciplinary approach of the synthesis process and design of the structures used to evidence the specific properties, adaptation of various structures for specific applications by multiscale modeling, engineering, and thermodynamic aspects.

Thematic is designed to be multidisciplinary and multidomain, in order to provide an open frame leading to the enlargement of ideas and approaches, and is based on the formation of advanced structures in well-defined conditions by combination of new structural and physicochemical concepts for designing and then investigation of the intelligent properties of polymeric materials. This research area presents a scientific interest—a better understanding of the nonlinear response to external stimuli—as well as a practical interest, elaboration, and characterization of new friendly materials, and potential development of commercial products. The struc-tures will be achieved by different mechanisms, considering the chains dynamics at the macro-molecular level and the interactions exhibited in well-established conditions. The efforts are focused on the investigation of polymer materials that are able to respond to the action of external stimuli in a predictable manner by modifying the physicochemical properties, such as the viscoelastic or conformational characteristics, hydrophilic–hydro-phobic balance, permeability, etc. All these challenges have become the strong driving forces for discovering more efficient processes and tech-nologies that offer new solutions for applications in advanced technologies. Therefore, much of works described in this book are motivated by pushing the performance limits of smart materials for precision processing.

This text is organized into 13 chapters that highlight interesting and innovative elements, relevant and up to date, insisting on how the polymer-based materials can influence the real world needs. The book considers the design of more materials, systems, and structures such as hybrid materials with complex architectures (micelles, vesicles, capsules, and gels), composite materials, and nanomaterials for sustainability, but the composition and/or microstructure is being designed to satisfy the required performance. This integrated system design approach may lead to enhanced properties and functions at different size scales, and to innovative applications which are also main subjects of this book.

The book debuts with an introduction concerning the smart structures—past, present, and future perspectives—and covers the key areas of the polymer-based smart materials for applications in advanced technologies (e.g., engineered environmental and medicine, healthcare as carriers for biomacromolecules and other substances, bone tissue engineering and regeneration, pharmaceutical applications, and interface that control of bacteria responses). After a basic introduction, the authors, that are specialists in the field of polymeric materials, including physicists, chemists, engineers, bioengineers, and biologists, explore topics that include major themes regarding the preparation, processing, design, properties, molecular technologies, thermodynamic aspects, nano- or bio-materials —involving biocompatibility, antimicrobial properties, and understand the biointerface (the interconnection between a synthetic or natural material and tissue, microorganism, cell, virus, or biomolecule), all discussed in the context of possible new applications. Some chapters are focused on the development of materials that can be tailored and combined with active materials, which allow a more integration of material and control system development. Combining different properties of the polymer-based materials with the technologies for obtaining structures with complex architectures for specific applications, this book examines the extraordinary potential for rapid implementation of the smart materials in a wide range of end user sectors and industries, including the biomedical, pharmaceutical, and environmental fields and provides real-world examples on use of the smart polymers. In addition, the book offers recent scientific information and can significantly enhance the basic knowledge of the students and researchers that are involved in smart material modeling including design or development, control of the systems behavior, and engineering approaches.

The concept of *"engineering"* materials and structures that respond to their environment is somehow an alien concept. Therefore, it is important because not only of the technological implications of these materials and structures addressed, but also of the associated issues with public understanding and acceptance. Such a general acceptance of smart materials and structures may be more difficult than some of the technological hurdles associated with their development. However, it is important to note that the present age has not left the engineering materials untouched, and that the fusion between the materials design and power of information storage and processing has led to a new family of engineering materials and structures, namely, *"smart materials."*

KEYWORDS

- smart materials
- polymer architectures
- modern technologies
- industrial applications
- future generations

REFERENCES

Addington, D. M.; Schodek, D. L. *Smart Materials and New Technologies for the Architecture and Design Professions*; Elsevier: Amsterdam, Netherlands, 2005.

Aguilar, M. R., et al. Smart Polymers and Their Applications as Biomaterials. In *E-book: Topics in Tissue Engineering*; Ashammakhi, N., Reis, R., Chiellini, E., Eds.; Oulu University: Oulu, Finland, 2007; Vol. 3, pp 1–27.

Drossel, W. G., et al. Smart³—Smart Materials for Smart Applications. *Procedia CIRP.* **2015,** *36,* 211–216.

Gandhi, M. V.; Thompson, B. D. *Smart Materials and Structures*; Springer: Netherlands, 1992.

Gupta, P.; Srivastava, R. K. Overview of Multi-functional Materials. In *New Trends in Technologies: Devices, Computer, Communication and Industrial Systems*; Meng, J. Er., Ed.; InTech: Rijeka, 2010; pp 1–14.

Ji, F. L., et al. Smart Polymer Fibers with Shape Memory Effect. *Smart Mater. Struct.* **2006,** *15* (6), 1547–1554.

Meléndez-Ortiz, H. I., et al. Smart Polymers and Coatings Obtained by Ionizing Radiation: Synthesis and Biomedical Applications. *Open J. Polym. Chem.* **2015,** *5,* 17–33.

Preumont, A. *Vibration Control of Active Structures*; Kluwer Academic Publisher: Dordrecht, 1997.

Pucci, A., et al. Colour Responsive Smart Polymers and Biopolymers Films Through Nanodispersion of Organic Chromophores and Metal Particles. *Prog. Org. Coat.* **2011,** *72* (1–2), 21–25.

Roy, I.; Gupta, M. N. Smart Polymeric Materials: Emerging Biochemical Applications. *Chem. Biol.* **2003,** *10,* 1161–1171.

Shailendra, K. B., et al. Smart Materials for Future. *Int. J. Adv. Res. Sci. Eng.* **2015,** *4* (2), 365–372.

Shivaji, S. A., et al. A Review on Smart Materials: Future Potentials in Engineering. *Int. J. Sci. Technol. Manag.* **2015,** *4* (10), 1–16.

CHAPTER 2

ENGINEERING SMART POLYMERIC MATERIALS WITH COMPLEX ARCHITECTURES FOR BIOMEDICAL APPLICATIONS

IONEL ADRIAN DINU[1*], MARIA VALENTINA DINU[1], MIHAI LOMORA[2], CORNELIA G. PALIVAN[2], and WOLFGANG MEIER[2]

[1]*"Mihai Dima" Department of Functional Polymers, "Petru Poni" Institute of Macromolecular Chemistry, Aleea Grigore Ghica Voda 41A, 700487 Iasi, Romania*

[2]*Department of Chemistry, University of Basel, Klingelbergstrasse 80, Basel 4056, Switzerland*

Corresponding author. E-mail: adinu@icmpp.ro

CONTENTS

ABSTRACT

In this chapter, a broad overview on recently published literature concerning the design, preparation, and applications of polymer-based smart systems and hybrid structured materials with complex architectures (micelles, vesicles, capsules, gels, and so on) and sizes ranging from nanometers toward several micrometers scale is provided. Aiming the fabrication of smart materials with enhanced properties for applications in biomedicine, the spatial organization and selective confinement of various molecules within the ordered domains of different polymer architectures were also considered. Moreover, the stimuli-responsiveness of these self-organized structures induces drastic changes in their morphology or properties when a specific stimulus is applied, leading to "on demand" release of the cargo or triggering a distinct in situ signal/reaction. Accordingly, results from the most relevant studies are briefly reviewed, more attention being paid to those polymeric systems with potential application in biomedical field, ranging from drug delivery to mimics of natural organelles or other cellular compartments.

2.1 INTRODUCTION

Artificial bio-inspired novel smart objects organized at nano- or/and microscale level in complex 3D polymeric self-assembled architectures are constantly under investigation as effect to the exigent demands of personalized medicine. Consequently, the field of polymeric self-structured materials has attracted an increasing interest, with a special focus on designing stimuli-responsive self-organizing polymers (Cabane et al., 2012; Cao and Wang, 2016; Che and van Hest, 2016; Guragain et al., 2015; Molina et al., 2015; Palivan et al., 2016; Tang et al., 2016; Wei et al., 2014). These polymers can act as building blocks for engineered smart materials, such as self-organized particulates with typical core–shell architectures (micelles), spherical hollow architectures in which the aqueous cavity is enclosed by a polymeric membrane/shell (vesicles/capsules), tight spherical supramolecular architectures [nanoparticles (NPs)], or cross-linked 3D polymer networks [nano- and microgels (NMGs)] (Fig. 2.1, Cao and Wang, 2016; Gaitzsch et al., 2016; Gunkel-Grabole et al., 2015; Joglekar eTrewyn, 2013; Palivan et al., 2016; Tang et al., 2016). The 3D polymeric self-assembled architectures are engineered, by tuning the chemical composition of polymer components, to accommodate and protect various payloads, such as drugs and therapeutic macromolecules (Balasubramanian et al., 2016; Car et al., 2014; Najer et al., 2016; Newland et al., 2013), proteins and enzymes (Gräfe et al., 2014; Sueyoshi et al., 2017; Vasquez et al., 2016), or fluorescent dyes and imaging agents (Cabane et al., 2012; Craciun et al., 2017; Lomora et al., 2015b, 2015c; Sigg et al., 2016a). Besides the protective role, their payload can be released "on demand" or trigger a distinct signal via physical or chemical changes (swelling, cleavage, disruption, fusion, dissociation, or disassembly) upon an applied stimulus, such as pH, temperature, light, redox agents, or other stimuli (Cabane et al., 2011; Cao and Wang, 2016; Car et al., 2014; Dinu et al., 2016; Najer et al., 2016; Palivan et al., 2016; Tang et al., 2016).

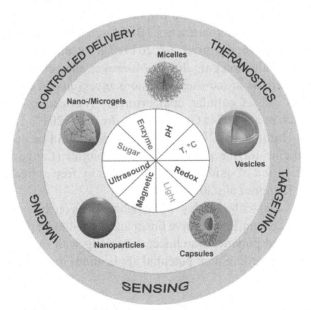

FIGURE 2.1   Schematic presentation of smart polymeric materials with 3D spherical complex architectures with applications in biomedicine.

In addition, their surface can be functionalized by conjugation with ligands to target specific receptors on the surface of different types of cells and tissues. This strategy, known as active targeting, increases the tissue-targeting capability and selectivity via ligand – receptor and antibody – antigen interactions, and thus improves the therapeutic outcome of the designed polymer-based self-assembled architectures (Chuard et al., 2017; Mohammadi et al., 2017; Palivan et al., 2016; Peer et al., 2007). Consequently, these preferential interactions result in accumulation of surface-functionalized 3D polymeric self-assembled architectures at the targeted site, where they could: (1) be uptaken into the cells, (2) release their payload into the close proximity of cells or identify the environmental changes, and/or (3) be attached on the cell membrane and release their cargo into the extracellular environment in a tailored manner (Egli et al., 2011; Palivan et al., 2016; Peer et al., 2007).

The confinement of active molecular and/or supramolecular species into the ordered domains of 3D polymeric architectures endows the complex smart hybrid systems with new functionalities offering the possibility of using these platforms in combined recognition/therapy strategies or in development of responsive devices that emulate the behavior of natural compartments (Buddingh' and van Hest, 2017; Gaitzsch et al., 2016; Garni et al., 2017; Liu et al., 2016; Palivan et al., 2016; Postupalenko et al., 2016). Moreover, they could be designed to produce reactions at nanoscale (nanoreactors) inside cells (artificial organelles) or act as cellular compartments (Fernandez-Trillo et al., 2017; Gaitzsch et al., 2016; Godoy-Gallardo et al., 2017; Tanner et al., 2011a; York-Duran et al., 2017). Similarly, they could be used in translational medicine for the production of antibiotics (Langowska et al., 2013) or mitigation of oxidative stress caused by free radicals (Richard et al., 2015; Tanner et al., 2011b).

Hence, this chapter highlights recent publications on the synthetic strategies toward the stimuli-responsive linear and cross-linked polymers, their 3D self-organized polymeric architectures, along with the requirements, and limitations regarding their potential applications in biomedicine.

2.2 DESIGNING POLYMERS AS BUILDING BLOCKS FOR ENGINEERED SMART MATERIALS

Polymers are very attractive for engineering smart materials for biomedical applications because they are available in a large diversity of molecular weights, functionalities, dispersities, defined sequences of repeating units, and specific chemical structures. Of particular interest for the development of self-organizing supramolecular architectures are the hydrophobic and hydrophilic monomers, which can be used in various combinations to design synthetic homopolymers and copolymers with well-defined properties (Hamidi et al., 2012; Ramasamy et al., 2017). When monomers with different functionalities are selected, copolymers with random, alternating, block, or graft conformations are obtained by changing the monomer ratio and position throughout the copolymer chain. The most common examples of hydrophilic and hydrophobic polymer building blocks involved in designing complex 3D architectures and the corresponding methods of synthesis are listed in Table 2.1. A complete list of abbreviations for all polymers and materials mentioned in this chapter can be found before the reference section.

TABLE 2.1 Common Hydrophilic and Hydrophobic Polymer Structures Used in Preparation of Smart Materials for Biomedical Applications.

Polymer name	Abbreviations	Structure	Methods of synthesis
		Hydrophilic	
Poly(ethylene glycol)/ poly(ethylene oxide)	PEG/PEO		ROP
Poly(2-(dimethylamino) ethyl methacrylate)	PDMAEMA		CRP
Poly(acrylic acid)	PAA		CRP
Poly(2-methyl-2-oxazoline)	PMOXA		ROP
Poly(aspartic acid)	PAsp	R = H or Na	ROP, polycondensation

TABLE 2.1   *(Continued)*

Polymer name	Abbreviations	Structure	Methods of synthesis
Poly(N-isopropylacrylamide)	PNIPAM	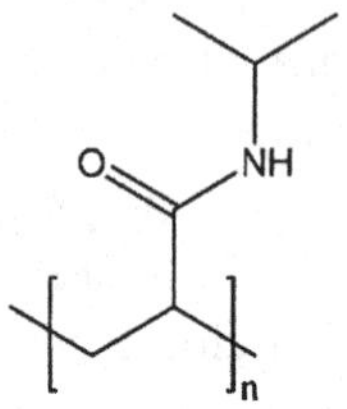	CRP
Poly(4-vinylpyridine)	P4VP	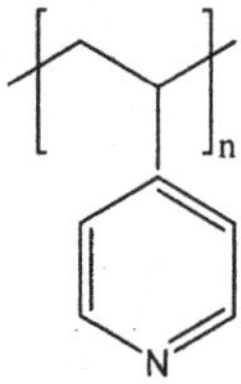	CRP, anionic
Poly(2-hydroxyethyl methacrylate)	PHEMA	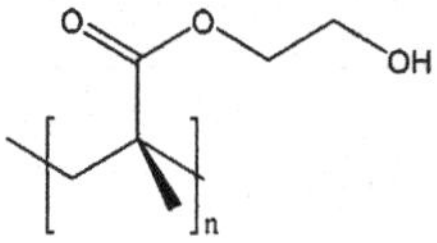	CRP
Polyethylenimine	PEI	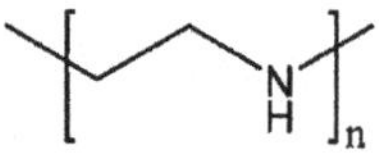	ROP

Hydrophobic

Polylactic acid	PLA	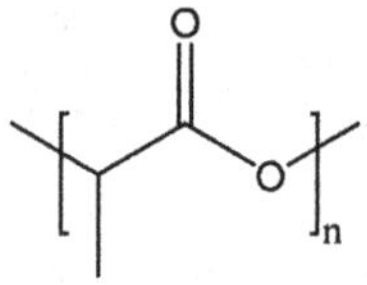	ROP
Poly(ε-caprolactone)	PCL	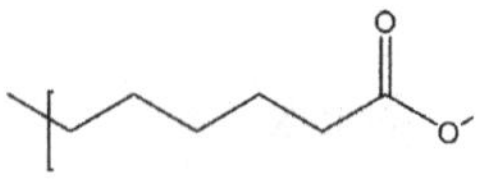	ROP
Poly(butyl methacrylate)	PBMA	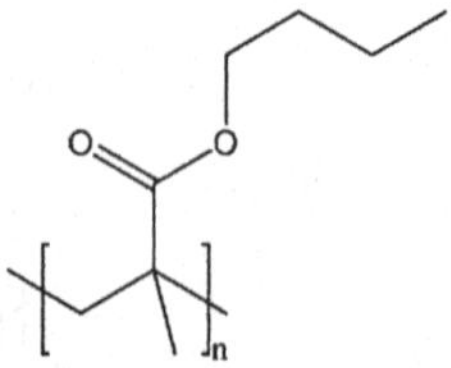	CRP

TABLE 2.1 *(Continued)*

Polymer name	Abbreviations	Structure	Methods of synthesis
Polydimethylsiloxane	PDMS		ROP, polycondensation
Poly(L-lysine)	PLLys		ROP, polycondensation
Polystyrene	PS		CRP; anionic
Poly(N-vinylcaprolactam)	PVCL		CRP

CRP: controlled radical polymerization; ROP: ring-opening polymerization.

Various synthetic strategies have been used to prepare both homopolymers and copolymers, such as living ionic polymerization, controlled radical polymerization (CRP), and ring-opening polymerization (ROP), which were recently reviewed in a comprehensive reference work (Hashmi, 2012). Depending on the monomer structure and the desired application, polymers with well-defined chemical compositions, molecular weights, and narrow polydispersity are synthesized by selecting the most appropriate polymerization method (Table 2.1).

Living ionic polymerizations, either anionic or cationic, are frequently used for the synthesis of polymers with certain molecular weight and tailored end-chain functional moieties if the reaction is performed in the absence of irreversible chain transfer and chain termination processes. In this regard, all polymer chains are almost instantaneously initiated and the highly reactive propagating chains are simultaneously growing in the presence of counterions which preserve the electroneutrality of the system. However, the propagating chain-ends must remain active for further chain extension or for functionalization with the appropriate terminal groups. Consequently, the ionic polymerization processes are strongly affected by the interaction strength between the "living" species and their counterions (Hashmi, 2012). Moreover, the solvent nature and the presence of water, carbon dioxide, and/or any other impurities are significantly influencing the polymerization mechanism, as well as the properties of the polymers fabricated using these methods. In addition, another serious drawback is the reduced number of polymers with functional side groups synthesized by ionic polymerization techniques. This limitation is dictated by the well-known intolerance of the polar propagating chains toward functional moieties, such as hydroxyl, amino, carboxyl, or mercapto groups, which can act as termination or chain transfer agents, quenching the polymerization. Nevertheless, this issue can be overcome using an approach involving the protection of functional group with a suitable moiety in order to convert it into a more stable form. In this manner, the protected functional monomer can be polymerized through an ionic mechanism, and finally, the original functional group is restored by removing the protective group.

Polymers with unique properties, as well as well-defined molecular weight and chemical composition, have been synthesized starting from cyclic monomers using ROP techniques (Dubois et al., 2009; Hashmi, 2012). Generally, most of the monomers are heterocycles and, due to their extremely polar nature, various electrophilic or nucleophilic reagents can promote an ionic ROP process. Thus, polyesters have been prepared from cyclic esters (lactones), polysiloxanes starting from cyclosiloxanes, or polyamides from cyclic amides (lactams) (see Table 2.1 for more examples), but a wide range of block copolymers can be also prepared by sequential ROP, usually those copolymers synthesized from different cyclic monomers having similar functional group, such as two lactones, two lactams, and so on. However, the ROP mechanism and the polymer polydispersity will drastically depend on the type of monomer and catalyst, as well as on their concentration. Moreover, these methods require an appropriate catalyst in order to proceed, which is typically based on heavy metal compounds; therefore, the complete removal of catalyst is challenging, when the synthesized polymers are designed to be used in biomedical and pharmaceutical applications.

CRP or reversible-deactivation radical polymerization is one of the most investigated strategies for synthesis of (co)polymers (Matyjaszewski et al., 2015). Typically, among all CRP techniques, atom transfer radical polymerization (ATRP), reversible addition-fragmentation chain transfer polymerization (RAFT), and nitroxide-mediated radical polymerization (NMP) are commonly used for synthesis of homopolymers and copolymers based on acrylates and acrylamides (Table 2.1). These methods have several advantages compared to ionic or free radical polymerization, including more suitable conditions for chain growth and less influence of impurities. Moreover, the growing chains in CRP are in a dormant state, and can be reactivated and functionalized, or can react with other monomers to obtain block copolymers. However, some limitations are inherent for these techniques involving radical intermediate species (Matyjaszewski et al., 2015). In this regard, the reduction of dead chain number or the increase of end chain functionality is still challenging, especially when polymers with high molecular weight are demanded.

One of the most interesting piece-from-the-puzzle of polymeric materials with high-end applications in biomedicine is, from the point of view of a polymer chemist, the development of smart polymer particulates, or structures able to accommodate and transport the payload to a specific target. Once the intelligent polymeric vehicle has reached its destination, it should either release its loaded content or trigger a precise signal as a response to internal or external stimuli (Table 2.2) such as pH, temperature, oxidizing or reducing agents, light, and so on (Che and van Hest, 2016; Li et al., 2014; Molina et al., 2015; Ramasamy et al., 2017; Thambi et al., 2016; Wang and Kohane, 2017). Consequently, specific characteristics of pathological tissues, such as low pH, hyperthermia, or overexpression of glutathione (GSH) were the main internal triggers taken into consideration in engineering smart self-structured vehicles with improved on-target properties and/or higher cellular uptake (Cabane et al., 2012; Che and van Hest, 2016; Hu et al., 2017; Joglekar and Trewyn, 2013; Karimi et al., 2016; Liu et al., 2017). Among the external stimuli, light is one of the most attractive and highly selective options used in biomedical applications, leading to the disassembly of self-organized structure or photocleavage of the chemically attached active molecules (Cabane et al., 2011; Dinu et al., 2016; Karimi et al., 2017; Marturano et al., 2017; Wajs et al., 2016).

TABLE 2.2 Examples of Single-responsive Polymers.

Stimulus	Architecture	Polymer building entity	Potential biomedical application	References
pH	Micelles	PDMS-*b*-PDMAEMA	DOX delivery	Car et al. (2014)
		PEG-*b*-PAsp	MG132 delivery	Quader et al. (2014)
		PEG-*b*-PDPA;	CAs for MRI	Zhu et al. (2016)
		PEG-*b*-PPDA		
	Vesicles	PMPC-*b*-PDMAEMA-*b*-PDPA;	DOX delivery	Du et al. (2012)
		PMPC-*b*-PDPA-*b*-PDMAEMA;		
		PMPC-*b*-P(DMAEMA-*stat*-DPA)		
	Capsules	Anis-PEG-PTTMA-PAA	Granzyme B delivery	Lu et al. (2015)
		PEO-*b*-PTTAMA	Drug delivery	Wang et al. (2015a)
		PDEAEMA-*co*-PPDSM	Nanoreactors and carriers	Huang et al. (2012)
	NPs	TA–PVPON	DOX delivery	Liu et al. (2014a)
		CS-Alg	Insulin delivery	Mukhopadhyay et al. (2015)
		PEG-PHis-PLLA	DOX delivery	Meng et al. (2016)
		PEG-*b*-PMCL-*b*-PDMAEMA	Protein delivery	Vasquez et al. (2016)
	NMGs	p(NIPAM-*co*-2VP)	Drug delivery	Lazim et al. (2012)
		dPG-AM-PEG-FPBA	MTX delivery	Zhang et al. (2015a)

TABLE 2.2 *(Continued)*

Stimulus	Architecture	Polymer building entity	Potential biomedical application	References
Temperature	Micelles	P(NIPAM-*co*-DMAM)-*b*-PLLA	Drug delivery	Akimoto et al. (2014)
		PEG-*b*-pHPMAmLac; PEG-*b*-pHEMAmLac	PTX, DEX, and DOX delivery	Talelli et al. (2015)
		PAMPA-*b*-PNIPAM;	Delivery of pDNA, siRNA, proteins, and peptides	Li et al. (2006);
		PDMAEMA-*b*-PNIPAM		Li et al. (2007)
	Vesicles	PVCL-*b*-PDMS-*b*-PVCL	DOX release	Liu et al. (2015a)
		PtBuA-*b*-PNIPAM	Drug delivery	Moughton and O'Reilly (2010)
	Capsules	CS/ELR	BSA delivery	Costa et al. (2013)
		Dx-GMA/GEL/PNIPAM gates	Chemokine (SDF-1α) delivery	Chen et al. (2013)
	NPs	Au-ssDNA/P(NIPAM-*co*-AM)	DOX delivery	Hamner et al. (2013)
		PBM-*b*-P(NIPAM-*co*-DMAM)	Theranostics	Hiruta et al. (2017)
		poly(NIPAM-*co*-AM)	BSA release	Sung et al. (2015)
	NMGs	PNIPAM-*co*-PAA	Drug delivery	Begum et al. (2016)
		P(NIPAM-*co*-NIPMAM)	DEX delivery	Fundueanu et al. (2016)

TABLE 2.2 *(Continued)*

Stimulus	Architecture	Polymer building entity	Potential biomedical application	References
Redox	Micelles	PTX-HA-SS-DOCA	PTX delivery	Li et al. (2012)
		PHEMA-*b*-(PBMA-SS-PBMA)-*b*-PHEMA	Drug delivery	Toughrai et al. (2014)
	Vesicles	PEtOXA-PLA-*g*-PEI-SS	DNA and DOX delivery	Gaspar et al. (2015)
		pPEGMA-PCL-SS-PCL-pPEGMA	DOX delivery	Kumar et al. (2015)
		PEG-SS-PAChol	Calcein release	Jia et al. (2014)
	Capsules	PEG–SS–Pasp; PEG–SS–P[Asp(DET)]	Drug and gene delivery	Dong et al. (2009)
		PHEMAPOSS-*b*-P(DMAEMA-*co*-CMA)	Photodynamic therapy	Zhang et al. (2016b)
	NPs	PMOXA-SS-PCL	Delivery of anticancer and antimalarial drugs	Najer et al. (2016)
		PHis-hepPDMS-Gd	MRI CAs	Sigg et al. (2016a)
	NMGs	P(Dex-MA/TMAEMA/PySS-MAM)	OVA delivery	Li et al. (2015)
	micelles	poly(DTPA-*co*-Cys)/PEG-*b*- PLLys)	Curcumin delivery	Lee et al. (2017)
Photo-/ light		PEG-*b*-PSpMA	Coumarin 102 delivery	Lee et al. (2007)
		pDNA/PEG-PAsp(DET)-PLLys	DNA delivery	Nomoto et al. (2014)

TABLE 2.2 *(Continued)*

Stimulus	Architecture	Polymer building entity	Potential biomedical application	References
		PMCL-ONB-PAA	Delivery of fluorescein, ATTO655, and eGFP	Cabane et al. (2011)
	Vesicles	PEO-b-PNBOC	Release of Nile red and DOX	Wang et al. (2014b)
		PBC-b-PDMA	CPT and DOX delivery	Liu et al. (2014b)
		PEO-b-PSPA	Drug delivery	Wang et al. (2015b)
	Capsules	Coumarin-modified microcapsules	Drug delivery	Jiang et al. (2017)
		Azobenzene/α-CD or Fc/β-CD-grafted dextran	Drug delivery	Wajs et al. (2016)
	NPs	ABQn	Drug delivery	Dinu et al. (2016)
		poly(2-nitroimidazole)/PVA	Cancer therapy	Qian et al. (2016)
		PE-PCL-b-PAA/Fe^{3+}	DOX delivery	Panja et al. (2016)
	NMGs	Au–AgNPs/PS-PEG	Hyperthermia	Wu et al. (2011)

Moreover, multiple stimuli-responsive triggering (for several examples see Table 2.3) was proved to be a more efficient strategy for achieving an improved therapeutic outcome (Cao and Wang, 2016; Guragain et al., 2015; Hu et al., 2017; Wei et al., 2014; Yuan et al., 2014a, 2014b), while targeted delivery was synergistically used to enhance the intracellular uptake or to accumulate the particulates in a specific region of the body (Dieu et al., 2014; Egli et al., 2011; Figueiredo et al., 2016; Najer et al., 2014).

In this respect, polymers with pH, temperature, redox, and light-responsiveness, and different compositions have been designed and evaluated as potential platforms for biomedical applications, some recent studies being discussed below.

2.2.1 *pH-RESPONSIVE POLYMERS*

The extracellular pH of tumors is generally slightly acidic (pH ~6.5−7.2) compared to that of normal tissues and blood (pH~7.4), while in the intracellular compartments, such as endosomes and lysosomes, the pH value is even much lower (pH ~4.5−5.5) (Hu et al., 2017). These pH gradients within the body were the main reason for designing a large number of pH-responsive polymer platforms as potential drug delivery carriers with various complex spherical architectures (Table 2.2). To engineer pH-sensitive self-organized systems, polymers incorporating acid-cleavable or ionizable groups are commonly used either as hydrophobic [polylactic acid (PLA), poly(ε-caprolactone) (PCL), or poly(L-lysine) (PLLys) or hydrophilic (poly(2-(dimethylamino)ethyl methacrylate) (PDMAEMA), poly(acrylic acid) (PAA), poly(aspartic acid) (PAsp), poly(4-vinyl pyridine) (P4VP), or polyethyleneimine (PEI)] building blocks (Table 2.1). When the pH decreases, the polymer architectures containing one of these polymers will exhibit a hydrolytic degradation or collapse and dissociation induced by the hydrophilic to hydrophobic conversion, which leads to a controlled release of the entrapped/encapsulated cargo. In addition, acid-labile bonds, such as hydrazone, imine, acetal, and so on, can be also involved in designing pH-responsive smart materials (Ramasamy et al., 2017; Zhang et al., 2015b). These moieties can be incorporated into the self-organized polymer architecture either as side or end chain groups for conjugation of drugs and other active biomolecules or as linkers connecting polymer blocks. Thus, amphiphilic PEO-*b*-PTTAMA diblock copolymers containing acetal side groups were synthesized by RAFT

TABLE 2.3 Examples of Dual- and Multi-responsive Polymers.

Polymeric system	Architecture	Responsiveness	References
PC-SPMA	Micelles	Redox/pH/light	Lee et al. (2014)
PEO-SS-PAPBA	Micelles	Glucose/pH/redox	Yuan et al. (2014a)
P(MEO$_2$MA-*co*-OEGMA)-SS-PCL-A–U–PEG	Micelles	Temperature/redox/pH/salt	Yuan et al. (2014b)
PNBM-SS-PDMAEMA	Micelles	Light/temperature/pH/redox	Cao et al. (2014)
PEG-b-PDEAEMA-*stat*-BMA	Vesicles	pH/light	Iyisan et al. (2016)
(OEG-STs)-b-PBOx	Vesicles	Temperature/pH	Jeong et al. (2015)
PiPrOx-b-CEtOxa	Vesicles	Temperature/pH	Zschoche et al. (2017)
PNIPAM-b-P(MA-co-BMA)/PAH	Capsules	Temperature/pH	Liu et al. (2017)
PNIPAM-co-PAA	NMGs	Temperature/pH	Begum et al. (2016)
Fe$_2$O$_3$/P(NIPAM-*co*-DMAEMAQ-*co*-MAA); Fe$_2$O$_3$/P(NIPAM-*co*-DMAEMAQ-*co*-MAA-*co*-HEMA)	NMGs	Temperature/pH/magnetic	Salehi et al. (2015)
PDMAEMA-SS-RhB cross-linked by Br-ONB-Br	NMGs	Temperature/pH/UV light/redox	Cao et al. (2016)
tPG/MNPs/BCN/Tf-PEG-azide	NMGs	Temperature/magnetic	Asadian-Birjand et al. (2016)
P(NIPAM-SS-AA)	NMGs	Temperature/pH/redox	Zhan et al. (2015) Yang et al. (2016)

polymerization of a pH-sensitive monomer (TTAMA) starting from PEO. These copolymers self-assembled into polymersomes whose membranes consist of pH-responsive hydrophobic bilayer, which was relatively stable under neutral pH, while upon exposure to acidic medium they underwent hydrolysis generating hydrophilic diol groups (Wang et al., 2015a). Moreover, the presence of ionizable moieties (e.g., carboxylic or amino groups) induces protonation or deprotonation as response to the pH variations. For example, PDMAEMA, which can be easily synthesized by ATRP, becomes ionized at a low pH, while PAA is ionized and dissolved more at high pH. The soluble to insoluble phase transition of pH-sensitive copolymers based on PDMAEMA was taken into account for engineering complex polymer structures, such as micelles or NPs that respond to the pH decrease and release their cargo (Car et al., 2014; Vasquez et al., 2016).

2.2.2 THERMO-SENSITIVE POLYMERS

Temperature is another common stimulus that has been intensively involved in the design of polymers with distinct responsiveness (Table 2.2). Implication of these polymers in biomedical applications was further supported by the local increase of temperature in inflamed tissues compared to normal tissues, which is also known as hyperthermia. In this context, polymers able to undergo hydrophilic to hydrophobic transitions when the local temperature is higher than their lower critical solution temperature (LCST) are typically selected to induce thermo-responsiveness (Akimoto et al., 2014; Hu et al., 2017). One of the most often used thermosensitive polymers is PNIPAM, which is particularly interesting because of its LCST of 32°C that is very close to the physiological temperature. Below this value, PNIPAM is hydrophilic in aqueous solution, due to the expansion of polymer chains, but changes to hydrophobic when the temperature is above the LCST (Hamner et al., 2013). This property can be exploited to thermally induce the assembly/disassembly of polymer architectures to release the entrapped/encapsulated payload (Table 2.2). A PtBuA-*b*-PNIPAM diblock copolymer end-functionalized at the terminal PNIPAM block with a permanently charged quaternary amine group has been synthesized by RAFT polymerization. This block copolymer exhibited a thermally induced transition from micelles to vesicles, which can be further investigated for entrapment and release of bioactive molecules (Moughton and O'Reilly, 2010). Moreover, when NIPAM is copolymerized with other monomers, the LCST and thermosensitivity of resulting copolymer can be modulated by changing the polymer concentration, the molecular weight, or the ratio between comonomers (Begum et al., 2016; Fundueanu et al., 2016; Sung et al., 2015). For example, LCST can be decreased by using hydrophobic comonomers or by increasing the molecular weight of copolymer, while the incorporation of hydrophilic comonomers able to form hydrogen bonds with NIPAM raises the copolymer LCST.

Temperature responsiveness can be also introduced by other N-alkylacrylamide derivatives, such as DMAM and DMAEMA, or by 2-isopropyl-2-oxazoline and N-vinylcaprolactam (Akimoto et al., 2014; Hiruta et al., 2017; Liu et al. 2015a; Zschoche et al., 2017). For example, PVCL-*b*-PDMS-*b*-PVCL triblock copolymers synthesized by RAFT polymerization starting from a bifunctional PDMS macroinitiator have been self-assembled into polymersomes with temperature-controlled permeability for sustained doxorubicin (DOX) delivery. The modification of temperature in the range of 37–42°C did not affect the polymer vesicles stability, while all morphological changes were reversible (Liu et al., 2015a).

2.2.3   REDOX-RESPONSIVE POLYMERS

The abnormal redox state of tumors and inflamed tissues leading to an overexpression of GSH drastically differentiates these pathological sites from the normal ones. Consequently, self-organized architectures with redox-responsiveness have been designed to transport contrast agents (CAs) or various therapeutics (Table 2.2). Usually, disulfide bonds, which can be reduced to thiol groups in the presence of GSH, are introduced during synthesis into the structure of amphiphilic block copolymers either inside of the hydrophobic block (Kumar et al., 2015; Toughrai et al. 2014) or between the hydrophilic and hydrophobic blocks (Jia et al., 2014; Najer et al., 2016). The incorporation of disulfide bonds into the hydrophobic block of amphiphilic block copolymers was taken into consideration to design NPs degradable into the reductive environment of the cell cytosol (Kumar et al., 2015). In this regard, poly(polyethylene glycol methacrylate)-poly(caprolactone)-SS-poly(caprolactone)-poly(polyethylene glycol methacrylate) (pPEGMA-PCL-SS-PCL-pPEGMA) block copolymers were synthesized by successive ROP of ε-caprolactone, and then the ATRP of PEGMA. Self-assembled NPs based on these block copolymers proved to be promising redox-sensitive carriers for delivery of DOX.

2.2.4 PHOTOSENSITIVE POLYMERS

Among the external triggers, light has attracted a great interest when smart polymeric materials are engineered (Table 2.2), because the responsiveness of self-organized structures can be simply and conveniently promoted at any location by applying the light on a specific body area or on a precise time range upon exposure to light (Che et al., 2016; Hu et al., 2017; Karimi et al., 2017; Marturano et al., 2017). The polymer photoresponsiveness is dependent on the nature of moieties absorbing light, inducing reversible or irreversible changes in the polymer structure. Thus, polymers containing photocleavable o-nitrobenzyl (Cabane et al., 2011; Dinu et al., 2016; Wang et al. 2014b) or coumarin (Jiang et al., 2017) linkers can undergo an irreversible light-induced degradation, while polymers comprising azobenzene (Wajs et al., 2016) or spiropyran moieties (Wang et al., 2015b) show reversible conformational changes. An interesting study reports a strategy that combines the photo-cleavable ability of o-nitrobenzyl groups with cross-linking amidation to induce a hydrophobic-to-hydrophilic transition, and simultaneously, stabilization and permeabilization of a self-assembled vesicular membrane (Wang et al., 2014b). To this end, amphiphilic diblock copolymers were synthesized by RAFT polymerization of 2-nitrobenzyloxycarbonylaminoethyl methacrylate (NBOC) starting from a PEO-based macroinitiator. The photo-cleavable NBOC units from the hydrophobic block are converted upon UV irradiation into hydrophilic 2-aminoethyl methacrylate repeating units, which can concurrently undergo extensive amidation reactions, leading to polymer cross-linking.

2.2.5 DUAL- AND MULTI-RESPONSIVE POLYMERS

The complexity of biological systems, which are able to adapt to multiple environmental changes, motivated the development of new multifunctional stimuli-responsive materials in order to offer a more flexible and efficient control over the response of engineered systems (Cao and Wang, 2016; Guragain et al., 2015). Consequently, dual- and multi-responsive polymers have been designed using modern methods of polymer chemistry by combining various single-responsive chemical entities into one polymeric structure (Table 2.3), which can simultaneously or sequentially respond to two or more different applied stimuli. Thus, polymers sensitive both to pH and temperature (Jeong et al., 2015; Liu et al., 2017; Zschoche et al.,

2017), or pH and light (Iyisan et al., 2016) have been reported. In addition, strategies to engineer triple- and multiple-stimuli responsive self-organized architectures have been also proposed (Cao et al., 2016; Lee et al., 2014; Yuan et al., 2014a, 2014b; Zhan et al., 2015). The stimuli-responsive building blocks capable of undergoing conformational changes upon stimulation have been mostly synthesized by ATRP or RAFT techniques due to the mild reaction conditions and high variety of available monomers. Using RAFT as polymerization technique, poly(N-isopropyl acrylamide)-*block*-poly(methacrylic acid-*co*-2-hydroxy-4-(methacryloyloxy) benzophenone) and polyallylamine (PNIPAM-*b*-P(MA-*co*-BMA) and PAH) have been prepared for engineering pH and temperature dual-sensitive, photo-cross-linked hollow 3D architectures as promising systems for mimicking cell functions (Liu et al., 2017). Thus, starting from a RAFT chain transfer agent and NIPAM, a thermo-responsive PNIPAM macroinitiator has been synthesized, and then MA and BMA have been statistically copolymerized to form the second block. The copolymers generated by this approach combine the advantage of comprising both thermo- and pH-responsive moieties (NIPAM and MA), and additionally, photo-cross-linkable groups (BMA). In contrast, thermo-, redox-, and pH-triple-responsive nanogels have been prepared by free radical polymerization of NIPAM and AA in the presence of sodium dodecyl sulfate (SDS) as a surfactant, using N,N'-bis(acryloyl)cystamine (BAC) as reduction-sensitive cross-linker (Zhan et al., 2015). These anionically charged self-organized systems hold a high potential for delivery of cationic therapeutics. Nevertheless, the design of multistimuli-responsive materials for biomedical applications is still challenging, due to the multiplicity of factors that are present in the physiological environment. Moreover, most of the studies investigating multiple-stimuli responsiveness focused on independent responses for individual triggers. However, evaluation of a synergistic response to combinations of different stimuli that are simultaneously applied is still lacking.

2.3 PREPARATION AND APPLICATIONS OF SELF-ORGANIZED POLYMER SYSTEMS WITH COMPLEX ARCHITECTURES

A plethora of supramolecular 3D self-structured spherical architectures, with sizes ranging from few nanometers toward several micrometers, have been designed starting from a large variety of natural and synthetic polymers (Cao and Wang, 2016; Guzman et al., 2017; He et al., 2013; Tang et al., 2016). Aiming the fabrication of smart materials with enhanced properties for applications in biomedicine, the spatial organization and selective confinement of various molecules within the ordered domains of these polymer architectures have been widely investigated (Che and van Hest, 2016; Guragain et al., 2015; Joglekar and Trewyn, 2013; Molina et al., 2015; Palivan et al., 2016). All self-organized systems can be designed to limit their size by (1) adjusting their chemical composition, (2) choosing the suitable preparation method, or (3) selecting a template with a specific size. In this respect, the main features, preparation strategies, as well as the promising or potential applications of such particulates in various biomedical applications have been extensively reviewed in the last decade; therefore, only a brief survey of the recent literature will be presented henceforth, highlighting the most relevant examples of 3D architectures and their biomedical applications.

2.3.1 MICELLES

Polymeric micelles are self-organized particulates with typical core–shell architectures, and obtained from amphiphilic block copolymers as result of self-assembly in aqueous media above the critical micelle concentration (CMC). Generally, the hydrophobic core is acting as a reservoir for hydrophobic entities, while the shell, also known as corona, comprises the hydrophilic part of amphiphilic macromolecules, whose major role is to provide steric stability to the micelle structures and good dispersibility in aqueous environments (Kulthe et al., 2012; Reddy et al., 2015; Tang et al. 2016). Segregation of hydrophobic core from the aqueous medium is generally controlled by attractive forces, including complexation, hydrophobic and/or ionic interactions, which are also governing the solubilization of drugs or other bioactive molecules within the micellar core to overcome either the solubility issues or to protect them from environmental degradation (Reddy et al., 2015). However, confining active molecules within the self-organized domains of micelles could result in a destabilization of the hydrophilic to hydrophobic balance, and consequently might decrease the stability of these polymeric vehicles.

In general, the polymeric micelles are built up from several hundred molecules, number which is usually referred as aggregation number, while their corresponding diameter is ranging from several nanometers up to

100 nm (Ahmad et al., 2014; Reddy et al., 2015). The size of micelles can be tuned by variation of some parameters, such as molecular weight of polymer, aggregation number or the methods, and conditions of preparation. Moreover, the balance between the hydrophilic and hydrophobic blocks of copolymers, defined as hydrophilic content, f, is also controlling the morphology of self-organizing architectures. Thus, spherical micelles are formed when amphiphilic block copolymers with a value of $f > 45\%$ are involved in preparation (Kulthe et al., 2012).

Even though the micelles are in a thermodynamic equilibrium with the individual solubilized polymeric chains, they are more stable toward dilution compared to the conventional surfactant micelles (Gunkel-Grabole et al., 2015; Kulthe et al., 2012; Tang et al., 2016). However, a further stabilization can be achieved by cross-linking after the micellar self-assembly to covalently bind the building blocks either in the corona or in the core (Talelli et al., 2015). Besides the enhanced stability, the shell-cross-linking preserves the micellar structure and increases the circulation time. In spite of these advantages, the cross-linking of corona is limited by the risk of intermicellar aggregation, if the reaction is performed at low dilutions. Moreover, the micelle hydrophilicity might be affected, and consequently, the circulation time within physiological environments. On the contrary, the stabilization by core-cross-linking should not have a negative impact on the surface properties, but it can hinder the effective release of the bioactive molecules.

Several methods have been used for preparation of polymeric micelles, including direct dissolution (Gong et al., 2009; Park et al., 2007), film casting or rehydration (Car et al., 2014; Seo et al., 2015), dialysis (Vangeyte et al., 2004; Wang et al., 2017), solvent evaporation (Danafar et al., 2017; Lee et al., 2014; Toughrai et al., 2015), and oil-in-water (O/W) emulsion (Danafar et al., 2017), but the selection of proper technique depends on the polymer solubility in water. The first method (i.e., direct dissolution) is most commonly used to prepare micelles from block copolymers with either a moderate solubility in water or low molecular weight and short insoluble block. This approach involves the addition of copolymers into the aqueous medium without using organic solvents or surfactants. In some cases, stirring, heating, or sonication have been used to facilitate dissolution and promote the formation of micelles (Kulthe et al., 2012; Tang et al., 2016). It can be used for fabrication of micelles based on Pluronics (block copolymers of ethylene oxide and propylene oxide) or poly(ethylene glycol)–poly(ε-caprolactone)–poly(ethylene   glycol)   (PEG–PCL–PEG)

(Kulthe et al., 2012; Reddy et al., 2015; Tang et al., 2016). This method was subsequently used for preparation assisted by sonication of drug-loaded micelles (Gong et al., 2009). Moreover, polyion complex micelles were obtained by mixing the solutions of poly(2-isopropyl-2-oxazoline)-*b*-poly(L-lysine) (PiPrOx-*b*-PLLys) and poly(2-isopropyl-2-oxazoline)-*b*-poly(aspartic acid) (PiPrOx-*b*-PAsp) prepared by direct dissolution of polymers in Tris HCl buffered solution (Park et al., 2007). Although direct dissolution is simple, it becomes inappropriate when the water solubility of copolymers is too low. Another method similar to direct dissolution is film casting or film rehydration, which consists in dissolution of copolymer or copolymer/drug mixtures in a volatile organic solvent, and consecutively, solvent evaporation to form a thin polymeric film, wherefore, the other names for this technique, as solution casting or drying-down method (Tang et al., 2016). Subsequently, empty or loaded micelles are formed by adding water or aqueous solutions containing the active species over the polymer film to rehydrate it. Using this method, anticancer drug-loaded micelles could be prepared by rehydration of poly(dimethylsiloxane)-*b*-poly(2-(dimethylamino)ethyl methacrylate) (PDMS-*b*-PDMAEMA) films with phosphate buffer saline solutions containing DOX (Car et al. 2014). In a similar manner, micelles loaded with either paclitaxel (PTX) or docetaxel (DTX) were prepared by rehydration with deionized water of mixed drug/polymer films containing amphiphilic diblock or triblock copolymers of 2-methyl-2-oxazoline (MOXA) and 2-alkyl-2-oxazoline (Seo et al., 2015).

When the overall water solubility of copolymer is very low, the dialysis method or solvent evaporation is recommended for micelle preparation. In both methods, the micelle-forming copolymer is first dissolved in a water-miscible organic solvent, such as ethanol (Sant et al., 2004), tetrahydrofuran (Toughrai et al., 2015; Wang et al., 2017), dimethylformamide (Hsiue et al., 2006), or N,N-dimethylacetamide (Quader et al., 2014). The solvent is subsequently removed either by dialysis against aqueous media or by evaporation. In some cases, the rapid addition of organic polymer solution in water or vice versa before dialysis induced the formation of micelles with a narrower size-distribution compared to the direct dialysis of organic polymer solution in the dialysis tube against water (Vangeyte et al., 2004). The O/W emulsion method is another strategy involving organic solvents for preparation of polymeric micelles. It is typically used for preparation of drug-loaded micelles, where the drug and copolymer are first dissolved in a volatile organic solvent, immiscible with water, such as chloroform, dichloromethane, or ethyl acetate (Ahmad et al., 2014, Danafar et al., 2017;

Tang et al., 2016). The O/W emulsion is formed thereafter by pouring the organic solution into an aqueous medium under stirring or sonication until complete evaporation of the organic solvent. In the O/W emulsion method, the drug is present inside of the organic droplets that are stabilized by copolymer. Upon solvent evaporation, the drug will be entrapped within the micellar core and thus, the drug loading efficiency increases compared to the dialysis method, where the hydrophobic drug may precipitate before its incorporation into the micelles (Sant et al., 2004). However, the dialysis is preferred over the emulsion technique, since less toxic solvents are used in the former method (Ahmad et al., 2014).

By tuning the type and strength of interactions between either the polymer constituents or polymers and drugs, the drug-loading ability can be considerably improved, while the sustained and controlled drug release from polymeric micelles is achieved through the modulation of their biodegradability or stimuli-responsiveness (Ahmad et al., 2014; Akimoto et al., 2014; Joglekar and Trewyn 2013; Kulthe et al., 2012; Li et al., 2014; Nakayama et al., 2014; Reddy et al., 2015; Talelli et al., 2015). Thus, based on the promising results from clinical trials, polymeric micelles might be considered as one of the most advanced self-organized polymer systems for drug delivery. For example, the biomedical formulation known as Genexol®, which is containing PTX-loaded micelles based on diblock copolymers containing PEG and poly(D, L-lactic acid) (PDLLA), has been approved for the treatment of nonsmall cell lung cancer, as well as for ovarian, breast and gastric cancers in Korea (Tang et al., 2016).

However, a performance which is still desired for polymeric micelles is the ability to deliver "on-demand" their payload (Tang et al., 2016). To achieve this milestone, intensive efforts have recently been made for developing smart micelles with stimuli-responsive properties, that is responsiveness to pH, temperature, redox agents, and light (Table 2.2), in order to increase the efficiency of intracellular drug delivery (Cao et al., 2014; Car et al., 2014; Gaspar et al., 2015; Lee et al., 2015; Talelli et al., 2015; Toughrai et al., 2015; Yuan et al., 2014a, 2014b; Zschoche et al., 2017). The most promising strategies involve either the use of polymers containing groups with variable ionization degrees depending on the environmental conditions or the introduction of stimuli-cleavable linkages within the polymer structure. Once the internal or external triggers are applied, the polymer architecture can be drastically changed leading to disintegration of self-organized assemblies, and subsequently, the release of payloads. In this respect, polymeric micelles based on PDMS-*b*-PDMAEMA block copolymers containing short hydrophilic blocks and showing excellent cell viability were proposed for intracellular delivery of DOX, a response to the decrease of pH from 7.4 to 5.5 (Fig. 2.2, Car et al., 2014). The DOX-loaded self-assembled micelles were taken up by endocytosis into the cells, while the acidic pH from endosomes/lysosomes caused the intracellular release of therapeutic cargo, and consequently the cell death.

Another interesting study reports the preparation of multifunctional triblock copolymer micelles based on poly(2-ethyl-2-oxazoline)–poly(L-lactide) copolymers grafted with bioreducible polyethyleneimine (PEtOXA–PLA-g–PEI-SS) for co-delivery of DOX and DNA (Gaspar et al., 2015). These polymeric vehicles combine the non-fouling properties of oxazolines with the complexation ability of PEI, the confining hydrophobic core provided by PLA, and the stimuli-cleavable disulfide linkage to confer biological stability, drug encapsulation, DNA complexation, and "on-demand" release. The DNA-loaded micelles penetrated in vitro into the tumor spheroid models and showed a higher gene expression compared to the non-reducible polymer systems. Moreover, the gene expression was detected up to 8 days after the intratumoral administration. Furthermore, both DOX and DNA were successfully co-encapsulated into these polymeric micelles with a high efficacy. The self-assembled systems showed a significant uptake and cytotoxicity in 2D cultures of cancer cells, indicating a high potential for future application in combinatorial DNA-drug therapy. Even more appealing is the development of multistimuli-responsive micelles (Table 2.3). Accordingly, triply responsive micelles for controlled release of insulin were reported as carriers sensitive to glucose, pH, and redox (Yuan et al., 2014a). The formation/dissociation of micelles containing the amphiphilic copolymer poly(ethylene oxide)-SS-poly(3-acrylamidophenylboronic acid) (PEO-SS-PAPBA) was favored by changes in the solution pH value, while the breakage of disulfide bonds was achieved after adding GSH. Moreover, the loaded insulin could be released by adding either glucose or GSH.

Polymeric micelles composed of amphiphilic block copolymers are also promising as diagnostic systems. The moieties involved in diagnostics can be covalently linked to the hydrophilic corona or non-covalently incorporated into the hydrophobic core. The resulting platforms can be used as particulate agents for diagnostic imaging, using scintigraphy, magnetic resonance imaging (MRI), or X-ray computed tomography (CT). Thus, the polymeric micelles can incorporate ^{111}In or ^{99m}Tc, Gd, or organic

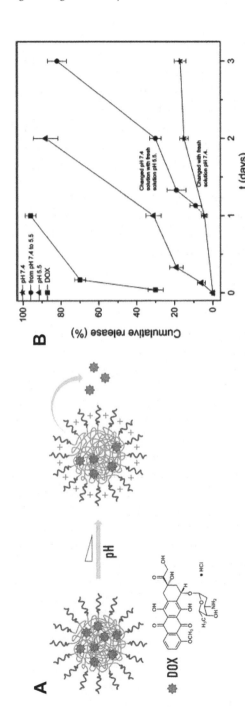

FIGURE 2.2 (A) Simplified representation of DOX release from PDMS-*b*-PDMAEMA micelles as response to the pH decrease. (B) Cumulative release of DOX from the self-assembled particulates based on block copolymers with the shortest hydrophilic block (five repeating units of DMAEMA). Reprinted with permission from Car, A., et al. pH-Responsive PDMS-b-PDMAEMA Micelles for Intracellular Anticancer Drug Delivery. *Biomacromolecules* **2014**, *15* (9), 3235–3245. © 2014 American Chemical Society.

iodine for use in scintigraphy, MRI, and CT, respectively (Reddy et al., 2015). Another interesting example is that of pH-responsive nanoparticle MRI contrast agents (CAs) for tumor detection (Craciun et al., 2017; Zhu et al., 2016). The lower extracellular pH that is surrounding the tumors can trigger the release of CA or amplify their signal.

Unfortunately, the dense structure and the absence of a confined aqueous space within their architecture impede the usage of polymeric micelles as advanced compartments for accommodation of biocatalysts, such as enzymes, and the further construction of nanoreactors. Moreover, despite the significant results obtained in the last decades, the therapeutic efficacy of polymeric micelle formulations is still an issue to be improved. Nevertheless, polymeric micelles can be mainly regarded as the most promising carriers for less soluble drugs and/or with serious side effects.

2.3.2 VESICLES

Polymer vesicles or polymersomes are also generated by self-assembly of amphiphilic copolymers in dilute solutions but, compared to micelles, have a spherical hollow architecture in which the aqueous cavity is enclosed by a polymeric membrane (Dinu et al., 2014; Discher and Ahmed, 2006). Similar to liposomes (vesicles formed by self-assembly of lipids), polymersome membrane consists of two hydrated hydrophilic layers situated on both the inside and outside of the hydrophobic middle part of the membrane. However, the membrane of a polymer vesicle is composed of species with higher molecular weight, leading to an enhanced mechanical stability compared to liposomes, but a reduced permeability (Discher and Ahmed, 2006; Guan et al., 2015). Consequently, this impermeable polymeric barrier is able to conceal the bioactive species encapsulated into their aqueous cavity and to protect them from environmental degradation. Segregation of the hydrophobic blocks of copolymers from the aqueous medium and their self-organization into a curved polymeric membrane is generally favored by the conical shape conformation of polymer chains. The vesicular morphology is controlled by the balance between hydrophobic and hydrophilic forces, and is supported by a hydrophilic content, f, between 10 and 35%, depending on the copolymer structure, while for other values of hydrophilic-to-hydrophobic ratios the formation of micelles, worms, or mixtures is favored (Discher and Ahmed, 2006; Guan et al., 2015; Meng and Zhong, 2011; Wu et al., 2014). A strong influence on morphology also has the molecular weight of each block. For example, when PDMS-*b*-PMOXA diblock copolymers with the same hydrophilic-to-hydrophobic ratio have been self-assembled in aqueous medium, the key parameter governing the final polymeric architecture was the molecular weight of hydrophobic block (Fig. 2.3, Wu et al., 2014). Thus, the morphology of self-assemblies changes from spherical micelles (Fig. 2.3C) to a combination of spherical micelles, worm-like micelles and small polymersomes (Fig. 2.3B), and subsequently to polymer vesicles (Fig. 2.3A), by increasing the length of hydrophobic block. Moreover, the membrane thickness of polymersomes has been shown to proportionally depend on the molecular weight of hydrophobic block (number of repeating units) (Itel et al., 2014).

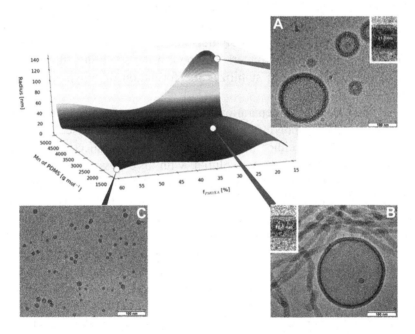

FIGURE 2.3 (See color insert.) 3D phase diagram of various amphiphilic diblock copolymers based on PDMS and PMOXA. The three white points with the corresponding cryo-TEM micrographs represent the supramolecular architectures formed by self-assembly of PDMS$_{65}$-*b*-PMOXA$_{14}$ (A), PDMS$_{39}$-*b*-PMOXA$_8$ (B), and PDMS$_{16}$-*b*-PMOXA$_7$ (C). As inset, the enlarged view of the membrane thickness for PDMS$_{65}$-*b*-PMOXA$_{14}$ (A; 21.3 nm), and PDMS$_{39}$-*b*-PMOXA$_8$, respectively (B; 16 nm), is inserted. Reprinted with permission from Wu, D., et al. Effect of Molecular Parameters on the Architecture and Membrane Properties of 3D Assemblies of Amphiphilic Copolymers. *Macromolecules* **2014**, *47* (15), 5060–5069. © 2014 American Chemical Society.

The aggregation number of polymersomes is typically higher than that of micelles, while their dimensions are spanning from nano- to microscale ranges (Meng and Zhong, 2011). However, the polymersome size can be easily controlled in a similar fashion to polymer micelles either by changing the macromolecular characteristics of polymers, such as composition, chemical structure, molecular weight, or by choosing the proper preparation methods and conditions (Bleul et al., 2015; Meng and Zhong, 2011; Warren and Armes, 2014; Wu et al., 2014; Zhang et al., 2012).

As mentioned above, in the previous section, the solubility of amphiphilic copolymers plays a key role in generation of supramolecular architectures, obtained by self-assembly, in particular for polymer vesicles. Therefore, the strategy for preparation of vesicles in aqueous conditions is specific for each polymer system. For example, the addition of small amounts of detergent (up to 1%) is commonly known to increase both the flexibility and permeability of polymersome membrane (Wu et al., 2014), while for amphiphilic copolymers with poor water solubility, the presence of detergents induces and stabilizes the formation of polymer vesicles (Marsden et al., 2010).

The main routes in preparation of vesicular architectures in aqueous media starting from polymer amphiphiles are generally based on dissolution of copolymer either in water or in a solvent, succeeded by the mixture of preformed polymer solution with an excess of water. Dissolution of copolymer in water avoids the use of potentially toxic organic solvents and can be achieved by direct dissolution, film rehydration, or electroformation. Direct dissolution is the easiest solvent-free method and involves the direct mixture of polymer with the aqueous solution, while the film rehydration technique consists first in dissolution of polymer in a volatile solvent that is afterwards removed upon evaporation resulting in a thin film, which is finally hydrated (Dinu et al., 2014; Gaitzsch et al., 2016; Tang et al., 2016). In both approaches, an external energy source is involved for a better diffusion, such as shaking, stirring, vortexing, or sonication, at desired temperature, but film rehydration offers more controlled conditions for generation of polymersomes compared to direct dissolution. A distinctive strategy is electroformation, which consists in hydration of the amphiphilic film under an oscillating electric field, leading to micrometer-sized polymer vesicles, also referred as giant vesicles (Wu et al., 2014). Nevertheless, a significant limitation of free-solvent methods is the broad size distribution of generated polymersomes, but this issue can be subsequently addressed by the repeated extrusion of vesicle dispersion through membrane filters with controlled pore sizes.

Another major route for polymersome preparation comprises the solvent-assisted methods, including kinetic and thermodynamic trapping, as well as double emulsion technique (Dinu et al., 2014; Gaitzsch et al., 2016). In these methods, polymeric vesicles are generated by dissolution of amphiphilic block copolymers in an organic solvent, which is successively hydrated with an aqueous medium. The hydration is done by either injecting the organic polymer solution into an excess of water, inducing phase inversion (kinetic trapping), or slowly adding the excess of water over the polymer solution (thermodynamic trapping), allowing the system to equilibrate (Dinu et al., 2014; Gaitzsch et al., 2016; Tang et al., 2016). Both of these techniques demand a solvent that is miscible with water, wherefore the name of "solvent-switch" or "phase inversion." Moreover, a controlled polymersome shaping can be promoted by careful selection of osmotic pressure and membrane permeability leading to predictable morphologies via out-of-equilibrium self-assembly (Rikken et al., 2016). This shape transformation was achieved by adding extra water to the polymersome dispersion, resulting in a series of shape transitions toward other vesicular architectures, such as discs, bowl-like vesicles, or stomatocytes, as result of deflation. These unusual morphologies were generated when amphiphilic block copolymers with a higher glass transition temperature (PEG-b-PS) were used (Rikken et al., 2016).

In a different approach, polymersomes are formed via "water-in-oil-in-water" (W/O/W) double emulsions in which the polymer is dissolved in a volatile immiscible organic solvent and used as a middle phase. Giant vesicles are fabricated using this method, which also offers the advantage of high encapsulation efficiency (Marguet et al., 2013). Despite the large variety of polymers that can be used in solvent-based techniques, their major drawback is the residual organic solvent that can be very difficult to be removed from the self-assembled structures. Since the solvent can interact with the polymersome membrane reducing its stability and denature biomolecules, such as proteins, enzymes, or siRNA, extensive purification is required to completely remove the remaining solvent traces through either evaporation or dialysis (Dinu et al., 2014; Gaitzsch et al., 2016; Marguet et al., 2013).

However, regardless the preparation method, the polymersome dispersions may have to be purified by size exclusion or dialysis to remove small molecules, such as detergents, non-encapsulated bioactive molecules, or even the disposal of undesirable polymeric 3D architectures, such as spherical and worm-like micelles, or larger aggregates (Bartenstein et al., 2016; Kita-Tokarczyk et al., 2005). Consequently, the specificity of each preparation method influences both the size of polymer vesicles and the encapsulation efficiency of bioactive molecules into their aqueous confined space.

Conventional methods for preparation of polymersomes are usually performed in dilute conditions (<1%), limiting the production of higher amounts of polymer vesicles. However, several techniques, including polymerization-induced self-assembly (PISA), centrifugation-induced self-assembly or microfluidics (Zhu et al., 2017), cannot only overcome this limitation, but they allow the formation of vesicular structures with more complicated architectures or monodispersed size distribution (Marguet et al., 2012a, 2012b). Recently, various polymeric nanostructures have been prepared through PISA at high solid contents and with a variety of surface functionalities (Warren and Armes, 2014). This method involves the use of hydrophilic monomers that can form hydrophobic polymers. In this manner, the morphology evolution of self-organizing objects from worm-like micelles to vesicles can be monitored and controlled by changing the hydrophilicity/hydrophobicity ratio during polymer synthesis. An interesting example is the self-assembly of poly(glycerol monomethacrylate)-*b*-poly(2-hydroxypropylmethacrylate), PGMA-*b*-PHPMA, during polymerization. Thus, PHPMA block becomes hydrophobic with the increase of its molecular weight, while PGMA block preserves its hydrophilicity (Warren and Armes, 2014). Moreover, a vesicle-to-worm (or vesicle-to-sphere) morphological transition was induced via dynamic covalent chemistry by the addition of 3-aminophenylboronic acid (APBA) to these polymersome dispersions (Deng et al., 2017a).

Polymer vesicles with a more complex morphology, such as vesosomes (polymersomes in polymersomes) can be prepared via simple centrifugation (Marguet et al., 2012a). This method holds a high potential for encapsulation of nano-sized objects in giant polymer vesicles or for preparation of artificial cells with "organelles" and "model cytoplasm" (Marguet et al., 2012b). A very effective approach for generation of giant vesicles with monodisperse size distribution is based on microfluidics (Zhu et al., 2017). Even though this method is mainly used for the preparation of liposomes, it might be used to form polymer vesicles with uniform size by tuning the size of droplet precursors.

A particular case of polymer vesicles is represented by PICsomes, also referred as PIC vesicles, which are hollow supramolecular structures, but their aqueous cavity is enclosed by a membrane formed by electrostatic interaction between either a pair of oppositely charged block copolymers or a charged block copolymer and an oppositely charged homoionomer (Kishimura, 2013; Koide et al., 2006; Sueyoshi et al., 2017; Takahashi et al., 2016). For example, vesicular architectures were obtained by mixing PEG-*b*-PAsp (anionic component) with PEG-*b*-PAsp(DET) (cationic component) in aqueous solution, where the hydrophobic domain of PICsome membrane is supported by the strong interaction between the oppositely charged polyelectrolyte counterparts (Koide et al., 2006). In this manner, the hydrophobic inter polyelectrolyte complex layer is sandwiched between two PEG hydrophilic layers. Recently, a very interesting self-organized PICsome system was formed by mixing a pair of anionic-neutral (AP) and cationic-neutral (MP) double-hydrophilic block copolymers (Takahashi et al., 2016). A reversible morphology transition between vesicles and spherical micelles can be induced by changing the mixing ratio between the polymer components. When both of the charged counterparts are nearly equimolar in solution, vesicular structures with a bilayer membrane are obtained, while the addition of either the anionic or cationic component into the PICsome dispersion induces electrostatic instability and shifts the morphology of nano-objects toward spherical micelles (Takahashi et al., 2016).

Potential application of PICsomes in biomedicine is promising due to their intrinsic semipermeable membrane and facile preparation strategy. Thus, the PICsome membrane is permeable to ions and small molecules but not to biomacromolecules, which can be encapsulated into their aqueous confined space. Consequently, these self-structured objects can be used either to design nanocarriers for transport and delivery of proteins and enzymes without the loss of their activity (Kishimura et al., 2007) and preparation of MRI CAs by encapsulation of commercial superparamagnetic iron oxide NPs (SPIONs) such as ferucarbotran (Resovist®) (Kokuryo et al., 2013), or to generate enzyme-loaded nanoreactors (Anraku et al., 2016; Chuanoi et al., 2014; Sueyoshi et al., 2017; Tang et al., 2017).

Nevertheless, polymersomes obtained by self-assembly of amphiphilic block copolymers are particularly appealing because of their ability to entrap both hydrophilic and hydrophobic species within their membrane and inside of the aqueous compartment (Balasubramanian et al., 2016;

Guan et al., 2015; Gunkel-Grabole et al., 2015; Hu et al., 2017; Martin et al., 2016; Mohammadi et al., 2017; Palivan et al., 2016). In this respect, polymer vesicles have been engineered as nanocarriers for delivery of anticancer drugs (Anajafi et al., 2016; Du et al., 2012; Figueiredo et al., 2016; Kumar et al., 2015; Liu et al., 2015a; Thambi et al., 2016), for cancer theranostics (Craciun et al., 2017; Liu et al., 2015b, Oliveira et al., 2012; Qin et al., 2015), for MRI by encapsulating quantum dots (Ye et al., 2014), for hyperthermia by loading magnetic NPs (Zhu et al., 2017), for gene therapy by delivering plasmid DNA (Li et al., 2006; Li et al., 2007), or for photodynamic therapy by generation of reactive oxygen species (Baumann et al., 2013; Baumann et al., 2014). Moreover, intensive efforts have been also focused on engineering smart stimuli-responsive polymer-somes (Table 2.2) that can react as response to an environmental stimulus and release their encapsulated cargo in a controlled manner (Cabane et al., 2011; Cabane et al., 2012; Cao and Wang, 2016; Che and van Hest, 2016; Deng et al., 2017a, 2017b; Hu et al., 2017). Once the stimulus is applied, the vesicular nanostructures can suffer various physical and/or chemical changes, including membrane fusion, swelling, disassembly, or polymer degradation, and finally leading to membrane disruption and release of the payload. In this respect, polymer vesicles containing hydrophobic self-immolative blocks, terminally modified with various stimuli-responsive capping groups, were used for combinational therapeutic delivery of camp-tothecin (CPT) and DOX (Liu et al., 2014b). Upon applying the proper stimulus, the end-caging moiety triggers the disintegration of polymer-somes resulting in drug release and controlled access of protons, oxygen, and enzymatic substrates. Cross-linked nanoreactors with pH-controlled enzymatic activity were formed using block copolymers based on PEG as hydrophilic block and a random hydrophobic block containing pH-respon-sive diethylaminoethyl methacrylate (DEAEMA) and photo-cross-linkable 4-(3,4-dimethylmaleimido) butyl methacrylate (DMIBM). The UV-irradiation of enzyme-loaded nanoreactors induced the cross-linking of DMIBM repeating units from the hydrophobic blocks, which allows the preservation of vesicular structure, while the membrane permeability is ensured by the protonation of pH-sensitive DEAEMA repeating units in acidic medium (Gräfe et al., 2014). However, their membrane perme-ability is nonspecific, and a strong pH change can significantly affect the enzyme activity. A very interesting strategy was proposed to induce simul-taneous membrane permeabilization and stabilization of polymer vesicles containing photo-labile carbamate-caged primary amines via the hydrophobicity-to-hydrophilicity transition of hydrophobic bilayer using the UV-initiated cross-linking process (Wang et al. 2014b). This UV-triggered cross-linking has been successfully used to accomplish the co-release of hydrophobic and hydrophilic species and the light-activated biocatalysis of enzyme-loaded nanoreactors (Wang et al. 2014b).

Polymersomes with promising potential for theranostic applications in MRI and targeted delivery of anticancer drugs have been recently engineered to improve the sensitivity of a T_1 MRI CA, as well as to enhance the efficiency of cancer chemotherapy (Liu et al., 2015b). The reported polymer vesicles consist of poly(L-glutamic acid)-b-poly(ε-caprolactone) block copolymers functionalized with folic acid (FA) or diethylenetriaminepentaacetic acid (DTPA) terminal groups. The DOX-loaded self-assemblies with a cancer-targeting outer corona and a Gd(III)-chelating inner corona exhibited an excellent antitumor activity, showing a slow release of their cargo in neutral conditions, but a burst release in acidic medium (Liu et al., 2015b). Another interesting example is that of polymersomes embedding ultrasmall SPIONs acting as negative T_2 MRI CAs and actuators for magnetic field induced DOX release (magneto chemotherapy) (Oliveira et al., 2012; Oliveira et al., 2013). Moreover, by surface functionalization with cell-specific targeting ligands (antibodies), the SPION-containing polymer vesicles were able to target cancer cells in vivo, in a bone metastasis model (Pourtau et al., 2013). These platforms can be used as particulate CAs in active targeting of tumors and maintain a high concentration at the targeted site, opening new perspectives in development of targeted tools for theranostics.

In addition to the intrinsic stimuli-responsive polymers, another approach for designing smart polymer vesicles is based on the selective confinement of naturally active biomolecules (proteins, enzymes, DNA, etc.), into these self-organizing architectures (Garni et al., 2017; Palivan et al., 2016). Such biomolecules act as specific gates or transporters for transferring the substrate and products inwards and outwards of the polymersome compartment. However, proper conditions must be selected for both encapsulated and reconstituted species to preserve their activity and simultaneously the functionality during encapsulation. Furthermore, supplementary requirements are necessary when more than one bioactive species are confined in the self-organized architectures due to the higher complexity of these systems (Klermund et al., 2017; Tanner et al., 2011b).

Even though the polymersome membrane is acting as a barrier with low permeability for water, ions, and neutral molecules, it can be tuned by insertion of biopores and ionophores (Lomora et al., 2015a, 2015b, 2015c), membrane proteins (Einfalt et al., 2015; Grzelakowski et al., 2015; Itel et al., 2015; Langowska et al., 2013; Palivan et al., 2016; Tanner et al., 2011b; Zhang et al., 2016a), and polysaccharide-based cell receptors (Najer et al., 2014; Najer et al., 2015). Insertion and functional reconstitution of all these biomolecules are supported by the high flexibility and fluidity of polymersome membrane as regard to the conformational freedom and mobility of polymer chains within the membrane (Itel et al., 2014). In this respect, the conformational changes offer both an increased stability to the polymer membrane and adaptability to allow the incorporation of biomolecules (Itel et al., 2014; Itel et al., 2015). However, the membrane dynamics also depends on the block copolymer structure and composition, as well as the length of each block. Thus, diblock copolymers showed a higher diffusion compared to triblock copolymers, forming a bilayer-like structure with slight interdigitation and weak entanglements, while the triblock copolymer chains rearrange in a combination of bent U-shape and stretched I-shape conformations, which reduces their overall mobility (Itel et al., 2014).

The permeability of polymersome membrane is crucial when these self-assembled structures are designed to obtain supramolecular architectures with complex functionalities, such as nanoreactors and nanodevices for biosensing and diagnostics or are aimed to mimic cell membrane and cellular compartments. Since biomolecules, such as membrane proteins, pores, and ion transporters play a key role in optimum cell activity, the biomimetic approach involves the reconstitution of these active species into the artificial membrane of polymersomes to achieve a similar functionality as in the biological membranes (Beales et al., 2017; Buddingh' and van Hest, 2017; Garni et al., 2017; Godoy-Gallardo et al., 2017; Palivan et al., 2016; York-Duran et al., 2017). In this respect, intelligent polymersomes confining active compounds have been engineered to act as nanoreactors (Baumann et al., 2013; Baumann et al., 2014; Dobrunz et al., 2012; Heinisch et al., 2013; Langowska et al., 2013; Klermund et al., 2017) or as artificial organelles (Tanner et al., 2011a; Tanner et al., 2013). Enzyme-loaded nanoreactors able to produce antibiotics "on demand" in solution are an interesting example of smart self-organized systems, which can locally synthesize an antibiotic by feeding them with substrates with no antibacterial activity (Langowska et al., 2013). Recently, nanoreactors with a pH-triggered activity have been engineered to transport the substrate molecules through the membrane in a stimuli-responsive manner using polymer membranes equipped with channel proteins, OmpF (Einfalt et al., 2015). The activity of enzyme encapsulated into nanoreactors based on amphiphilic triblock copolymers containing PMOXA as hydrophilic blocks and PDMS as hydrophobic block, $PMOXA_6$–b–$PDMS_{44}$–b–$PMOXA_6$, was successfully tuned by reconstitution of chemically modified OmpF biomolecules, which can act as pH-controlled "gates" for the enzyme-loaded polymersomes (Fig. 2.4). The pH-responsive molecular cap that was chemically attached to the channel porin opens the protein "gates" when the pH decreases, allowing the in and out passage of substrates and products of the reaction catalyzed by horseradish peroxidase (HRP) (Einfalt et al., 2015).

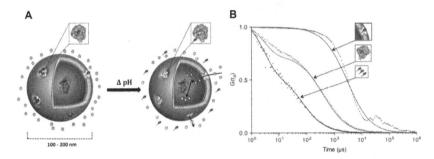

FIGURE 2.4 (See color insert.) Schematic representation of a nanoreactor with a pH-triggered enzymatic activity controlled by chemically modified porin "gates" reconstituted in a polymersome membrane based on $PMOXA_6$–b–$PDMS_{44}$–b–$PMOXA_6$ triblock copolymer. (A) The pH change induces the cleavage of molecular cap (green dots) opening the "gate" for both the entrance of substrates (red dots) and release of the products resulted from enzymatic reaction (yellow dots). (B) Autocorrelation curves of pH-sensitive fluorescent cap, cyanine5-hydrazide (Cy5-hydrazide), in phosphate buffer (black), Cy5-modified OmpF in 3% octyl-glucopyranoside micelles (red), and polymersomes with reconstituted Cy5-modified OmpF (blue). The significant increase of diffusion time for micelles and polymersomes containing the capped porin compared to the freely diffusing dye indicates the chemical modification of channel protein and its insertion into the polymer membrane. Reprinted with permission from Einfalt, T., et al. Stimuli-triggered Activity of Nanoreactors by Biomimetic Engineering Polymer Membranes. *Nano Lett.* **2015,** *15* (11), 7596–7603.© 2015, American Chemical Society.

Polymer membranes based on PMOXA-b-PDMS-b-PMOXA were successfully involved in another interesting biomimetic approach comprising the insertion of a small bio pore, gramicidin, which allows the diffusion of protons, or Na^+ and K^+ ions through it, while the polymersome architecture remains unaffected (Fig. 2.5A, Lomora et al., 2015c). To this end, pH-, Na^+-, and K^+- sensitive dyes were encapsulated into the aqueous cavity of polymersomes, and the proper insertion and functionality of gramicidin were assessed by the changes in fluorescence of encapsulated dyes resulting from the exchange of proton or ions across the membrane upon biopore insertion. Thus, there were no changes in fluorescence intensity of the encapsulated dye in the absence of biopore, while the insertion of gramicidin resulted in an increase of fluorescence upon addition of monovalent cations, which confirms the passage of both Na^+ and K^+ ions across the polymersome membrane (Fig. 2.5B,C, blue curves).

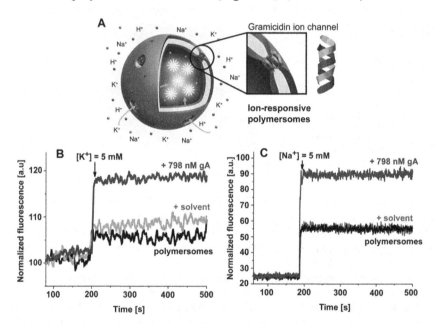

FIGURE 2.5 (A) Polymersomes designed for selective permeability of protons and monovalent cations upon insertion of gramicidin into polymersome membranes formed by self-assembly of PMOXA–*b*–PDMS–*b*–PMOXA amphiphilic block copolymers, (B) fluorescence intensity of polymersomes encapsulating dyes sensitive for K^+ ions, and (C) Na^+ ions, respectively, upon addition of cations (black curves), upon addition of cations in the presence of DMSO/EtOH mixture used as negative control (grey curves), and upon addition of cations in the presence of DMSO/EtOH mixture, after insertion of gramicidin (blue curves). Reprinted with permission from Lomora, M., et al. Polymersomes with Engineered Ion Selective Permeability as Stimuli-responsive Nanocompartments with Preserved Architecture. *Biomaterials* **2015c**, *53*, 406–414. © 2015 Elsevier.

Furthermore, polymersome membranes engineered to allow the transport only for Ca^{2+} ions were obtained by insertion of an ionophore, ionomycin, into the artificial membranes also based on PMOXA-b-PDMS-b-PMOXA block copolymers (Fig. 2.6A, Lomora et al., 2015a, 2015b). In a similar fashion with the gramicidin approach, specific fluorescence markers were encapsulated into the inner cavity of polymersomes to assess the selective membrane permeability toward Ca^{2+} ions after insertion of ionomycin. In this respect, stable calcium-sensitive supramolecular systems were designed by insertion of the ion transporter with a size of only about 2 nm into artificial membranes with thicknesses up to about nine times larger compared to the size of ionophore (Lomora et al., 2015a, 2015b). The dye-loaded 3D supramolecular architectures formed by self-assembly have spherical structures and no changes in their morphology or size have been observed upon addition of ionomycin or/and Ca^{2+} ions (Fig. 2.6B,C). Moreover, there were no differences in fluorescence intensity of the encapsulated dye in the absence of ionophore and only when ionomycin was inserted into the membrane, the fluorescence intensity significantly increased (Fig. 2.6D, blue curve), which demonstrates the proper insertion and functionality of ion transporter.

These strategies showed the possibility to properly insert small biopores and to assess their functionality in artificial membranes with thicknesses much larger than the size of biomolecule if the polymersome membrane has the suitable properties. However, the rearrangement of copolymer chains and the adjustment of membrane thickness to the size of inserted biomolecule are limited to a certain value. Thus, gramicidin was inserted and remained functional in PMOXA-b-PDMS-b-PMOXA polymer membranes with thicknesses up to 12.1 nm, while the ion transport by ionomycin was not affected when inserted in up to 13.4 nm thick membranes (Lomora et al., 2015a, 2015b, 2015c). Consequently, the biomimetic approach of inserting biopores with specific selectivity into polymer membranes offers an interesting opportunity to engineer and develop biomembrane mimics, biosensors, or artificial organelles.

Up to now the design of smart biomimetic systems with 3D architectures has been focused on generation of self-organized hybrid supramolecular structures with specific properties (Palivan et al., 2016). However, the development of synthetic compartments with a more complex functionality and spatial organization is even much more appealing for fabrication

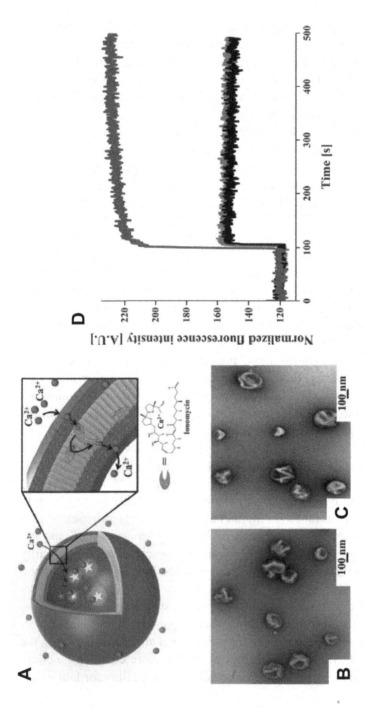

FIGURE 2.6 (A) Polymer vesicles with Ca²⁺-selective permeability engineered by insertion of ionomycin into polymersome membranes, (B) TEM micrographs of polymersomes encapsulating the Ca²⁻ sensitive dye in the absence, and (C) in the presence of inserted ionomycin and added CaCl₂; fluorescence intensity of dye-loaded polymersomes upon the addition of Ca²⁺ ions (black curve), in the presence of EtOH, used as negative control (red curve), and upon addition of Ca²⁺ ions in the presence of EtOH, after insertion of ionomycin (blue curve). Reprinted from Lomora, M., et al. Selective Ion-permeable Membranes by Insertion of Biopores into Polymersomes. *Phys. Chem. Chem. Phys.* **2015b,** *17* (24), 15538–15546.Open access: http://pubs.rsc.org/en/Content/ArticleLanding/2015/CP/c4cp05879h#!divAbstract.

of artificial platforms able to mimic the sophisticated structures existent in cells. One of these fascinating approaches is to mimic the connection of specific organelles by membrane contact sites, which play a key role in cell functionality. In this respect, a biomimetic strategy to self-organize polymersomes into clusters with a controlled spatial topology has been designed (Liu et al., 2016). Thus, polymersomes generated by self-assembly of amphiphilic block copolymers were selected as mimics of cell compartments and functionalized on their surface with single-stranded DNA (ssDNA) using "click" chemistry. Taking the advantage of polymersome mechanical stability and using the DNA hybridization, self-organized clusters with spatial supramolecular topologies were generated by specific recognition between complementary ssDNA-functionalized polymersomes. Moreover, the double-stranded DNA rigid bridge between polymer vesicles serves as spacer between the polymersomes, preventing their fusion. The interconnected supramolecular structures hold a high potential in translational medicine, favoring the cascade reactions between different compartments, and consequently, sustaining the cell signaling or the intracellular interactions.

Another exciting and challenging approach in designing smart 3D architectures with potential applications in biomedicine has recently attracted a great interest. This strategy consists of engineering synthetic micro/nanomotors with "on demand" self-propulsion by converting the energy of various physical or chemical triggers into motion (Wu et al., 2016). Biodegradable stomatocyte nanomotors formed by self-assembly of PEG-b-PCL and PEG-b-PS, entrapping Pt NPs, were designed to act as potential vehicles for DOX delivery. Moreover, the self-assembled motors can move in a unidirectional manner by conversion of hydrogen peroxide, used as chemical fuel, into motion (Tu et al., 2017a). A better control over the movement of these self-propelled vehicles was achieved using a temperature-responsive controlling mechanism. To this end, polymer brushes based on PNIPAM were chemically grown onto the surface of stomatocytes, which can control the access of hydrogen peroxide into the stomatocyte opening by changing the temperature, and accordingly, can tune the vehicle movement (Tu et al., 2017b).

In conclusion, these smart hybrid architectures are expected to offer new perspectives on designing and developing active biomimetic complex systems able to respond "on demand" to the physiological changes.

2.3.3 LbL CAPSULES

Capsules are hollow spherical self-organized structures with relatively large sizes (typically >1 μm), which consist of a polymer shell and a hollow inner space (Costa et al., 2015). Numerous methods have been developed for designing polymer capsules, but the most popular technique to generate the polymeric shell is through the electrostatic interaction between oppositely charged polyelectrolytes (Costa et al., 2015; Shutava et al., 2012; Guzman et al., 2017). However, in order to distinguish these 3D architectures from the PIC vesicles, this section presents only polymer capsules formed by consecutive adsorption of positively and negatively charged polyelectrolytes onto sacrificial colloidal templates using the layer-by-layer (LbL) deposition technique. In this approach, the polymer shell is formed in aqueous media around a pre-formed template particle that is subsequently removed, resulting in an empty polymeric shell (Tang et al., 2016; Zhao et al., 2017). Generally, the thickness of multilayered polyelectrolyte shell can be controlled by the number of adsorbed polyelectrolyte layers. In this respect, the solid core is removed by chemical dissolution when the desired thickness is reached, using an appropriate agent that is not affecting the shell integrity. Common templates involved in fabrication of polymer capsules include $CaCO_3$ particles, which are dissolved in ethylenediaminetetraacetic acid (EDTA), PS latex particles removed by THF, silica particles leached by HF, and melamine formaldehyde resin particles dissolved by HCl (Guzman et al., 2017; Liu et al., 2017; Zhang et al., 2016b). The classical approach involves several intermediate steps of cleaning and centrifugation for the removal of nonadsorbed polyelectrolyte chains, in order to avoid the formation of inter polyelectrolyte complexes. However, this technique is challenging when the template microparticles are replaced by NPs. This issue can be overcome using other methods for separation of polyelectrolyte in excess, such as filtration-based methodologies (Voigt et al., 1999). Another interesting technique used for preparation of LbL capsules is based on microfluidic chips. However, the use of expensive instrumentation and the limitations regarding the optimization of LbL process for each particular system are the major drawbacks of this method (Elizarova and Luckham, 2016). Nevertheless, the LBL technique is one of the most promising strategies for an effective and cheap preparation of multifunctional capsules as carriers for drugs, enzymes, or cells (De Crozals et al., 2016; Huang et al., 2012;

Jiang et al., 2017; Liu et al., 2014a; Marturano et al., 2017; Shutava et al., 2012). Moreover, the facile fabrication strategy, a permeability controlled by adjusting the multilayer thickness and porosity, and the broad variety of polyelectrolytes are several advantages which endow the LbL capsules with a great potential for delivery of peptide and protein-based drugs (De Crozals et al., 2016; Guzman et al., 2017). In addition, generation of LbL capsules that can adjust their shell permeability as response to an applied stimulus (Table 2.2), and consequently control the release of encapsulated cargo as response to various triggers, such as pH, temperature, salt concentration, and so forth, was also possible by selecting the appropriate pairs of polymers (Costa et al., 2013; Wuytens et al., 2014).

An interesting strategy to obtain nano-sized capsules with controllable shell structure and thickness is to design core–shell hybrid NPs by surface-initiated ATRP (SI-ATRP) (Tang et al., 2016). In this approach, silica NPs are commonly used as templates, while the polymer shells that are formed via polymerization on the surface of NPs are usually stabilized by cross-linking before the template etching. RAFT polymerization is another polymerization process that can be used in addition to ATRP for grafting polymers on silica NPs. Thus, nanocapsules with sizes ranging from 450 to 900 nm have been obtained using the RAFT polymerization approach, whose shell thickness is governed by controlling the molecular weight of the grafted copolymer (Huang et al., 2011). The same route was used to synthesize smart nanocapsules based on DEAEMA as pH-sensitive monomer and disulfide containing monomer to cross-link the shell membrane (Huang et al., 2012). The membrane permeability was tuned by adjusting the pH value and using dithiol cross-linkers with different lengths. Thus, the nanocapsules showed a significant swelling from 560 nm at pH 8.0 to 780 nm at pH 4.0 when the longest cross-linker was used. Moreover, the presence of disulfide bonds allowed the degradation of capsules with GSH as reducing agent. In addition, myoglobin-loaded nanocapsules with a pH-controlled enzymatic activity were prepared by enzyme post-encapsulation at pH 4.0 and tuning the membrane permeability by changing the pH between 6.0 and 8.0. Therefore, the capsules generated by this robust approach are expected to have a high potential in the biomedical field, especially for delivery of bioactive substances, reducing the loss of the payload and increasing the efficiency. Furthermore, the surface of orthopedic devices can be modified with smart LbL nanocapsules encapsulating antibiotics or anti-inflammatory drugs, which can release their cargo "on demand" to prevent the implant failure and rejection (Costa et al., 2015). Additionally, functionalization of their surface with suitable markers can be very useful in specific targeting or biosensing applications, such as cancer theranostics, while keeping the healthy cells not affected.

2.3.4 NANOPARTICLES

NPs are the most stable and tight spherical supramolecular architectures compared to other self-organized structures and are more easily to be prepared (Tang et al., 2016). The self-assembly method is one of the most extensively used techniques for fabrication of NPs. This method consists of controlled aggregation of hydrophilic or amphiphilic polymers by several physicochemical interactions, such as hydrophobic or electrostatic interactions, hydrogen bonding, stereo complexation, or supramolecular chemistry (Li et al., 2017). In this context, a smart NP platform based on self-assembled, reduction-responsive amphiphilic graft copolymers entrapping hydrophobic guest molecules were recently reported (Najer et al., 2016). The reduction-responsive nanoparticles with mean diameters of about 30–50 nm were successfully prepared using the thiol-disulfide exchange reaction between the thiolated hydrophilic block (PMOXA) and pyridyl disulfide (PySS) functionalized hydrophobic block (PCL) (Fig. 2.7A–C, Najer et al., 2016).

The self-assembled NPs were stable in cell media at body temperature and proved to be non-toxic up to a dosage of 1 mg/mL, while a fast and complete disassembly was observed in a physiological reducing environment. Moreover, the reduction-sensitive NPs delivered DOX to HeLa cells, and a metabolically unstable antimalarial drug (serine hydroxymethyltransferase inhibitor) to red blood cells infected with *Plasmodium falciparum* parasites, with a higher efficiency compared to the non-sensitive drug-loaded NPs. The encouraging results obtained with these responsive self-assembled NPs, as well as the biodegradability of PCL block and the protein-repellant PMOXA block, recommend them as promising smart drug nanocarriers (Najer et al., 2016). A similar strategy was used to design peptide-based NPs for co-delivery of DOX and plasmid DNA (Sigg et al., 2016b). In contrast to another previously reported peptide-based co-delivery system (Han et al., 2013), the sizes of these reductive-sensitive peptide NPs correspond to the values required for intravenous application, while their highly reduction sensitive architecture allows the release co-loaded payloads into the cytosol with a great efficacy (Sigg et al., 2016b). Another interesting example that supports the use of NPs as efficient stimuli-responsive platforms consist of supramolecular assemblies based on asymmetric PEG-*b*-PMCL-*b*-PDMAEMA copolymers (Vasquez et al., 2016). Key parameters, such as the hydrophilic to hydrophobic ratio, NP size, and the nature of loaded proteins, were shown to control the localization and attachment efficiency of bovine serum albumin (BSA) and acid sphingomyelinase (ASM). The outer shell containing predominantly PDMAEMA blocks allowed the pH-triggered release of both proteins in acidic conditions with a high efficiency.

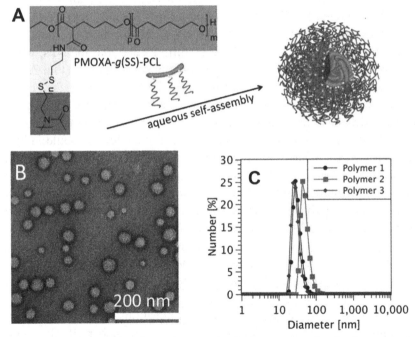

FIGURE 2.7 (See color insert.) (A) Chemical structure of reduction-responsive amphiphilic graft copolymer, PMOXA-*g*(SS)-PCL, and the schematic representation of NPs self-assembled in aqueous medium; (B) TEM micrographs of NPs based on PMOXA-*g*(SS)-PCL copolymers; and (C) hydrodynamic diameter of NPs and their number distribution in phosphate buffer determined by DLS. Reprinted with permission from Najer, A., et al. An Amphiphilic Graft Copolymer-based Nanoparticle Platform for Reduction-responsive Anticancer and Antimalarial Drug Delivery. *Nanoscale* **2016,** *8* (31), 14858–14869. © 2016 Royal Society of Chemistry.

The type and density of charged moieties are also important factors affecting the encapsulation efficiency and stability of polyelectrolyte-based NPs engineered for biomedical applications. However, the high toxicity of positively charged polymers and the burst release of the payload are the major limitations of these systems. Nevertheless, a versatile strategy has been recently proposed, which is based on preparation of photo-responsive NPs with low cytotoxicity, high stability, and slow release profile (Dinu et al., 2016). To this end, photo-cleavable 2-nitrobenzyl moieties were attached as pendant groups onto PDMS-b-PDMAEMA diblock copolymers (ABQn), and then self-assembled into NPs. The positively charged photo-responsive ABQn copolymers could undergo a cationic to zwitterionic structural transformation when exposed to UV light. The self-assembled NPs based on ABQn were nontoxic for HeLa cells both before and after UV irradiation up to a dose of 300 μg mL^{-1} (Fig. 2.8A), compared to the non-sensitive propyl-quaternized NPs (ABQPn). Sulforhodamine B (SRB), used as model drug, was successfully loaded into the NPs, while its slow release was controlled in a photo-responsive manner (Fig. 2.8B), being governed only by the dye diffusion (Fig. 2.8C).

Besides drug delivery, stimuli-responsive NPs can be very good candidates as MRI CAs for tumor detection (Craciun et al., 2017). The lower pH of the extracellular environment surrounding the tumors could induce the release of CA loaded into pH-responsive NPs or amplify their signal. A relevant example is that of Ca$_3$(PO$_4$)$_2$-containing NPs comprising a PEG shell used for modulating the "off–on" contrast in the microenvironment of tumor tissues (Mi et al., 2016). The Mn^{2+} species that are confined into the NPs are released by disintegration of NPs at a low pH and can bind to the proteins causing the contrast enhancement. In this manner, the engineered NPs can selectively mark the solid tumors, sensing the hypoxic regions, and detecting the metastatic tumors. The reductive potential of extracellular medium surrounding the tumors can be considered when CA-loaded NPs are designed for MRI applications. In this regard, an appealing strategy was developed for co-assembly of hep-PDMS copolymers into NPs with entrapped Gd^{3+} species and covered by reduction-sensitive amphiphilic peptides (Sigg et al., 2016a). The increase of CA relaxivity in reductive conditions supports these polymeric NPs as effective MRI agents for the selective targeting of solid tumors.

Even though significant steps have been made in development of stimuli-responsive NPs, the design of self-organized polymeric architectures showing both multifunctionality and multi-stimuli responsiveness is still challenging especially for emerging applications such as theranostics, which combines the capabilities for targeting therapy and diagnostics in a single polymer platform.

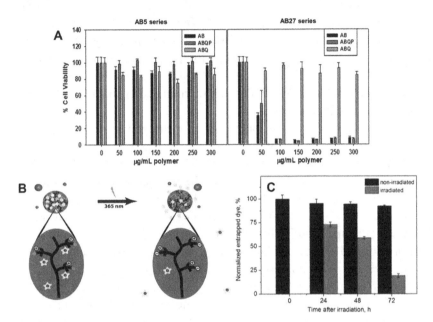

FIGURE 2.8 **(See color insert.)** (A) HeLa cell viability vs. polymer concentration after 24 h for both quaternized and non-quaternized NPs; (B) schematic representation of the photo-triggered release mechanism for the quaternized self-assembled NPs loaded with anionic payload; and (C) the content of entrapped dye (%), evaluated by fluorimetry, at various time points of dye release from the irradiated ABQ27 NPs compared to non-irradiated NPs. Reprinted with permission from Dinu, I. A., et al. Engineered Non-toxic Cationic Nanocarriers with Photo-triggered Slow-release Properties. *Polym. Chem.* **2016**, *7* (20), 3451–3464. © 2016 Royal Society of Chemistry.

2.3.5 NANO- AND MICROGELS

Spherical nano-/microgels (NMGs) are physical or chemical cross-linked 3D polymer networks able to absorb large quantities of water (Molina et al. 2015; Tahara and Akiyoshi 2015; Zhang et al. 2015b). A broad variety of techniques, including emulsion or inverse mini-emulsion polymerization, precipitation polymerization, and "one step" polymerization have been developed to prepare spherical NMGs (Tang et al., 2016; Wani et al., 2014). The inverse mini-emulsion approach is based on the water-in-oil (W/O) heterogeneous polymerization process, where template nano- or microdroplets are formed by oil-soluble surfactants (mainly SDS) in a continuous organic phase (Fundueanu et al., 2016; Li et al., 2015; Sarika et al., 2015). According to this method, polymeric NMGs can be prepared by free radical copolymerization of hydrophilic monomers in the presence of di- or multi-functional cross-linkers. In this context, thermo-responsive microgels based on P(NIPAM-*co*-NIPMAM) were generated for pulsatile release of dexamethasone (DEX) under thermal cycling operation between 32 and 38°C (Fundueanu et al., 2016).

Another commonly used approach for the preparation of NMGs, especially for thermo-sensitive ones (e.g., NIPAM-based nanogels) is the precipitation technique (Begum et al., 2016; Plamper and Richtering, 2017; Molina et al., 2015; Tang et al., 2016; Yang et al., 2016). In this method, the growing NIPAM chains collapse when they reach a critical length and the polymerization temperature is above the LCST of resulting polymers (Plamper and Richtering, 2017). Consequently, precursor particles are formed during nucleation step, which is continuously growing by (1) addition of monomers; (2) adsorption of polymer chains onto the surface of existing particles; or (3) aggregation of precursor particles. By adding an appropriate cross-linker, the assembled polymer chains are stabilized in colloidally stable supramolecular structures, which can considerably swell when the temperature is reduced below LCST. For example, dual-responsive hybrid nanogels, consisting of hydroxypropyl cellulose and PAA, and loaded with CdSe quantum dots were obtained using the abovementioned strategy (Wu et al., 2010). These nanogels hold a high potential as polymeric carriers for optical sensing of pH changes, cancer cell imaging, and pH-triggered drug delivery, which can open new opportunities for theranostics.

"One-step" polymerization in homogeneous aqueous solution has been also used in preparation of NMGs (Sanson and Rieger, 2010). However, highly diluted solutions of di- or multi-functional monomers are required in this approach to avoid the formation of macroscopic gels. By introduction of –SS– linkages into difunctional phosphate monomers, reduction-responsive NMGs have been obtained (Xiong et al., 2012). The release of an entrapped antibiotic, vancomycin, from these nanogels was significantly accelerated by the presence of alkaline phosphatase or DTT. pH-responsive nanogels with high drug loading capacity and efficiency for DOX have been also obtained by copolymerization of amino acids with ionizable side groups, followed by loading of ionizable drug into the nanogel core by electrostatic interactions (Shi et al., 2012).

Dual- and multistimuli-responsive NMGs have recently received an increasing attention for their potential application in biomedicine and pharmaceutics (Table 2.3). In this regard, the introduction of pH-cleavable bonds into the network of NMGs resulted in a large variety of degradable NMGs for applications in drug delivery (Jiang et al., 2014; Molina et al., 2015; Yang et al., 2016; Yu et al., 2006; Zhan et al., 2015; Zhang et al., 2015a). An interesting study was focused on preparation of redox-sensitive/degradable nanogels based on NIPAM and AA cross-linked by N,N'-bis(acryloyl)cystamine (Yang et al., 2016). These nanogels showed a dual pH/redox-triggered delivery of DOX in vitro and in tumor cells, where the lysosomal pH (pH 4.5) and cytosolic reduction (GSH) conditions favored the release of anti-cancer drug. Another approach based on adenosine-5'-triphosphate (ATP) and pH dual-responsive nanogels have been proposed for the intracellular delivery of anticancer drugs (Zhang et al., 2015a), in which MTX rapidly released in the presence of ATP and in acidic pH conditions. Multistimuli-responsive NMGs with increased stability have been also engineered and used for the selective release of co-encapsulated hydrophobic and hydrophilic payloads (Cao et al., 2016). In this respect, nanogels based on pH- and thermo-responsive hydrophilic PDMAEMA, and a hydrophobic photo-cleavable ONB-based cross-linker, were synthesized. The hydrophobic cargo, Coumarin 102, was physically entrapped within the nanogels, while the hydrophilic Rhodamine B was chemically attached to PDMAEMA via redox-cleavable disulfide bonds. These dual-loaded multi-responsive nanogels released their hydrophobic payload in response to temperature and pH changes, or UV light exposure, while the hydrophilic molecules were released in the presence of DTT (Cao et al., 2016).

Besides the drug delivery applications, NMGs also showed a great potential for bioimaging (Asadian-Birjand et al., 2016; Molina et al., 2015; Wang et al., 2014a), and tissue engineering (Jiang et al., 2014). A relevant example is that of poly(NIPAM-AM) nanogels entrapping fluorescent carbon NPs (Wang et al., 2014a). The resultant hybrid nanogels showed multiple functionalities combined into one polymeric platform, providing them abilities for light/thermo-responsive drug delivery, cell imaging, and fluorescent temperature sensing. These multifunctional hybrid nanogels endowed with unique responsive properties hold a promising potential for the development of new multifunctional polymeric materials for biomedical applications.

2.4 REQUIREMENTS AND LIMITATIONS FOR USING POLYMER MATERIALS IN BIOMEDICAL APPLICATIONS

A complex, but comprehensive, set of requirements are necessary to be fulfilled when designing a polymer material for biomedical applications. Generally, the criteria used in engineering polymeric architectures with applications in biomedicine must take into account the quality, efficacy, and safety of these materials to protect public health, according to the protocols indicated by the International Council for Harmonization of Technical Requirements for Pharmaceuticals for Human Use (ICH, http://www.ich.org/home.html). Moreover, the final products and formulations must be approved by national and/or international regulations and fulfill the standards demanded by US Food and Drug Administration (FDA), Health Canada (HC), European Medicines Agency (EMA) or Japanese Pharmaceuticals, and Medical Devices Agency.

In any respect, the overall prerequisites for an efficient polymer material, which is intended to be used in biomedical applications, should include: (1) an efficient loading of the bioactive molecules within the polymeric self-organized architecture, (2) the stability of self-assembled objects in aqueous environment, as well as the protection of the confined biomolecules from degradation, (3) a good accessibility of substrate-containing fluids to the bioactive molecules that are entrapped/encapsulated into the polymeric particulates, in order to ensure their proper functionality, and (4) a prolonged blood circulation time in order to avoid their rapid clearance, to allow the accommodated active agent reaching the targeted site, and to accumulate at a certain level for an efficient drug release or better signaling of the changes in local environment. Additionally, there are supplementary requisites that may have to be achieved in respect to both the polymer properties (such as biocompatibility, biodegradability, toxicity, flexibility, and functionality) (Markovsky et al., 2012), and the features of self-organized structures, such as size (Brinkhuis et al., 2012; Dash et al., 2010), morphology (Venkataraman et al., 2011), surface charge and chemistry (Chuard et al., 2017; Dash et al., 2010), stimuli-responsiveness (Cao and Wang, 2016; Tang et al., 2016), and mechanical parameters (Markovsky et al., 2012). Furthermore, when a biologically active molecule is entrapped/ encapsulated within the ordered domains of the polymeric 3D architecture to form hybrid-structured materials, the criteria of designing structured smart materials may also be adjusted depending on the specifications of loaded active molecule.

Biocompatibility and biodegradability of particulate-forming components play an essential role in designing polymeric materials for biomedical applications. Moreover, the products of degradation should have not at all or only insignificant side effects, and do not accumulate into the body (Balasubramanian et al., 2010; Balasubramanian et al., 2011; Tanner et al., 2011a). Nevertheless, the polymer biodegradability needs to be delayed in order to ensure the protection of bioactive agents concealed within their organized structure, even after the cellular uptake. A special concern may still have to be addressed to the cell toxicity, especially after their loaded cargo is released, even though a plenty of synthetic amine-containing polymers, including PEI, PLLys, poly(amidoamine) (PAMAM), and PDMAEMA were extensively studied for delivery of bioactive molecules (Kim et al., 2016).

In addition, functionality is another key parameter to be taken into account and consists in a distinct response triggered by the presence of a specific stimulus. To address the requirement, composition of the supramolecular architecture is adjusted depending on the biomedical application and the expected outcome. This could be tuned either by synthesizing or selecting polymer constituents with a suitable stimuli-responsiveness (Cao and Wang, 2016; Car et al., 2014; Dinu et al., 2016; Najer et al., 2016; Tang et al., 2016, Vasquez et al., 2016), or by selecting proper confining/gating approaches for transport and permeabilization (Dinu et al., 2015; Einfalt et al., 2015; Garni et al., 2017; Lomora et al., 2015a, 2015b, 2015c).

The size and morphology of self-organized polymeric structures also play a major role on biodistribution, circulation time, and interaction of particulates with plasma proteins and cell membranes (Duan and Li, 2013; Gao and He, 2014; von Roemeling et al., 2017). Furthermore, the minimization of interactions with serum and tissue proteins, known as "stealth" behavior, which is required to prevent the fast degradation, prolong the stability in biological media, and improve the efficacy was achieved by surface functionalization with PEG or other polymers with protein repellent properties (Cavadas et al., 2011; Wattendorf and Merkle, 2008; Wilson et al., 2017). Moreover, surface modification with a large variety of functional groups allows the conjugation of different biomolecules, such as growth factors, peptides, antibodies, and carbohydrates. This strategy is essential when active targeting is demanded for an effective personalized therapy, or for a selective sensing, in order to minimize the nonspecific uptake and to favor the recognition of certain molecules or cells (Dieu et al., 2014; Egli et al., 2011; Figueiredo et al., 2016; Mohammadi et al., 2017; Palivan et al., 2016; Pourtau et al., 2013; Zhang and Zhang, 2017). However, a nonspecific or insufficient targeting, an inadequate release at the target site or the multidrug resistance is still challenged to be overcome when designing novel biomedical polymeric platforms.

2.5 CONCLUDING REMARKS

The new developments in biotechnology can definitely help and beneficially improve the life quality of many individuals with various acute or chronic diseases, such as cancer, infectious or autoimmune diseases, and so on. To better address the specific demands of biomedicine, various smart polymeric materials with complex spherical 3D architectures, and increased functionalities were engineered and intensively investigated for their potential biomedical applications, such as biodetection and biosensing, diagnosis, and/or treatment. In this respect, the introduction of stimuli-triggered responsiveness allowed the designed polymer platforms to recognize the variations of physiological conditions and induce a suitable response. Thus, the progress made in polymer chemistry affords a continuously increasing library of building blocks that were used to design stimuli-responsive linear or cross-linked polymers with tailorable properties. Consequently, the most relevant synthetic strategies toward stimuli-responsive polymers were first highlighted in our chapter. Furthermore, the fabrication strategies involved in the development of intelligent platforms with complex spherical self-organized architectures (micelles, vesicles, capsules, nanoparticles, and NMGs), which can be implemented in various fields of biomedicine were also presented. In conclusion, different polymer platforms overcoming the critical drawbacks, such as biocompatibility, controlled release, specific targeting, or stealth properties, have been developed. However, the selection of suitable conditions and techniques for preparation of self-organized polymer structures should take into consideration the intrinsic properties of polymer building blocks, such as solubility, chain flexibility, molecular weight, interactions between polymer and payload, charge and composition, and so on. Nevertheless, the advance in polymer systems for clinical applications is still challenging, these systems being expected to play a key role in further development of smart materials for personalized medicine.

So far, the main application of self-organized polymer materials with spherical 3D architectures is oriented toward therapeutics and theranostics based on their ability to reply "on demand." Moreover, another approach consisting of the selective confinement of naturally active biomolecules (proteins, enzymes, DNA, etc.) into the self-organizing polymer architectures have been introduced for designing smart polymer vehicles. Such biomolecules, acting as specific gates, transporters or catalysts, open new emerging perspectives for the bioengineered architectures. The complex hybrid systems endowed with bio-inspired functions, such as transport, compartmentalization, or catalysis, have a high potential for mimicking the function of natural organelles or even cells. Even though promising results have been obtained in this challenging field, further development is still required to increase their application range in biomedicine.

KEYWORDS

- stimuli-responsive
- micelles
- polymersomes
- PICsomes
- capsules
- nano- and microgels
- biomedical applications

REFERENCES

Ahmad, Z., et al. Polymeric Micelles as Drug Delivery Vehicles. *RSC Adv.* **2014**, *4* (33), 17028–17038.

Akimoto, J.; Nakayama, M.; Okano, T. Temperature-responsive Polymeric Micelles for Optimizing Drug Targeting to Solid Tumors. *J. Controll. Release* **2014**, *193*, 2–8.

Anajafi, T., et al. Acridine Orange Conjugated Polymersomes for Simultaneous Nuclear Delivery of Gemcitabine and Doxorubicin to Pancreatic Cancer Cells. *Bioconjugate Chem.* **2016,** *27* (3), 762–771.

Anraku, Y., et al. Systemically Injectable Enzyme-loaded Polyion Complex Vesicles as In Vivo Nanoreactors Functioning in Tumors. *Angew. Chem. Int. Ed.* **2016,** *55* (2), 560–565.

Asadian-Birjand, M., et al. Transferrin Decorated Thermo-responsive Nanogels as Magnetic Trap Devices for Circulating Tumor Cells. *Macromol. Rapid Commun.* **2016,** *37* (5), 439–445.

Balasubramanian, V., et al. Protein Delivery: From Conventional Drug Delivery Carriers to Polymeric Nanoreactors. *Exp. Opin. Drug Del.* **2010,** *7* (1), 63–78.

Balasubramanian, V., et al. A Surprising System: Polymeric Nanoreactors Containing a Mimic with Dual-enzyme Activity. *Soft Matter.* **2011,** *7* (12), 5595–5603.

Balasubramanian, V., et al. Multifaceted Polymersome Platforms: Spanning from Self-assembly to Drug Delivery and Protocells. *Prog. Polym. Sci.* **2016,** *60,* 51–85.

Bartenstein, J. E., et al. Stability of Polymersomes Prepared by Size Exclusion Chroma-tography and Extrusion. *Colloids Surf. A Physicochem. Eng. Asp.* **2016,** *506,* 739–746.

Baumann, P., et al. Light-responsive Polymer Nanoreactors: A Source of Reactive Oxygen Species on Demand. *Nanoscale* **2013,** *5* (1), 217–224.

Baumann, P., et al. Cellular Trojan Horse Based Polymer Nanoreactors with Light-sensi-tive Activity. *J. Phys. Chem. B.* **2014,** *118* (31), 9361–9370.

Beales, P. A., et al. Durable Vesicles for Reconstitution of Membrane Proteins in Biotech-nology. *Biochem. Soc. Trans.* **2017,** *45* (1), 15–26.

Bedard, M. F., et al. Polymeric Microcapsules with Light Responsive Properties for Encap-sulation and Release. *Adv. Colloid Interface Sci.* **2010,** *158* (1–2), 2–14.

Begum, R.; Farooqi, Z. H.; Khan, S. R. Poly(N-isopropylacrylamide-acrylic Acid) Copo-lymer Microgels for Various Applications: A Review. *Int. J. Polym. Mater. Polym. Biomater.* **2016,** *65* (16), 841–852.

Bleul, R.; Thiermann, R.; Maskos, M. Techniques to Control Polymersome Size. *Macro-molecules* **2015,** *48* (20), 7396–7409.

Brinkhuis, R. P., et al. Size Dependent Biodistribution and SPECT Imaging of In-111-labeled Polymersomes. *Bioconj. Chem.* **2012,** *23* (5), 958–965.

Buddingh', B. C.; van Hest, J. C. M. Artificial Cells: Synthetic Compartments with Life-like Functionality and Adaptivity. *Acc. Chem. Res.* **2017,** *50* (4), 769–777.

Cabane, E., et al. Photo-responsive Polymersomes as Smart, Triggerable Nanocarriers. *Soft Matter* **2011,** *7,* 9167–9176.

Cabane, E., et al. Stimuli-responsive Polymers and Their Applications in Nanomedicine. *Biointerphases* **2012,** *7* (9), 1–27.

Cao, Z. Q.; Wang, G. J. Multi-stimuli-responsive Polymer Materials: Particles, Films, and Bulk Gels. *Chem. Rec.* **2016,** *16* (3), 1398–1435.

Cao, Z.; Zhou, X.; Wang, G. Selective Release of Hydrophobic and Hydrophilic Cargos from Multi-stimuli-responsive Nanogels. *ACS Appl. Mater. Interfaces* **2016,** *8* (42), 28888–28896.

Car, A., et al. pH-Responsive PDMS-*b*-PDMAEMA Micelles for Intracellular Anticancer Drug Delivery. *Biomacromolecules* **2014,** *15* (9), 3235–3245.

Cavadas, M., et al. Pathogen-mimetic Stealth Nanocarriers for Drug Delivery: A Future Possibility. *Nanomed. Nanotechnol. Biol. Med.* **2011,** *7* (6), 730–743.

Che, H.; van Hest, J. C. M. Stimuli-responsive Polymersomes and Nanoreactors. *J. Mater. Chem. B.* **2016,** *4* (27), 4632–4647.

Chen, F. M., et al. Surface-engineering of Glycidyl Methacrylated Dextran/Gelatin Micro-capsules with Thermo-responsive Poly(N-isopropylacrylamide) Gates for Controlled Delivery of Stromal Cell-derived Factor-1α. *Biomaterials* **2013,** *34* (27), 6515–6527.

Chuanoi, S., et al. Fabrication of Polyion Complex Vesicles with Enhanced Salt and Temperature Resistance and Their Potential Applications as Enzymatic Nanoreactors. *Biomacromolecules* **2014,** *15* (7), 2389–2397.

Chuard, N., et al. Strain-promoted Thiol-mediated Cellular Uptake of Giant Substrates: Liposomes and Polymersomes. *Angew. Chem. Int. Ed.* **2017,** *56* (11), 2947 –2950.

Costa, R. R., et al. Nanostructured and Thermoresponsive Recombinant Biopolymer-based Microcapsules for the Delivery of Active Molecules. *Nanomed. Nanotechnol. Biol. Med.* **2013,** *9* (7), 895–902.

Costa, R. R.; Alatorre-Meda, M.; Mano, J. F. Drug Nano-reservoirs Synthesized Using Layer-by-layer Technologies. *Biotechnol. Adv.* **2015,** *33* (3), 1310–1326.

Craciun, I., et al. Expanding the Potential of MRI Contrast Agents through Multifunctional Polymeric Nanocarriers. *Nanomedicine (Lond.)* **2017,** *12* (7), 811–817.

Danafar, H.; Rostamizadeh, K.; Hamidi, M. Polylactide/poly(Ethylene Glycol)/Polylactide Triblock Copolymer Micelles as Carrier for Delivery of Hydrophilic and Hydrophobic Drugs: A Comparison Study. *J. Pharm. Investig.* **2017,** DOI:10.1007/s40005-017-0334-8.

Dash, B. C., et al. The Influence of Size and Charge of Chitosan/polyglutamic Acid Hollow Spheres on Cellular Internalization, Viability and Blood Compatibility. *Biomaterials* **2010,** *31* (32), 8188–8197.

De Crozals, G., et al. Nanoparticles with Multiple Properties for Biomedical Applications: A Strategic Guide. *Nano Today* **2016,** *11* (4), 435–463.

Deng, R., et al. Using Dynamic Covalent Chemistry to Drive Morphological Transitions: Controlled Release of Encapsulated Nanoparticles from Block Copolymer Vesicles. *J. Am. Chem. Soc.* **2017a,** *139* (22), 7616–7623.

Deng, Z.; Hu, J.; Liu, S. Reactive Oxygen, Nitrogen, and Sulfur Species (RONSS)-respon-sive Polymersomes for Triggered Drug Release. *Macromol. Rapid Commun.* **2017b,** *38* (11), 1600685.

Dieu, L. H., et al. Polymersomes Conjugated to 83–14 Monoclonal Antibodies: In Vitro Targeting of Brain Capillary Endothelial Cells. *Eur. J. Pharm. Biopharm.* **2014,** *88* (2), 316–324.

Dinu, I. A., et al. Polymer Vesicles. In *Encyclopedia of Polymeric Nanomaterials*; Kobayashi, S., Müllen, K., Eds.; Springer: Berlin, Heidelberg, 2014; Vol. 1, pp 1–11.

Dinu, I. A., et al. Engineered Non-toxic Cationic Nanocarriers with Photo-triggered Slow-release Properties. *Polym. Chem.* **2016,** *7* (20), 3451–3464.

Dinu, M. V., et al. Filling Polymersomes with Polymers by Peroxidase-catalyzed Atom Transfer Radical Polymerization. *Macromol. Rapid Commun.* **2015,** *36* (6), 507–514.

Discher, D. E.; Ahmed, F. Polymersomes. *Annu. Rev. Biomed. Eng.* **2006,** *8,* 323–341.

Dobrunz, D., et al. Polymer Nanoreactors with Dual Functionality: Simultaneous Detoxi-fication of Peroxynitrite and Oxygen Transport. *Langmuir* **2012,** *28* (45), 15889–15899.

Dong, W. F., et al. Monodispersed Polymeric Nanocapsules: Spontaneous Evolution and Morphology Transition from Reducible Hetero-PEG PICmicelles by Controlled Degra-dation. *J. Am. Chem. Soc.* **2009,** *131* (11), 3804–3805.

Du, J. Z.; Fan, L.; Liu, Q. M. pH-sensitive Block Copolymer Vesicles with Variable Trigger Points for Drug Delivery. *Macromolecules* **2012,** *45* (20), 8275–8283.

Duan, X.; Li, Y. Physicochemical Characteristics of Nanoparticles Affect Circulation, Biodistribution, Cellular Internalization, and Trafficking. *Small* **2013,** *9* (9–10), 1521–1532.

Dubois, P.; Coulombier, O.; Raquez, J. M. *Handbook of Ring-Opening Polymerization*; Wiley-VCH, Verlag GmbH & Co. KGaA: Weinheim, Germany, 2009.

Egli, S., et al. Biocompatible Functionalization of Polymersome Surfaces: A New Approach to Surface Immobilization and Cell Targeting Using Polymersomes. *J. Am. Chem. Soc.* **2011,** *133* (12), 4476–4483.

Einfalt, T., et al. Stimuli-triggered Activity of Nanoreactors by Biomimetic Engineering Polymer Membranes. *Nano Lett.* **2015,** *15* (11), 7596–7603.

Elizarova, I. S.; Luckham, P. F. Fabrication of Polyelectrolyte Multilayered Nanocapsules Using a Continuous Layer-by-layer Approach. *J. Colloid Interface Sci.* **2016,** *470,* 92–99.

Fernandez-Trillo, F., et al. Vesicles and Their Multiple Facets: Underpinning Biological and Synthetic Progress. *Angew. Chem. Int. Ed.* **2017,** *56* (12), 3142–3160.

Figueiredo, P., et al. Angiopep2-functionalized Polymersomes for Targeted Doxorubicin Delivery to Glioblastoma Cells. *Int. J. Pharm.* **2016,** *511* (2), 794–803.

Fundueanu, G., et al. Poly(N-isopropylacrylamide-co-N-isopropylmethacrylamide) Thermo-responsive Microgels as Self-regulated Drug Delivery System. *Macromol. Chem. Phys.* **2016,** *217* (22), 2525–2533.

Gaitzsch, J.; Huang, X.; Voit, B. Engineering Functional Polymer Capsules Toward Smart Nanoreactors. *Chem. Rev.* **2016,** *116* (3), 1053–1093.

Garni, M., et al. Biopores/Membrane Proteins in Synthetic Polymer Membranes. *Biochim. Biophys. Acta.* **2017,** *1859* (4), 619–638.

Gao, H.; He, Q. The Interaction of Nanoparticles with Plasma Proteins and the Consequent Influence on Nanoparticles Behavior. *Expert Opin. Drug Deliv.* **2014,** *11* (3), 409–420.

Gaspar, V. M., et al. Bioreducible Poly(2-ethyl-2-oxazoline)–PLA–PEI-SS Triblock Copolymer Micelles for Co-delivery of DNA Minicircles and Doxorubicin. *J. Controlled Release* **2015,** *213,* 175–191.

Gong, C., et al. Novel Composite Drug Delivery System for Honokiol Delivery: Self-assembled Poly(Ethylene Glycol)-poly(ε-caprolactone)-poly(Ethylene Glycol) Micelles in Thermosensitive Poly(Ethylene Glycol)–poly(ε-caprolactone)–poly(Ethylene Glycol) Hydrogel. *J. Phys. Chem. B.* **2009,** *113* (30), 10183–10188.

Godoy-Gallardo, M., et al. Multicompartment Artificial Organelles Conducting Enzymatic Cascade Reactions Inside Cells. *ACS Appl. Mater. Interfaces* **2017,** *9* (19), 15907–15921.

Gräfe, D., et al. Cross-linked Polymersomes as Nanoreactors for Controlled and Stabilized Single and Cascade Enzymatic Reactions. *Nanoscale* **2014,** *6* (18), 10752–10761.

Grzelakowski, M., et al. A Framework for Accurate Evaluation of the Promise of Aquaporin Based Biomimetic Membranes. *J. Membr. Sci.* **2015,** *479,* 223–231.

Guan, L., et al. Polymersomes and Their Applications in Cancer Delivery and Therapy. *Nanomedicine (Lond.)* **2015,** *10* (17), 2757–2780.

Gunkel-Grabole, G., et al. Polymeric 3D Nano-architectures for Transport and Delivery of Therapeutically Relevant Biomacromolecules. *Biomater. Sci.* **2015,** *3* (1), 25–40.

Guragain, S., et al. Multi-stimuli-responsive Polymeric Materials. *Chem. Eur. J.* **2015,** *21* (38), 13164–13174.

Guzmán, E., et al. Layer-by-layer Polyelectrolyte Assemblies for Encapsulation and Release of Active Compounds. *Adv. Colloid Interface. Sci.* **2017**, *249*, 290–307.

Hamidi, M.; Shahbazi, M. A.; Rostamizadeh, K. Copolymers: Efficient Carriers for Intelligent Nanoparticulate Drug Targeting and Gene Therapy. *Macromol. Biosci.* **2012**, *12* (2), 144–164.

Hamner, K. L., et al. Using Temperature-sensitive Smart Polymers to Regulate DNA-mediated Nanoassembly and Encoded Nanocarrier Drug Release. *ACS Nano.* **2013**, *7* (8), 7011–7020.

Han, K., et al. Synergistic Gene and Drug Tumor Therapy Using a Chimeric Peptide. *Biomaterials* **2013**, *34* (19), 4680–4689.

Hashmi, S. *Reference Module in Materials Science and Materials Engineering*; Elsevier: Amsterdam, Netherlands; 2012.

He, W., et al. Atom Transfer Radical Polymerization of Hydrophilic Monomers and Its Applications. *Polym. Chem.* **2013**, *4* (10), 2919–2938.

Heinisch, T., et al. Fluorescence-based Assay for the Optimization of the Activity of Artificial Transfer Hydrogenase Within a Biocompatible Compartment. *Chem. Cat. Chem.* **2013**, *5* (3), 720–723.

Hiruta, Y.; Nemoto, R.; Kanazawa, H. Design and Synthesis of Temperature-responsive Polymer/silica Hybrid Nanoparticles and Application to Thermally Controlled Cellular Uptake. *Colloids Surf. B Biointerface* **2017**, *153*, 2–9.

Hsiue, G. H., et al. Environmental-sensitive Micelles Based on Poly(2-ethyl-2-oxazoline)-b-poly(l-lactide) Diblock Copolymer for Application in Drug Delivery. *Int. J. Pharm.* **2006**, *317* (1), 69–75.

Hu, X., et al. Stimuli-responsive Polymersomes for Biomedical Applications. *Biomacromolecules* **2017**, *18* (3), 649–673.

Huang, X., et al. Synthesis of Well-defined Photo-cross-linked Polymeric Nanocapsules by Surface-initiated RAFT Polymerization. *Macromolecules* **2011**, *44* (21), 8351–8360.

Huang, X., et al. Tailored Synthesis of Intelligent Polymer Nanocapsules: An Investigation of Controlled Permeability and pH-dependent Degradability. *ACS Nano* **2012**, *6* (11), 9718–9726.

Itel, F., et al. Molecular Organization and Dynamics in Polymersome Membranes: A Lateral Diffusion Study. *Macromolecules* **2014**, *47* (21), 7588–7596.

Itel, F., et al. Dynamics of Membrane Proteins Within Synthetic Polymer Membranes with Large Hydrophobic Mismatch. *Nano Lett.* **2015**, *15* (6), 3871–3878.

Iyisan, B., et al. Multifunctional and Dual-responsive Polymersomes as Robust Nanocontainers: Design, Formation by Sequential Post-conjugations, and pH-controlled Drug Release. *Chem. Mater.* **2016**, *28* (5), 1513–1525.

Jeong, E. S.; Park, C.; Kim, K. T. Doubly Responsive Polymersomes Towards Monosaccharides and Temperature Under Physiologically Relevant Conditions. *Polym. Chem.* **2015**, *6* (22), 4080–4088.

Jia, L., et al. Reduction-responsive Cholesterol-based Block Copolymer Vesicles for Drug Delivery. *Biomacromolecules* **2014**, *15* (6), 2206–2217.

Jiang, Y., et al. Click Hydrogels, Microgels and Nanogels: Emerging Platforms for Drug Delivery and Tissue Engineering. *Biomaterials* **2014**, *35* (18), 4969–4985.

Jiang, N.; Cheng, Y.; Wei, J. Coumarin-modified Fluorescent Microcapsules and Their Photo-switchable Release Property. *Colloid Surf. A Physicochem. Eng. Aspects.* **2017,** *522,* 28–37.

Joglekar, M.; Trewyn, B. G. Polymer-based Stimuli-responsive Nanosystems for Biomedical Applications. *Biotechnol. J.* **2013,** *8* (8), 931–945.

Karimi, M. et al. Smart Micro/nanoparticles in Stimulus-responsive Drug/gene Delivery Systems. *Chem. Soc. Rev.* **2016,** *45* (5), 1457–1501.

Karimi, M., et al. Smart Nanostructures for Cargo Delivery: Uncaging and Activating by Light. *J. Am. Chem. Soc.* **2017,** *139* (13), 4584−4610.

Kim, K., et al. Polycations and Their Biomedical Applications. *Prog. Polym. Sci.* **2016,** *60,* 18–50.

Kishimura, A. Development of Polyion Complex Vesicles (PICsomes) from Block Copolymers for Biomedical Applications. *Polym. J.* **2013,** *45* (9), 892–897.

Kishimura, A., et al. Encapsulation of Myoglobin in PEGylated Polyion Complex Vesicles Made from a Pair of Oppositely Charged Block Ionomers: A Physiologically Available Oxygen Carrier. *Angew. Chem. Int. Ed.* **2007,** *46* (32), 6085–6088.

Kita-Tokarczyk, K., et al. Block Copolymer Vesicles—Using Concepts from Polymer Chemistry to Mimic Biomembranes. *Polymer* **2005,** *46* (11), 3540–3563.

Klermund, L.; Poschenrieder, S. T.; Castiglione, K. Biocatalysis in Polymersomes: Improving Multienzyme Cascades with Incompatible Reaction Steps by Compartmentalization. *ACS Catal.* **2017,** *7* (6), 3900−3904.

Koide, A., et al. Semipermeable Polymer Vesicle (PICsome) Self-assembled in Aqueous Medium from a Pair of Oppositely Charged Block Copolymers: Physiologically Stable Micro-/Nanocontainers of Water-soluble Macromolecules. *J. Am. Chem. Soc.* **2006,** *128* (18), 5988–5989.

Kokuryo, D., et al. SPIO-PICsome: Development of a Highly Sensitive and Stealth-capable MRI Nano-agent for Tumor Detection Using SPIO-loaded Unilamellar Polyion Complex Vesicles (PICsomes). *J. Controlled Release* **2013,** *169* (3), 220–227.

Kulthe, S. S., et al. Polymeric Micelles: Authoritative Aspects for Drug Delivery. *Des. Monomers Polym.* **2012,** *15* (5), 465–521.

Kumar, A., et al. ROP and ATRP Fabricated Dual Targeted Redox Sensitive Polymersomes Based on pPEGMA-PCL-ss-PCL-pPEGMA Triblock Copolymers for Breast Cancer Therapeutics. *ACS Appl. Mat. Interfaces* **2015,** *7* (17), 9211–9227.

Langowska, K.; Palivan, C. G.; Meier, W. Polymer Nanoreactors Shown to Produce and Release Antibiotics Locally. *Chem. Commun.* **2013,** *49* (2), 128–130.

Lazim, A.; Julian, E.; Melanie, B. Incorporation of Gold Nanoparticles into pH Responsive Mixed Microgel Systems. *Mediterr. J. Chem.* **2012,** *1* (5), 259–272.

Lee, H. I., et al. Light-induced Reversible Formation of Polymeric Micelles. *Angew. Chem.* **2007,** *46* (14), 2453–2457.

Lee, J. Y., et al. Nanoparticle-loaded Protein–polymer Nanodroplets for Improved Stability and Conversion Efficiency in Ultrasound Imaging and Drug Delivery. *Adv. Mater.* **2015,** *27* (37), 5484–5492.

Lee, P. Y., et al. Nanogels Comprising Reduction-cleavable Polymers for Glutathione-induced Intracellular Curcumin Delivery. *J. Polym. Res.* **2017,** *24* (66), 1–10.

Lee, S. Y., et al. pH/redox/photo Responsive Polymeric Micelle Via Boronate Ester and Disulfide Bonds with Spiropyran-based Photochromic Polymer for Cell Imaging and Anticancer Drug Delivery. *Eur. Polym. J.* **2014,** *57,* 1–10.

Li, D., et al. Reduction-Sensitive Dextran Nanogels Aimed for Intracellular Delivery of Antigens. *Adv. Funct. Mater.* **2015**, *25* (20), 2993–3003.

Li, J., et al. Redox-sensitive Micelles Self-assembled from Amphiphilic Hyaluronic Acid Deoxycholic Acid Conjugates for Targeted Intracellular Delivery of Paclitaxel. *Biomaterials* **2012**, *33* (7), 2310–2320.

Li, Y., et al. In Situ Formation of Gold-"decorated" Vesicles from a RAFT-synthesized, Thermally Responsive Block Copolymer. *Macromolecules* **2007**, *40* (24), 8524–8526.

Li, Y., et al. Stimuli-responsive Cross-linked Micelles for On-demand Drug Delivery Against Cancers. *Adv. Drug Deliv. Rev.* **2014**, *66*, 58–73.

Li, Y.; Lokitz, B. S.; McCormick, C. L. Thermally Responsive Vesicles and Their Structural "Locking" Through Polyelectrolyte Complex Formation. *Angew. Chem. Int. Ed.* **2006**, *45* (35), 5792–5795.

Liu, F., et al. Encapsulation of Anticancer Drug by Hydrogen-bonded Multilayers of Tannic Acid. *Soft Matter.* **2014a**, *10* (46), 9237–9247.

Liu, F., et al. Temperature-sensitive Polymersomes for Controlled Delivery of Anticancer Drugs. *Chem. Mater.* **2015a**, *27* (23), 7945–7956.

Liu, G., et al. Self-immolative Polymersomes for High-efficiency Triggered Release and Programmed Enzymatic Reactions. *J. Am. Chem. Soc.* **2014b**, *136* (20), 7492–7497.

Liu, J., et al. DNA-mediated Self-organization of Polymeric Nanocompartments Leads to Interconnected Artificial Organelles. *Nano Lett.* **2016**, *16* (11), 7128–7136.

Liu, M., et al. Internal Stimuli-responsive Nanocarriers for Drug Delivery: Design Strategies and Applications. *Mater. Sci. Eng. C.* **2017**, *71*, 1267–1280.

Liu, Q. M., et al. An Asymmetrical Polymer Vesicle Strategy for Significantly Improving T1 MRI Sensitivity and Cancer Targeted Drug Delivery. *Macromolecules* **2015b**, *48* (3), 739–749.

Lomora, M., et al. Does Membrane Thickness Affect the Transport of Selective Ions Mediated by Ionophores in Synthetic Membranes? *Macromol. Rapid Commun.* **2015a**, *36* (21), 1929–1934.

Lomora, M., et al. Selective Ion-permeable Membranes by Insertion of Biopores into Polymersomes. *Phys. Chem. Chem. Phys.* **2015b**, *17* (24), 15538–15546.

Lomora, M., et al. Polymersomes with Engineered Ion Selective Permeability as Stimuli-responsive Nanocompartments with Preserved Architecture. *Biomaterials* **2015c**, *53*, 406–414.

Lu, L., et al. Anisamide-decorated pH-sensitive Degradable Chimaeric Polymersomes Mediate Potent and Targeted Protein Delivery to Lung Cancer Cells. *Biomacromolecules* **2015**, *16* (6), 1726–1735.

Marguet, M.; Bonduelle, C.; Lecommandoux, S. Multicompartmentalized Polymeric Systems: Towards Biomimetic Cellular Structure and Function. *Chem. Soc. Rev.* **2013**, *42* (2), 512–529.

Marguet, M.; Edembe, L.; Lecommandoux, S. Polymersomes in Polymersomes: Multiple Loading and Permeability Control. *Angew. Chem. Int. Ed.* **2012a**, *51* (5), 1173–1176.

Marguet, M.; Sandre, O.; Lecommandoux, S. Polymersomes in "Gelly" Polymersomes: Toward Structural Cell Mimicry. *Langmuir* **2012b**, *28* (4), 2035–2043.

Markovsky, E., et al. Administration, Distribution, Metabolism and Elimination of Polymer Therapeutics. *J. Controll. Release* **2012**, *161* (2), 446–460.

Marsden, H. R., et al. Detergent-aided Polymersome Preparation. *Biomacromolecules* **2010**, *11* (4), 833–838.

Martin, C., et al. Recent Advances in Amphiphilic Polymers for Simultaneous Delivery of Hydrophobic and Hydrophilic Drugs. *Ther. Deliv.* **2016**, *7* (1), 15–31.

Marturano, V., et al. Light-responsive Polymer Micro- and Nano-capsules. *Polymers.* **2017**, *9* (1), 8–19.

Matyjaszewski, K., et al. *Controlled Radical Polymerization: Mechanisms*; American Chemical Society: Washington, DC, 2015.

Meng, F.; Zhong, Z. Polymersomes Spanning from Nano- to Microscales: Advanced Vehicles for Controlled Drug Delivery and Robust Vesicles for Virus and Cell Mimicking. *J. Phys. Chem. Lett.* **2011**, *2* (13), 1533–1539.

Meng, Z., et al. NIR-laser-triggered Smart Full-polymer Nanogels for Synergic Photothermal-/chemo-therapy of Tumors. *RSC Adv.* **2016**, *6* (93), 90111–90119.

Mi, P., et al. A pH-activatable Nanoparticle with Signal-amplification Capabilities for Noninvasive Imaging of Tumour Malignancy. *Nat. Nanotechnol.* **2016**, *11* (8), 724–730.

Mohammadi, M., et al. Biocompatible Polymersomes-based Cancer Theranostics: Towards Multifunctional Nanomedicine. *Int. J. Pharm.* **2017**, *519* (1–2), 287–303.

Molina, M., et al. Stimuli-responsive Nanogel Composites and Their Application in Nanomedicine. *Chem. Soc. Rev.* **2015**, *44* (17), 6161–6186.

Moughton, A. O.; O'Reilly, R. K. Thermally Induced Micelle to Vesicle Morphology Transition for a Charged Chain End Diblock Copolymer. *Chem. Commun.* **2010**, *46* (7), 1091–1093.

Mukhopadhyay, P., et al. pH-sensitive Chitosan/alginate Core-shell Nanoparticles for Efficient and Safe Oral Insulin Delivery. *Int. J. Biol. Macromol.* **2015**, *72*, 640–648.

Najer, A., et al. Nanomimics of Host Cell Membranes Block Invasion and Expose Invasive Malaria Parasites. *ACS Nano* **2014**, *8* (12), 12560–12571.

Najer, A., et al. Analysis of Molecular Parameters Determining the Antimalarial Activity of Polymer-based Nanomimics. *Macromol. Rapid Commun.* **2015**, *36* (21), 1923–1928.

Najer, A., et al. An Amphiphilic Graft Copolymer-based Nanoparticle Platform for Reduction-responsive Anticancer and Antimalarial Drug Delivery. *Nanoscale* **2016**, *8* (31), 14858–14869.

Nakayama, M.; Akimoto, J.; Okano, T. Polymeric Micelles with Stimuli-triggering Systems for Advanced Cancer Drug Targeting. *J. Drug Target.* **2014**, *22* (7), 584–599.

Newland, B., et al. The Neurotoxicity of Gene Vectors and Its Amelioration by Packaging with Collagen Hollow Spheres. *Biomaterials* **2013**, *34* (8), 2130–2141.

Nomoto, T., et al. Three-layered Polyplex Micelle as a Multifunctional Nanocarrier Platform for Light-induced Systemic Gene Transfer. *Nat. Commun.* **2014**, *5*, 3545.

Oliveira, H.; Thevenot, J.; Lecommandoux, S. Smart Polymersomes for Therapy and Diagnosis: Fast Progress Toward Multifunctional Biomimetic Nanomedicines. *Rev. Nanomed. Nanobiotechnol.* **2012**, *4* (5), 525–546.

Oliveira, H., et al. Magnetic Field Triggered Drug Release from Polymersomes for Cancer Therapeutics. *J. Control. Release* **2013**, *169* (3), 165–170.

Palivan, C. G., et al. Bioinspired Polymer Vesicles and Membranes for Biological and Medical Applications. *Chem. Soc. Rev.* **2016**, *45* (2), 377–411.

Panja, S., et al. Metal Ion Ornamented Ultra-fast Light-sensitive Nanogel for Potential In Vivo Cancer Therapy. *Chem. Mater.* **2016**, *28* (23), 8598–8610.

Park, J. S., et al. Preparation and Characterization of Polyion Complex Micelles with a Novel Thermosensitive Poly(2-isopropyl-2-oxazoline) Shell Via the Complexation of Oppositely Charged Block Ionomers. *Langmuir* **2007**, *23* (1), 138–146.

Peer, D., et al. Nanocarriers as an Emerging Platform for Cancer Therapy. *Nat. Nanotechnol.* **2007**, *2*, 751–760.

Plamper, F. A.; Richtering, W. Functional Microgels and Microgel Systems. *Acc. Chem. Res.* **2017**, *50* (2), 131–140.

Postupalenko, V., et al. Bionanoreactors: From Confined Reaction Spaces to Artificial Organelles. In *"Organic Nanoreactors from Molecular to Supramolecular Organic Compounds"*; Sadjadi, S., Ed.; Academic Press: London, 2016; Vol. 1, pp 341–371.

Pourtau, L., et al. Antibody-functionalized Magnetic Polymersomes: In Vivo Targeting and Imaging of Bone Metastases Using High Resolution MRI. *Adv. Healthcare Mater.* **2013**, *2* (11), 1420–1424.

Ramasamy, T., et al. Smart Chemistry-based Nanosized Drug Delivery Systems for Systemic Applications: A Comprehensive Review. *J. Control Release* **2017**, *258*, 226–253.

Reddy, B. P. K., et al. Polymeric Micelles as Novel Carriers for Poorly Soluble Drugs—A Review. *J. Nanosci. Nanotechnol.* **2015**, *15* (6), 4009–4018.

Richard, P. U., et al. New Concepts to Fight Oxidative Stress: Nanosized Three-dimensional Supramolecular Antioxidant Assemblies. *Expert Opin. Drug Deliv.* **2015**, *12* (9), 1527–1545.

Rikken, R. S. M., et al. Shaping Polymersomes into Predictable Morphologies *via* Out-of-equilibrium Self-assembly. *Nat. Commun.* **2016**, *7* (August), 12606.

Qian, C., et al. Light-activated Hypoxia-responsive Nanocarriers for Enhanced Anticancer Therapy. *Adv. Mater.* **2016**, *28* (17), 3313–3320.

Qin, J. Y., et al. Rationally Separating the Corona and Membrane Functions of Polymer Vesicles for Enhanced T2 MRI and Drug Delivery. *ACS Appl. Mater. Interfaces* **2015**, *7* (25), 14043–14052.

Quader, S., et al. Selective Intracellular Delivery of Proteasome Inhibitors Through pH-sensitive Polymeric Micelles Directed to Efficient Antitumor Therapy. *J. Control Release* **2014**, *188*, 67–77.

Salehi, R.; Rasouli, S.; Hamishehkar, H. Smart Thermo/pH Responsive Magnetic Nanogels for the Simultaneous Delivery of Doxorubicin and Methotrexate. *Int. J. Pharm.* **2015**, *487* (1–2), 274–284.

Sanson, N.; Rieger, J. Synthesis of Nanogels/Microgels by Conventional and Controlled Radical Crosslinking Copolymerization. *Polym. Chem.* **2010**, *1* (7), 965–977.

Sant, V. P.; Smith, D.; Leroux, J. C. Novel pH-sensitive Supramolecular Assemblies for Oral Delivery of Poorly Water Soluble Drugs: Preparation and Characterization. *J. Control Release* **2004**, *97* (2), 301–312.

Sarika, P. R., et al. Nanogels Based on Alginic Aldehyde and Gelatin by Inverse Miniemulsion Technique: Synthesis and Characterization. *Carbohydr. Polym.* **2015**, *119*, 118–125.

Seo, Y., et al. Poly(2-oxazoline) Block Copolymer Based Formulations of Taxanes: Effect of Copolymer and Drug Structure, Concentration, and Environmental Factors. *Polym. Adv. Technol.* **2015**, *26* (7), 837–850.

Shi, F., et al. Intracellular Microenvironment Responsive PEGylated Polypeptide Nanogels with Ionizable Cores for Efficient Doxorubicin Loading and Triggered Release. *J. Mater. Chem.* **2012**, *22* (28), 14168–14179.

Sigg, S. J., et al. Nanoparticle-based Highly Sensitive MRI Contrast Agents with Enhanced Relaxivity in Reductive Milieu. *Chem. Commun.* **2016a**, *52* (64), 9937–9940.

Sigg, S. J., et al. Stimuli-responsive Co-delivery of Oligonucleotides and Drugs by Self-assembled Peptide Nanoparticles. *Biomacromolecules* **2016b**, *17* (3), 935–945.

Sueyoshi, D., et al. Enzyme-loaded Polyion Complex Vesicles as In Vivo Nanoreactors Working Sustainably Under the Blood Circulation: Characterization and Functional Evaluation. *Biomacromolecules* **2017**, *18* (4), 1189–1196.

Sung, B.; Kim, C.; Kim, M. H. Biodegradable Colloidal Microgels with Tunable Thermosensitive Volume Phase Transitions for Controllable Drug Delivery. *J. Colloid Interface. Sci.* **2015**, *450*, 26–33.

Shutava, T. G., et al. Architectural Layer-by-layer Assembly of Drug Nanocapsules with PEGylated Polyelectrolytes. *Soft Matter* **2012**, *8* (36), 9418–9427.

Tahara, Y.; Akiyoshi, K. Current Advances in Self-assembled Nanogel Delivery Systems for Immunotherapy. *Adv. Drug Deliv. Rev.* **2015**, *95*, 65–76.

Talelli, M., et al. Core-crosslinked Polymeric Micelles: Principles, Preparation, Biomedical Applications and Clinical Translation. *Nano Today* **2015**, *10* (1), 93–117.

Takahashi, R., et al. Reversible Vesicle–spherical Micelle Transition in a Polyion Complex Micellar System Induced by Changing the Mixing Ratio of Copolymer Components. *Macromolecules* **2016**, *49* (8), 3091–3099.

Tang, H., et al. Development of Enzyme Loaded Polyion Complex Vesicle (PICsome): Thermal Stability of Enzyme in PICsome Compartment and Effect of Coencapsulation of Dextran on Enzyme Activity. *Macromol. Biosci.* **2017**, *17* (8), 1600542.

Tang, Z., et al. Polymeric Nanostructured Materials for Biomedical Applications. *Prog. Polym. Sci.* **2016**, *60*, 86–128.

Tanner, P., et al. Polymeric Vesicles: From Drug Carriers to Nanoreactors and Artificial Organelles. *Acc. Chem. Res.* **2011a**, *44* (10), 1039–1049.

Tanner, P., et al. Enzymatic Cascade Reactions Inside Polymeric Nanocontainers: A Means to Combat Oxidative Stress. *Chem. Eur. J.* **2011b**, *17* (16), 4552–4560.

Tanner, P.; Balasubramanian, V.; Palivan, C. G. Aiding Nature's Organelles: Artificial Peroxisomes Play Their Role. *Nano Lett.* **2013**, *13* (6), 2875–2883.

Thambi, T.; Park, J. H.; Lee, D. S. Stimuli-responsive Polymersomes for Cancer Therapy. *Biomater. Sci.* **2016**, *4* (1), 55–69.

Toughrai, S., et al. Reduction-sensitive Amphiphilic Triblock Copolymers Self-assemble into Stimuli-responsive Micelles for Drug Delivery. *Macromol. Biosci.* **2014**, *15* (4), 481–489.

Tu, Y., et al. Biodegradable Hybrid Stomatocyte Nanomotors for Drug Delivery. *ACS Nano.* **2017a**, *11* (2), 1957–1963.

Tu, Y., et al. Self-propelled Supramolecular Nanomotors with Temperature-responsive Speed Regulation. *Nat. Chem.* **2017b**, *9* (5), 480–486.

Vangeyte, P., et al. About the Methods of Preparation of Poly(Ethylene Oxide)-b-poly(ε-caprolactone) Nanoparticles in Water. Analysis by Dynamic Light Scattering. *Colloid Surf. A Physicochem. Eng. Aspects* **2004**, *242* (1–3), 203–211.

Vasquez, D., et al. Asymmetric Triblock Copolymer Nanocarriers for Controlled Localization and pH-sensitive Release of Proteins. *Langmuir* **2016**, *32* (40), 10235–10243.

Venkataraman, S., et al. The Effects of Polymeric Nanostructure Shape on Drug Delivery. *Adv. Drug Del. Rev.* **2011**, *63* (14–15), 1228–1246.

von Roemeling, C., et al. Breaking Down the Barriers to Precision Cancer Nanomedicine. *Trend Biotechnol.* **2017**, *35* (2), 159–171.

Voigt, A., et al. Membrane Filtration for Microencapsulation and Microcapsules Fabrication by Layer-by-layer Polyelectrolyte Adsorption. *Ind. Eng. Chem. Res.* **1999**, *38* (10), 4037–4043.

Wang, H., et al. Responsive Polymer-fluorescent Carbon Nanoparticle Hybrid Nanogels for Optical Temperature Sensing, Near-infrared Light-responsive Drug Release, and Tumor Cell Imaging. *Nanoscale* **2014a**, *6* (13), 7443–7452.

Wang, L., et al. Acid-disintegratable Polymersomes of pH-responsive Amphiphilic Diblock Copolymers for Intracellular Drug Delivery. *Macromolecules* **2015a**, *48* (19), 7262–7272.

Wang, L., et al. Reversibly Switching Bilayer Permeability and Release Modules of Photochromic Polymersomes Stabilized by Cooperative Noncovalent Interactions. *J. Am. Chem. Soc.* **2015b**, *137* (48), 15262–15275.

Wang, X., et al. Concurrent Block Copolymer Polymersome Stabilization and Bilayer Permeabilization by Stimuli-regulated "Traceless" Crosslinking. *Angew. Chem. Int. Ed.* **2014b**, *53* (12), 3138–3142.

Wang, Y.; Kohane, D. S. External Triggering and Triggered Targeting Strategies for Drug Delivery. *Nat. Rev. Mater.* **2017**, *2*, 1–14.

Wang, Z., et al. Cationic Amphiphilic Copolymers: Synthesis, Characterization, Self-assembly and Drug-loading Capacity. *Polym. Int.* **2017**, *66*, 1199–1205.

Wani, T. U., et al. Targeting Aspects of Nanogels: An Overview. *Int. J. Pharm. Sci. Nanotechnol.* **2014**, *7* (4), 2612–2630.

Warren, N. J.; Armes, S. P. Polymerization-induced Self-assembly of Block Copolymer Nano-objects Via RAFT Aqueous Dispersion Polymerization. *J. Am. Chem. Soc.* **2014**, *136* (29), 10174–10185.

Wajs, E., et al. Preparation of Stimuli-responsive Nano-sized Capsules Based on Cyclodextrin Polymers with Redox or Light Switching Properties. *Nano Res.* **2016**, *9* (7), 2070–2078.

Wattendorf, U.; Merkle, H. P. PEGylation as a Tool for the Biomedical Engineering of Surface Modified Microparticles. *J. Pharm. Sci.* **2008**, *97* (11), 4655–4669.

Wei, J., et al. Multi-stimuli-responsive Microcapsules for Adjustable Controlled-release. *Adv. Funct. Mater.* **2014**, *24* (22), 3312–3323.

Wilson, P., et al. Poly(2-oxazoline)-based Micro- and Nanoparticles: A Review. *Eur. Polym. J.* **2017**, *88*, 486–515.

Wu, D., et al. Effect of Molecular Parameters on the Architecture and Membrane Properties of 3D Assemblies of Amphiphilic Copolymers. *Macromolecules* **2014**, *47* (15), 5060–5069.

Wu, W., et al. In-situ Immobilization of Quantum Dots in Polysaccharide-based Nanogels for Integration of Optical pH-sensing, Tumor Cell Imaging, and Drug Delivery. *Biomaterials* **2010**, *31* (11), 3023–3031.

Wu, W., et al. Water-dispersible Multifunctional Hybrid Nanogels for Combined Curcumin and Photothermal Therapy. *Biomaterials* **2011**, *32* (2), 598–609.

Wu, Z., et al. Recent Progress on Bioinspired Self-propelled Micro/nanomotors Via Controlled Molecular Self-assembly. *Small* **2016**, *12* (23), 3080–3093.

Xiong, M. H., et al. Bacteria-responsive Multifunctional Nanogel for Targeted Antibiotic Delivery. *Adv. Mater.* **2012**, *24* (46), 6175–6180.

Yang, H., et al. Smart pH/Redox Dual-responsive Nanogels for On-demand Intracellular Anticancer Drug Release. *ACS Appl. Mater. Interfaces* **2016,** *8* (12), 7729–7738.

Ye, F., et al. Biodegradable Polymeric Vesicles Containing Magnetic Nanoparticles, Quantum Dots and Anticancer Drugs for Drug Delivery and Imaging. *Biomaterials* **2014,** *35* (12), 3885–3894.

York-Duran, M. J., et al. Recent Advances in Compartmentalized Synthetic Architectures as Drug Carriers, Cell Mimics and Artificial Organelles. *Colloid Surf. B Biointerface* **2017,** *152,* 199–213.

Yuan, W., et al. Formation–dissociation of Glucose, pH and Redox Triply Responsive Micelles and Controlled Release of Insulin *Polym. Chem.* **2014a,** *5* (13), 3968–3971.

Yuan, W., et al. Synthesis, Self-assembly, and Multi-stimuli Responses of a Supramolec-ular Block Copolymer. *Macromol. Rapid Commun.* **2014b,** *35* (20), 1776–1781.

Zhan, Y., et al. Thermo/redox/pH- Triple Sensitive Poly(N-isopropylacrylamide-co-acrylic Acid) Nanogels for Anticancer Drug Delivery. *J. Mater. Chem. B.* **2015,** *3* (20), 4221–4230.

Zhang, X., et al. Mimicking the Cell Membrane with Block Copolymer Membranes. *J. Polym. Sci. A Polym. Chem.* **2012,** *50* (12), 2293–2318.

Zhang, X.; Achazi, K.; Haag, R.Boronate Cross-linked ATP- and pH-responsive Nanogels for Intracellular Delivery of Anticancer Drugs. *Adv. Healthcare Mater.* **2015a,** *4* (4), 585–592.

Zhang, X., et al. Micro- and Nanogels with Labile Crosslinks—From Synthesis to Biomed-ical Applications. *Chem. Soc. Rev.* **2015b,** *44* (7), 1948–1973.

Zhang, X., et al. Active Surfaces Engineered by Immobilizing Protein–Polymer Nanoreac-tors for Selectively Detecting Sugar Alcohols. *Biomaterials* **2016a,** *89,* 79–88.

Zhang, X. Y.; Zhang, P. Y. Polymersomes in Nanomedicine—A Review. *Curr. Nanosci.* **2017,** *13* (2), 124–129.

Zhang, Z., et al. Hollow Polymeric Capsules from POSS-based Block Copolymer for Photodynamic Therapy. *Macromolecules* **2016b,** *49* (22), 8440–8448.

Zhao, L., et al. Engineering and Delivery of Nanocolloids of Hydrophobic Drugs. *Adv. Colloid Interface. Sci.* **2017,** *249* (November), 308–320.

Zhu, L., et al. Surface Modification of Gd Nanoparticles with pH-responsive Block Copo-lymers for Use as Smart MRI Contrast Agents. *ACS Appl. Mater. Interfaces* **2016,** *8* (7), 5040–5050.

Zhu, Y., et al. Polymer Vesicles: Mechanism, Preparation, Application, and Responsive Behavior. *Prog. Polym. Sci.* **2017,** *64,* 1–22.

Zschoche, S., et al. Temperature- and pH-dependent Aggregation Behavior of Hydrophilic Dual-sensitive Poly(2-oxazoline)s Block Copolymers as Latent Amphiphilic Macromol-ecules. *Eur. Polym. J.* **2017,** *88,* 623–635.

CHAPTER 3

STRATEGIES TO IMPROVE THE XANTHAN PROPERTIES FOR SPECIFIC APPLICATIONS

CRISTINA-ELIZA BRUNCHI*, SIMONA MORARIU, and MARIA BERCEA

Department of Electroactive Polymers and Plasmochemistry, "Petru Poni" Institute of Macromolecular Chemistry, 41A Grigore Ghica Voda Alley, 700487 Iasi, Romania

*Corresponding author. E-mail: brunchic@icmpp.ro

CONTENTS

ABSTRACT

Xanthan is a polysaccharide produced by the plant-pathogenic bacterium *Xanthomonas campestris*. It presents some remarkable properties such as good solubility in hot or cold water, stability over a wide range of temperatures and pH values, thickening properties (giving high viscous solutions even at very low concentrations), the ability to modify the surface activity or the rheological properties of the materials which contain it, etc. These make the use of xanthan interesting in various kinds of systems from food, pharmaceutical, cosmetic formulations to agricultural, textile, ceramic, petroleum industries, etc. The chemical structure (existence of a large number of functional groups on the polymer chain) and helical conformation of xanthan chains (influenced by temperature, ionic strength, and pH changes) are versatile characteristics which can be efficiently exploited in order to create new behaviors of this unique polysaccharide.

This chapter provides an overview of the strategies used to enhance the properties of xanthan for specific applications. The discussion is mainly focused on two topics: (1) strategies which are applied during the production process changing the properties of native xanthan; and (2) strategies involving structural modifications of macromolecular chains by degradation or chemical functionalization.

3.1 INTRODUCTION

Polysaccharides are considered the most abundant carbohydrate polymers available in nature. Naturally, polysaccharides are obtained from biosynthesis processes in plants (cellulose, hemicellulose, starch, pectin, etc.), algae (agar, carrageenan, alginate, etc.), animals (chitin, glycogen, hyaluronic acid, chondroitin sulfate, etc.), or through action of different microorganisms (pullulan, xanthan, gellan, dextran, levan, welan, etc.). In living organisms, they perform various biological functions such as ensure the structural support and mechanical strength of tissues by forming a three dimensional network (cellulose, chitin, etc.), serve as an energy reservoir (starch for plants and glycogen for animals), or together with proteins and polynucleotides are implied in intercellular communication, cell adhesion, or molecular recognition in the immune system (Dwek, 1996; Yang and Zhang, 2009). The biological properties (biocompatibility, biodegradability, and bioactivity) and specific physi-cochemical characteristics of polysaccharides are exploited in food and nonfood industrial formulations where they act as thickeners, gelling, emulsifying/suspending stabilization agents, or control the active ingre-dient release from drugs, pesticides, flavors, etc. Although some poly-saccharides possess antioxidative, anti-inflammatory, anti-HIV immune regulatory, antimutagenic, antitumor, or anticoagulant properties, their bioactivity is not always satisfying for different requirements (Li et al., 2016). A way to solve such problem is to improve the properties by tailoring and diversification the polysaccharide structures by using chem-ical, physical, or biological methods.

This chapter is focused on xanthan and tries to provide an overview of the main strategies reported in literature in order to modify and function-alize its structure with impact on the final properties required in different applications.

Since its discovery (in the late 1950s by US scientists), xanthan has gained a high interest from the scientists and industrial producers due to its distinct properties in solution, which are superior to any commercially available plant hydrocolloids. Based on the extensive biochemical and toxicological studies, xanthan received in 1969 the approval from the US Food and Drug Administration (FDA) to be used in food and pharmaceu-tical industry as being harmless to living organism. In Europe, xanthan was introduced in the food emulsifier/stabilizer list by European Economic Community from 1982 and was appointed E415. The major producers of xanthan are United States by Kelco (started in the early 1960s by the Kelco Company), Merck and Pfizer, Austria by Jungbunzlauer AG (from 1986) and Solvay, France by Rhône Poulenc, Mero-Rousselot-Santi, and Sanofi-Elf, and Saidy Chemical in China (García-Ochoa et al., 2000; Faria et al., 2011; Petri, 2015).

3.2 CHEMICAL STRUCTURE AND CONFORMATION OF XANTHAN CHAINS

Xanthan is an anionic heteropolysaccharide with molecular weight ranging from 2 million to 50 million Da (Rosalam and England, 2006) and branched structure composed of repeated pentasaccharide units as shown in Figure 3.1.

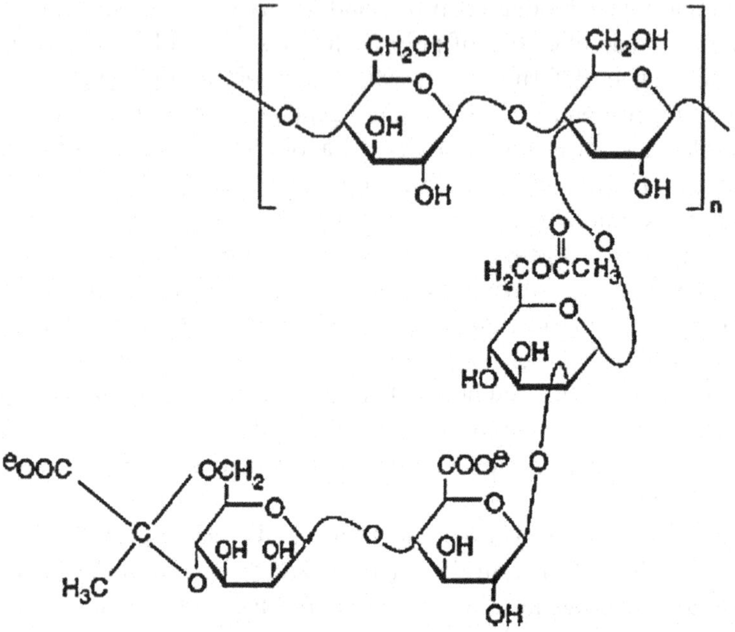

FIGURE 3.1 Primary structure of xanthan. Reprinted with permission from Brunchi, C. E.; Morariu, S.; Bercea, M. Intrinsic Viscosity and Conformational Parameters of Xanthan in Aqueous Solutions: Salt Addition Effect. *Colloids Surf. B* **2014**, *122*, 512–519. © 2014 Elsevier.

The repeated units consist of two glucose residues (as backbone), two mannose residues (namely, α-D-mannose and β-D-mannose), and one β-D-glucuronic acid residue (as a side chain linked at every second glucose residue of the backbone). The α-D-mannose (inner mannose) is linked to the backbone and it may contain an O-acetyl group at the C-6 position, while the β-D-mannose (outer mannose) may contain a pyruvate ketal at the C-4 and C-6 positions (Hassler and Doherty, 1990). The glucuronic acid residue is located between α-D- and β-D-mannose. Some studies suggest that the content and location of functional groups in the side chain structure may undergo various changes (Kulicke and van Eikern, 1992; Kool et al., 2013, 2014; Bilanovic et al., 2016). The various types of xanthan were reported as a function of the acetate and pyruvate content from the repeated units: (1) *low-pyruvate xanthan* in which one pyruvyl group is found at 20 repeated units, (2) *high-pyruvate xanthan* with one pyruvyl group to each repeated unit, (3) *high-acetate xanthan* that has one acetyl group to each repeated unit, and (4) *low-acetate xanthan* which contains one acetyl group to 20 repeated units (Bilanovic et al., 2016). In many cases, xanthan is presented with a structure in which the inner mannose is nearly stoichiometric acetylated (about 70–85%) and the outer one is nonstoichiometric pyruvate (about 30–50%) (Rocherfort and Middleman, 1987; Cadmus et al., 1976; Orentas et al., 1963; Sutherland, 1981; Bilanovic et al., 2015). Also, it was hypothesized (Hassler and Doherty, 1990; Kool et al., 2014) the existence of six-sided chain containing only acetyl or pyruvyl groups, both types of groups or none of them.

If pyruvyl group is located exclusively at the outer mannose, the acetyl group may be found at the inner mannose unit or at the outer one (in this position the acetylation is about 24%) or at both mannose units (Hassler and Doherty, 1990). Generally, it is assumed that the acetyl group from inner mannose is involved (by its methyl groups) in hydrogen bonding with adjacent hemi-acetal oxygen atom of the backbone promoting thus the intra-molecular association and stabilizing the ordered rigid rod-like conformation (structure) of xanthan (Tako and Nakamura,1989; Fitzpatrick et al., 2013). As it turns, the pyruvate group has a destabilizing effect of ordered structure due to the increase of the electrostatic repulsion interactions (between COO⁻ groups) (Viebke, 2005; Fitzpatrick et al., 2013), but by methyl group may contribute to the intermolecular association of xanthan macromolecules into quaternary structure (Tako and Nakamura, 1989).

Thus, the acetate and pyruvate content can be seen as an indicator of the xanthan rheological quality, high pyruvate xanthan is more viscous than that with low pyruvate content (Borges et al., 2009). By elimination of terminal mannose residue, the thickening ability is reduced compared with standard xanthan, whereas an opposite effect was observed by elimination of both terminal mannose and glucuronic acid residues (Hassler and Doherty, 1990). The ordered xanthan structure with stiffness (Tinland and Rinaudo,1989; Brunchi et al., 2013) intermediate that of double-stranded DNA (Smidsrød and Haug, 1971) and triple-stranded collagen or schizophyllan (Zirnsak et al., 1999) is generally accepted, but the question if the ordered xanthan is formed by a single (Norton et al., 1984; Milas and Rinaudo, 1986) or double (Lui and Norisuye, 1988; Camesano and Wilkinson, 2001; Holzwarth and Prestridge, 1977) helix or as dimmers obtained by association of single chains in solution (Foss et al., 1987; Gamini et al., 1991) is still debated. However, it was established that,

80   Smart Materials

depending on solute nature, xanthan exists both in single- and double-stranded conformation in solution. Dintzis et al. (1970) reported that single-stranded xanthan is dissolved in urea, while Southwick et al. (1980) suggested the self-association of xanthan chains in aqueous solutions is induced by the presence of urea. This uncertainness related to the xanthan conformation may be due, on the one hand, to the limitations of analytical techniques used in solution as a result of the formed aggregates that mask the individual behavior of macromolecules and, on the other hand, to the lack of evidences of smaller fractions implied in deviation from the average behavior (the properties being determined in bulk solution) (Camesano and Wilkinson, 2001). Atomic force microscopy (AFM) studies have shown that, at equilibrium, the double-helix configuration of xanthan molecules is based on intramolecular (antiparallel) or inter-molecular (antiparallel or parallel) associations (Moffat et al., 2016) (Fig. 3.2).

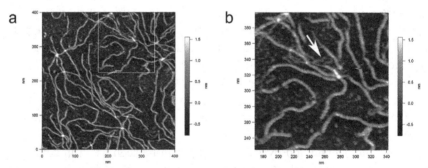

FIGURE 3.2 AFM images of xanthan chains (a) with unraveled ends and (b) electronic zoom of the marked region. Reprinted with permission from Moffat, J., et al. Visualization of Xanthan Conformation by Atomic Force Microscopy. *Carbohydr. Polym.* **2016,** *148,* 380–389. https://creativecommons.org/licenses/by/4.0/

However, it is unanimously accepted that xanthan exists in two ordered structures (i.e., native and re-natured) and one disordered (de-natured) (Milas et al., 1996; Capron et al., 1998; Matsuda et al., 2009).

3.3 STRATEGIES FOR CHANGING XANTHAN PROPERTIES THROUGH PHYSICOCHEMICAL PROCESSES

Increasing performance of xanthan, and therefore its applicability in larger activity domains, can be achieved by modifying the chain structure. Thus, xanthan chains can undergo conformational changes and degradations (by cleavage of backbone and side chains) during exposure to different environmental conditions such as temperature, salt concentration and type, pH, enzyme, radiation, ultrasound, high pressure, etc. as well as be functionalized with different organic groups.

3.3.1 STRUCTURAL MODIFICATIONS INDUCED FROM PRODUCTION PROCESS

Xanthan structure with regard to molecular weight, acetate and pyruvate content in the side chain, and chain conformation can be tailored even from the production process. So, it is important to highlight the production conditions that lead not only at maximal production but also at a desired composition. A general outline of the production process of xanthan from the selected bacterial strain to the final product is presented in Figure 3.3. It is well known that xanthan production is influenced by operation mode of bioreactor (batch or continuous processes), medium composition, cultivation and fermentation conditions (temperature, pH, stirrer speed, airflow rate, fermentation time, and strain of the microorganisms), and post-fermentation conditions (heat treatment, recovery, and purification).

Molar mass is strongly influenced by thermal treatment of fermentation broth that allows chains aggregation (Casas et al., 2000). Besides temperature, obtaining of xanthan with high molar mass is influenced by pH and stirrer speed. Likewise, homologous series of xanthan can be obtained by xanthan removal from fermentation broth at different intervals.

Concerning the *content of the side chain*, literature shows (Cadmus et al., 1978; Thonart et al., 1985) that for xanthan sample with low pyruvate content the culture temperature varying between 27 and 31°C, while temperatures of 31 and 33°C favor the xanthan production. Xanthan is a microbial exopolysaccharide produced by bacterium *Xanthomonas* (*X*) (a plant-pathogenic bacterium), and the type of bacterial strain (such as: *X. arboricola, X. axonopodis, X. campestris, X. carotae, X. citri, X. fragariae, X. gummisudas, X. juglandis, X. malvacearum, X. phaseoli, X. vasculorum*, etc.) influences the content of the side chain. Among bacterial strains, *X. campestris* is the most used in industrial production of xanthan with high pyruvate and acetate content and *X. arboricola* vs. *pruni* leads to xanthan with low concentration of pyruvate due to the presence of rhamnose (Borges et al., 2008; Moreira et al., 2001). Rhamnose is a natural

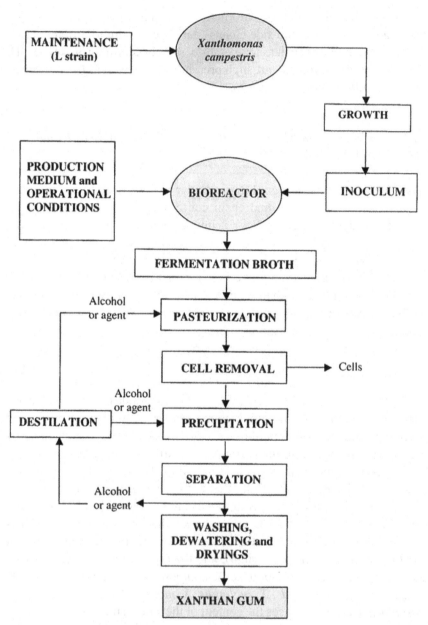

FIGURE 3.3   Outline of the xanthan production process. Reprinted with permission from García-Ochoa, F., et al. Xanthan Gum: Production, Recovery and Properties. *Biotechnol. Adv.* **2000,** *18* (7), 549–579. © 2000 Elsevier.

deoxy sugar commonly bound to other sugars in nature and is believed that it can replace some molecule of outer mannose.

Callet et al. (1989) reported that temperature in combination with pH value of the broth led to different composition of the side chain. So, acid pH favors the hydrolysis of pyruvate groups while under basic pH the acetates were released. Casas et al. (2000) and Papagianni et al. (2001) have proved the relation between the stirrer speed of fermentation broth and the acetate and pyruvate content in the side chain. Thus, it was shown that, as the stirrer speed increases from 100 to 500 rpm, the acetate concentration also increases from 1.53 to 4.44% (Casas et al., 2000), whereas Papagianni et al. (2001) found that at 100 rpm the pyruvation degree is 1.54% and it increases to 3.49% at 600 rpm. However, it was observed that the stirrer speed has no significant effect on the final molar mass of xanthan. Another factor that stimulated xanthan production and high pyruvate content is citric acid. Acid regulates the incorporation of pyruvate groups in the xanthan molecule, and enhances the solubilization of calcium and iron salts (from culture medium components) in the presence of phosphates and stimulates the activity of the Krebs cycle (Kennedy and Bradshaw, 1984).

3.3.2 CONFORMATIONAL TRANSITIONS

3.3.2.1 CONFORMATIONAL TRANSITIONS INDUCED BY SALT ADDITION AND pH CHANGE

In aqueous solutions at 25°C, the backbone of xanthan is disordered or partially ordered in the form of a randomly broken helix but extended, due to the electrostatic repulsions interactions between COO^- groups on the side chains, allowing them to align and associate by hydrogen bonding and to form a weakly structured material. By salt addition, counter ions shield the intramolecular charge–charge interactions and the side chains fold down on the backbone forming a fivefold ordered helical structure (Rochefort and Middleman, 1987). The conformational changes occurring in the presence of salts diminish the chain rigidity reflected in the decrease of the persistence length and intrinsic viscosity. The intrinsic viscosity decreases by salt addition due to the enhancement of the non-Coulombic interactions (Brunchi et al., 2014). The increase of environmental pH determines a conformational change of xanthan chains from double helix to coils in disordered state (Brunchi et al., 2016a). At low (acidic) pH value the protonation of the anionic carboxylic groups occurs and the chains become uncharged and partially flexible. In weak acid environment, the carboxylic groups become partially protonated (around pH = 4) or completely deprotonated (around pH = 6) (Bueno et al., 2013; Bueno and Petri, 2014; Shiroodi and Lo, 2015).

3.3.2.2   CONFORMATIONAL TRANSITIONS INDUCED BY TEMPERATURE

Besides salt and pH solution, conformational transition can be induced by changing temperature and the content of acetyl and pyruvyl groups (Rochefort and Middleman, 1987; Smith et al., 1981).

Temperature at which the transition occurs is known as melting temperature (T_m). The value of T_m depends on salt concentration (Milas and Rinaudo, 1979; Liu et al., 1987), and also pyruvate substitution of the outer mannose (Holzwarth and Ogletree, 1979; Smith et al., 1981) increases of pyruvate substitution lead to a decrease of transition temperature. For a xanthan sample with the ratio of pyruvyl/acetyl substitution of 0.9, T_m was found to be 66°C and for a ratio of 0.4, T_m was 74°C (Kierulf, 1988). Considering the xanthan existence in two ordered structures (i.e., native and renatured) and one disordered (denatured) (Milas et al., 1996; Capron et al., 1998; Matsuda et al., 2009), the heating at a temperature above T_m (80–90°C) can destroy reversibly or irreversibly the ordered conformation of xanthan. By heating the solution of ordered xanthan for few minutes, the side chains become free to rotate and macromolecules adopt a disordered coiled structure. Fast cooling (at room temperature or below) of hot solution avoid the chains degradation and so, the renatured ordered structure is obtained (Rinaudo et al., 1999; Milas et al., 1996). The ordered renatured structure of the xanthan is similar to the native one only at the local level and it is more flexible (Capron et al., 1997; Oviatt and Bran, 1994). The investigation related to thermal behavior of double-helical xanthan in aqueous solution reports the effect of polymer and salt solution concentration on the denaturation, renaturation, and aggregation process (Matsuda et al., 2009). The denaturation of xanthan in dilute solution $(c_p \leq 1$ mg/cm^3) occurs by dissociation of double-helical structure in two single chains. At sufficiently low ionic strength each single chain reconstructed the double-helical structure with the hairpin loops at one end by renaturation. In the case of concentrated xanthan solution ($c_p = 10$ mg/cm³), the denaturation leads to a double-helical structure unwounded at both ends stabilized by the ionic strength of xanthan itself. Renaturation process occurs by multiple and mismatched molecule aggregations that led to apparent molecular weights higher than in native state. In the most cases, the commercial samples of xanthan available on market are found in the renatured state (Milas et al., 1996).

Heating xanthan solution ($T > T_m$) more than 2 h the structural degradation takes place (Capron et al., 1997; Nishinari et al., 1996; Fitzpatrick et al., 2013). Taking into account that a high ionic strength stabilizes the ordered double-stranded xanthan conformation, it is expected that the degradation rate of its chains (measured as loss viscosity) to be lower in salt solutions. In addition, it was found that highly pyruvated samples show a greater thermal stability than lower pyruvated ones (Sandford et al., 1997).

Annealing is, also, a thermal treatment which consists of heating xanthan solution at a temperature higher than that of the gel–sol transition (temperature of gel–sol transition is lower than T_m) followed by cooling. Thus, xanthan chains undergo conformational transitions through intramolecular rearrangements which lead to a decrease of the shear viscosity (Milas and Rinaudo, 1984). Temperature of the gel–sol transition is influenced by the concentration of xanthan solution, its molar mass and the nature of the existing cations. By annealing in the sol state, as other non-gelling polysaccharides, xanthan may form hydrogels (Fujiwara et al., 2000a, 2000b; Iseki et al., 2001; Quinn and Hatakeyama, 1994). Some reports show that solutions with $c_p > 0.5$ wt% form gels through the junction zone composed of oriented bundles of xanthan double helix. On the other hand, the gelling is favored by temperature and the time of annealing process, as well as by the polymer concentration.

3.3.2.3 CONFORMATIONAL TRANSITIONS INDUCED BY THE PRESENCE OF OTHERS POLYMERS

Mixtures of natural and synthetic polymers can give polymeric materials with improved mechanical properties and biological performances. The viscometric and Fourier transform infrared spectroscopy (FTIR) investigations of xanthan/poly(vinyl alcohol) (PVA) mixtures revealed the formation of intermolecular associates through hydrogen bonds interactions (Brunchi et al., 2016b). In solid state, the carboxylate groups from xanthan chains interact preferentially with OH groups of PVA through hydrogen bonding interactions. In aqueous solutions, associations with less or more compact structures between the carboxylate groups from the pyruvic moieties and the hydrophilic units from PVA as well as intermolecular associations of glucuronate units of the side chain with water molecules were evidenced.

A synergistic combination of xanthan with other polysaccharides changes considerably the rheological characteristics and this aspect was extensively investigated in order to obtain physical gels. The feature was attributed to different chain topologies and stiffness of two polysaccharides in the mixture especially if one component is xanthan. Thus, the addition of a small amount of xanthan to a guar gum homogeneous solution induced gel-like properties. The increase of the molecular weight of xanthan or the presence of an electrolyte determines a stronger synergistic effect (Schorsch et al., 1997).

Mixtures of xanthan and galactomannans or glucomannan give thermally reversible gelation attributed by many researchers to chain associations. In the case of galactomannans, the gel strength depends on the mannose to galactose ratio (the interaction decreases as galactose content decreases), increasing from guar to tara and locust bean gum. Associations were attributed to interactions between ordered helix of xanthan and unsubstituted regions of galactomannan chains (Dea et al., 1977). Other explanations discussed by Annable et al. (1994) are briefly given here:

- Mixing above transition temperature, thus xanthan interacts in disordered state.
- Deacetylated xanthan increases the interactions and gives stronger gels due to enhanced side chain mobility.
- The ionic strength controls the gelation for xanthan/Konjac mannan mixtures, the synergistic interaction appears in pure water above the transition temperature (which is located at approx. 82°C), whereas in 0.04 M NaCl around 42°C.
- Xanthan and guar gum interact favorably in both disordered and ordered states.

The charge density of polysaccharide chains and the nature of counter ions play an important role in synergistic interaction and gel formation (Annable et al., 1994). The interaction between the xanthan side chain backbone and glucomannan (Konjac mannan) in aqueous solutions occurs nearly instantaneously bellow the conformational transition temperature of xanthan, determining the associated formation which is thermodynamically favored, the driving force being the decrease of the number of xanthan/water contacts. The presence of electrolyte shifts the conformational change of xanthan to higher temperatures. Divalent cations present a more pronounced effect than monovalent ones because they are more effective for promoting the aggregation or ordering of xanthan chains. The polymer–polymer heterocontacts involve both ordered and disordered xanthan sequences. The gelation temperature decreases in the presence of electrolytes because the self-association of xanthan macromolecules is favored in a greater extent than xanthan/Konjac mannan associations.

3.3.3 DEGRADATION OF XANTHAN CHAINS

Xanthan degradation is strongly related to its application in different areas schematically presented in Section 3.4. During degradation, xanthan chains can undergo a depolymerization by cleavage of noncovalent bonds implied in stabilization of double-helical structure (by hydrolytic processes acid pH and radiolysis assisted, thermal treatments) or glycoside bonds from backbone and side chain (by acid and enzymatic hydrolysis). Also, a long treatment at high temperatures in seawater produces the side chain degradation evidenced by a decrease of mannose/glucose ratio and pyruvate content (Christensen and Smidsrød, 1991). The rate of degradation is closely related to xanthan conformation, the ordered double-stranded conformation has a low rate of degradation by free-radical de-polymerization or acid-catalyzed hydrolysis, but at temperatures around T_m or even higher, the xanthan conformation is partly or total disordered and the degradation rate is high.

In this section, some aspects concerning the xanthan degradation through ultrasound, radiation, high-pressure treatments or pH, and enzyme action are briefly presented.

3.3.3.1 ULTRASOUND DEGRADATION

Comparatively with classical methods of polymer degradation (e.g., thermal, mechanical, and chemical degradation), many authors (Kulicke et al., 1996; Schittenhelm and Kulicke, 2000; Pfefferkorn et al., 2003; Goodwin et al., 2011) agreed that ultrasonic degradation is an adequate method to control molecular weight and to produce fragments of definite molecular size. The simplicity, inexpensively, environment friendly (especially because requires a short processing time), and lack of purification steps of degraded samples (does not need the addition of reagents) are few advantages of this method. During ultrasound degradation the polymer chain scission occurs close to gravity center of macromolecule (Kulicke et al., 1996) without major change in the chemical structure of xanthan. Milas et al. (1986) confirm the lack of changes in chemical structure of xanthan degraded by sonication (the substituent content in the xanthan side chain remains unchanged). They observed that by sonication a xanthan sample with low polydispersity index (around 1.4) was obtained; generally, the polydispersity ranges from 1.4 to 2.8. The ultrasound degradation rate depends on the frequency and intensity of ultrasound (Mark et al., 1998; Czechowska-Biskup et al., 2005; Koda et al., 2011; Price and Smith, 1993a), nature of the solvent (Price and Smith, 1991; Malhotra, 1982; Li and Feke, 2005a, 2005b), concentration of polymer solution and temperature (Price and Smith, 1993a, 1993b), and sonication time (e.g., before reaching a constant minimum value, the molecular weight decreases exponentially in time). A comprehensive study about the effect of temperature, salts, and pyruvate groups on ultrasonic degradation of xanthan in aqueous solution (c_p = 3.0 g/L) has been conducted by Li and Feke (2015a, 2015b). With regard to temperature, they observed a faster degradation rate or better degradation at 25°C than at 35°C. The increase of temperature favors both the disordered conformation of xanthan (that is the most flexible and the reactive structure with high degradation potential (Lambert and Rinaudo, 1985; Christensen and Smidsrød, 1991)) and the increase of the solvent vapor pressure. The vapor pressure slows down the movement of solvent molecules, the shock waves are diminished and so, the chains degradation is annihilated (Price and Smith, 1993a). The salt effect on xanthan degradation is complex and depends on specific behavior of salt in aqueous solution (e.g., salting-out or salting-in behavior according to the Hofmeister theory). Salts with salting-out behavior inhibit the dissolution of polymers and so, decrease the efficiency of ultrasonic degradation, whereas the salts with salting-in behavior have opposite effect (Li and Feke, 2015a, 2015b). As mentioned above, the pyruvate groups decrease the stability of ordered conformation of xanthan and consequently decrease the degradation efficiency of xanthan.

3.3.3.2 DEGRADATION INDUCED BY RADIATION

Irradiation represents a conventional method by which the polymer structure can be modified by degradation, grafting, and cross-linking. Sometimes, even a few cross-links or scission sites in the molecular structure can dramatically affect the strength or solubility of a polymer.

As a function of the radiation energy, ionizing and non-ionizing radiations exist. Ionizing radiation is made up of energetic subatomic particles (include alpha and beta particles and neutrons), ions, or atoms moving at high speeds (usually greater than 1% of the speed of light) and electromagnetic waves on the high-energy end of the electromagnetic spectrum (gamma-ray, X-ray, and the high ultraviolet) while the low ultraviolet part of the electromagnetic spectrum as well as the part of the spectrum below UV including visible light, infrared, microwaves, and radio waves are considered non-ionizing radiations. Among ionizing radiation, gamma-ray enjoys great attention because can lead to polysaccharide degradation (El-Mohdy, 2017; Charlesby, 1997; Li et al., 2011; Binh et al., 2016) without introducing chemical reagents or control temperature, environment, and additives; the degradation of the polymer chain in aqueous solution is produced through the free radicals resulted from water radiolysis (i.e., hydroxyl radical, hydrate electron, and hydrogen atom) (Al-Assaf et al., 1995).

The most important effect of polysaccharide degradation induced by gamma-ray is the scission of the C—O bond connecting the glycoside groups on the main chain (Şen et al., 2016) accompanied by the appearance of low molecular weight oligomers (Hayashi and Aoki, 1985; Ghali et al., 1979; Raffi et al., 1981). Such oligosaccharide with new functional characteristics (such as improved bioactivity and antioxidant and bioadhesive activity) can be used to obtain biomedical products (Clough, 2001; Rosiak et al., 1995) or in agriculture to enhance the crop yields and defense system of the plants, to reduce the loss of chemical fertilizers, etc. (Luan et al., 2003; Binh et al., 2016).

Degradation by gamma irradiation was conducted on xanthan both in dry form (Şen et al., 2016) and in solution (Li et al., 2011; Binh et al., 2016). Şen et al. (2016) irradiated xanthan in solid state at different doses (2, 5, 5.0, 10, 20, 30, and 50 kGy) and dose rates (0.1, 3.3 and 7.0 kGy/h), and found that: (1) the chain scission is more effective at low dose rates and (2) irrespective of the dose rate and dose used the non-Newtonian and shear thinning behavior of xanthan solution (1% concentration) has not changed. Another effect of irradiation treatment in solid state is related to xanthan solubility; the radiation dose reduces the time required for dissolution and thus, increases the xanthan solubility (Binh et al., 2016). Li et al. (2010) subjected the irradiation of 2% w/v xanthan solution (at different radiation doses and a dose rate of 0.5 kGy) and concluded that the degradation was caused by simultaneous action of gamma-ray on crystal-lization and amorphous area of xanthan. Therefore, the molecular weight (M_w) and the polydispersity (I) of xanthan decrease with gamma-ray dose from M_w = 5755.5 kDa and I = 3.3 at 0 kGy to M_w = 7.3 kDa and I = 2.4 at 120 kGy.

Comparing the molecular weight of xanthan obtained by irradiation (at the same dose and dose rate), it was observed that the xanthan degrada-tion is more efficient in solution than that in solid state (Binh et al., 2016; Yan-Jie et al., 2010; Şen et al., 2016).

3.3.3.3   AUTOCLAVING EFFECT ON XANTHAN STRUCTURE

Lagoueyte and Paquin (1998) have observed that the mechanical stress during high-pressure treatment can degraded xanthan chains; the degrada-tion was evidenced by continuous reduction of xanthan molecular weight as well as solution shear viscosity.

Later, Gurlez et al. (2012) confirmed the degradation of macromolec-ular chains at high pressures when the xanthan solution was submitted to different treatments (such as heating, high-pressure homogenization, auto-claving, and irradiation) and they reported a similar effect of autoclaving in the case of oat-glycan. The effects of these treatments are presented in Figure 3.4.

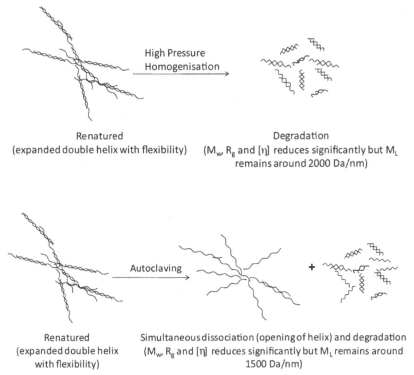

FIGURE 3.4 Schematic illustration of structural and conformational changes of xanthan in aqueous solution due to the effect of high pressure and autoclaving. Reprinted with permission from Gulrez, S. K. H., et al. Revisiting the Conformation of Xanthan and the Effect of Industrially Relevant Treatments. *Carbohydr. Polym.* **2012**, *90* (3), 1235–1243. © 2012 Elsevier.

3.3.3.4 CHEMICAL DEGRADATION

Chemical degradation of xanthan can be induced by acid, alkaline, and enzymatic processes.

Christensen et al. (1991, 1993, 1996) studied the xanthan degradation by acid hydrolysis (e.g., pH = 1–4) at 80°C when the xanthan conformation was varied from fully ordered to partially disordered. Thus, depolymerization of double-stranded xanthan by partial acid hydrolysis can lead to xanthan fractions with conformational properties more or less analogs to undegraded xanthan (depending on ionic strength and degradation time), various chemical compositions of the side chains and reduced molecular

weights. In the side chain the terminal β-mannose hydrolysis occurs preferentially and leads to modified xanthan fractions from the intact "polypen-
tamer" to "polytetramer"; the inner α-mannose acidic hydrolysis is slow
while the glycosidic bond between glucuronic acid and α-mannose is resistant to hydrolysis. "Polytetramer" fractions obtained from xanthan hydrolysis present low viscosity, while the lack of the two terminal units from
the side chains (β-mannose and glucuronic acid) leads to a "polytrimer"
with higher viscosity (resulted by enzymatic hydrolysis). The acetyl group
has a negative effect on the "polytetramer" viscosity and does not affect the
"polytrimer" xanthan viscosity (Hassler and Doherty, 1990).

   Enzymes are substances that act as catalysts in living organisms, regulating the rate at which chemical reactions take place without their alteration. They are constituted by linear chains of amino acids that fold to
produce a three-dimensional structure.

   Even if xanthan is a polysaccharide highly stable to degradation induced
by most of microorganisms, it can be degraded by cellulase (Rinaudo and
Milas, 1980), as well as xanthanase (endo-1,4-β-D-glucanase) and xanthan
lyase (4,5-transeliminase) found in microbes. Some xanthanases have
been categorized as the cellulase family members. Taking into account
that the biodegradation processes described in literature (Ahlgren, 1991;
Shatwell et al., 1990; Sutherland, 1982, 1987; Hashimoto et al., 1998) are
associated with those of hydrolysis, xanthanase catalyzes the hydrolysis of
the xanthan backbone, whereas xanthan lyase catalyzes the cleavages of
the link between the terminal mannosyl and glucuronyl residues from the
side chain of xanthan.

   The degradation process produced by enzyme is followed by changes
in molecular mass, viscosity and morphology. Thus, Rinaudo and Milas
(1980) demonstrated that in salt-free solution the rate of enzymatic hydrolysis of xanthan is reduced by the presence of aggregations or microgel
structure; the xanthan chains in unordered conformation present a higher
ability to degrade because small side chains protect the glycosidic bond
against hydrolysis.

   Another way to degrade the xanthan structure is based on chemical
reactivity of acetyl group making possible the xanthan deacetylation by
alkaline treatment.

   The literature shows that, irrespective of deacetylation conditions (i.e.,
0.025 M NaOH solution for 3 h (Covielo et al., 1986; Dentini, 1984),
0.01 M KOH solution for 10 h (Tako and Nakamura, 1984, 1989; Tako,

1991) or for 2.5 h (Khouryieh et al., 2007)), the process occurs without further modification of xanthan backbone. The deacetylation degree and the viscosity values at 10 s^{-1} of xanthan solutions in the two types of alkaline solutions were close (1.3%, 410 mPa·s in 0.01 M NaOH and 1.4%, 420 mPa·s in 0.01 M KOH) (Pinto et al., 2011).

3.3.4 CHEMICAL FUNCTIONALIZATION OF XANTHAN

Chemical modification of xanthan can be realized by grafting different groups on the macromolecular chains. Grafting of xanthan is based on free radical mechanism microwave and plasma assisted.

In grafting process of xanthan with acrylamide (Behari et al., 2001), methacrylic acid (Kumar et al., 2007), acrylic acid (Pandey et al., 2003), 2-acrylamido-2-methyl-1-propanesulfonic acid (Srivastava and Behari, 2009), 2-acrylamidoglycolic acid (Sand et al., 2010) and 4-vinyl pyridine (Kumar et al., 2009a, 2009b) different redox systems were used and properties such as flocculation (Brunchi and Ghimici, 2013), solubility, thermal stability, binding strength, water retention, and resistance to biodegradation were improved.

Drugs release from xanthan matrix was improved if acrylamide grafting on xanthan chains was initiated by $Fe^{2+}/BrO3^-$ redox system in aqueous medium under a nitrogen atmosphere (Behari et al., 2001) or ceric ion (Mundargi et al., 2007) or was microwave assisted (Kumar et al., 2009a, 2009b). Based on FTIR spectra of pure components (xanthan and polyethylacrylate) and graft copolymer based on xanthan and polyethylacrylate, it was supposed that grafting might have taken place on OH sites of xanthan (Pandey and Mishra, 2011; Gils et al., 2009), whereas in case of hydrophobically modified xanthan by grafting with octylamine has reported the contribution of carboxylic acid groups of both the glucuronic acid and the pyruvyl group (Roy et al., 2014).

Another practical application of grafting xanthan is to improve the gelling ability. Cross-linked hydrogels with N,N'-methylenebisacrylamide (MBA) (Gils et al., 2009) or glutaraldehyde (Jampala et al., 2005) were synthesized. Usually, the cross-linking agent reacts with amines, amides, and thiol groups in proteins and lack of such groups in xanthan structure requires its functionalization. Thus, Jampala et al. have functionalized xanthan by ethylenediamine grafting in cold plasma; in plasma environment was implanted SiH_xCl_y in order to anchor ethylenediamine in situ conditions (Fig. 3.5) (Jampala et al., 2005). Authors reported that stabilization of xanthan gels formed by cross-linking with glutaraldehyde increases with polymer concentration and time of plasma treatment.

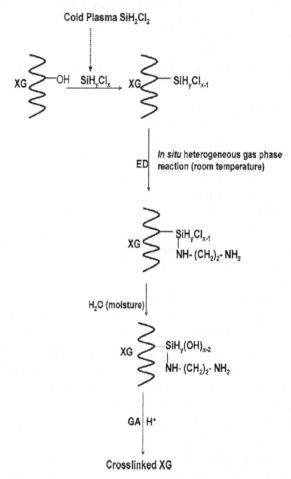

FIGURE 3.5   Schematic representation of modification reaction for xanthan. Reprinted with permission from Jampala, S. N., et al. Plasma-enhanced Modification of Xanthan Gum and Its Effect on Rheological Properties. *J. Agric. Food Chem.* **2005,** *53* (9), 3618–3625.© 2005 American Chemical Society.

In an attempt to avoid toxic cross-linking agents (such as epichlorohydrin, glutaraldehyde, MBA, etc.) in obtaining of hydrogels, Hamcerencu et al. (2007) have prepared unsaturated xanthan derivatives through esterification with acrylic acid (under homogenous condition) or acryloyl chloride and maleic anhydride (under heterogeneous conditions). Strategies for xanthan modification are presented in Figure 3.6.

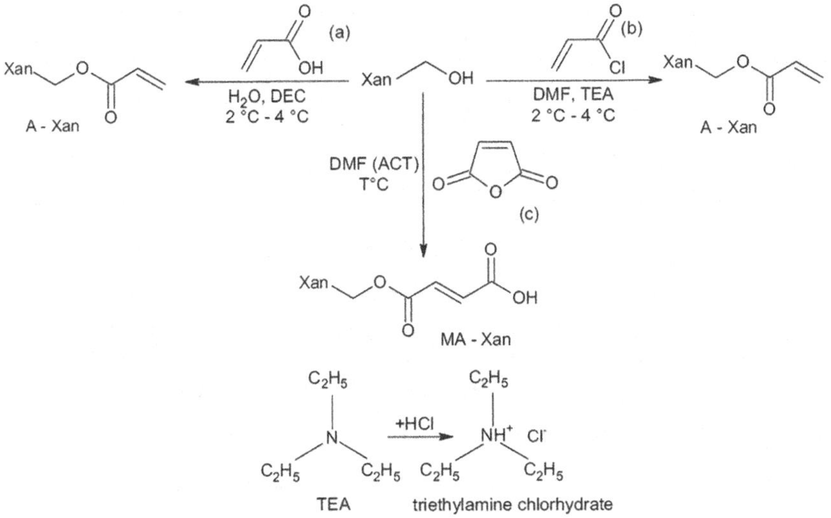

FIGURE 3.6   Strategies for xanthan modification: (a) acrylic acid; (b) acryloyl chloride; and (c) maleic anhydride. Reprinted with permission from Hamcerencu, M., et al. New Unsaturated Derivatives of Xanthan Gum: Synthesis and Characterization. *Polymer* **2007,** *48* (7), 1921–1929. © 2007 Elsevier.

Taking into account that unsaturated esters of xanthan can be used in synthesis of grafted/cross-linked structure with hydrogel properties, the degrees of substitution (DS) of xanthan is important parameter influenced by time and temperature reaction, chemical nature of esterification agent or molar ratio; high DS of xanthan was obtained by using maleic acid and so, it is able to form hydrogels with dense network.

3.4   APPLICATIONS OF XANTHAN

Xanthan receives a great attention from researchers due to its special properties (high viscosity even at low concentration, pseudoplasticity, biocompatibility and biodegradability, and nontoxicity) which make it useful in various applications from the food, cosmetic and pharmaceutical industries, etc. (Fig. 3.7).

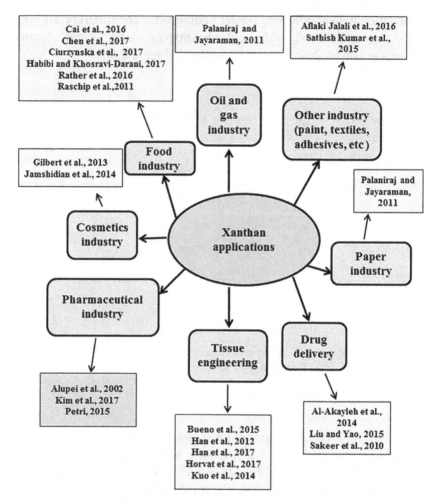

FIGURE 3.7 Potential applications of xanthan.

The most important application of xanthan is in the food industry as additive which, in small quantities, confers on products the stability during storage, acts as an emulsifier for oil–water systems and it can have the role of thickener for aqueous solutions (Saha and Bhattacharya, 2010). In addition, the xanthan improves the texture and the viscoelastic properties of different food products (i.e., gluten-free dough (Cai et al., 2016), sauce, mayonnaise, dressings (Aslanzadeh et al., 2012; Wang et al., 2016), etc.). Added to cosmetics and personal care products, it improves the shear thinning flow properties for the tube packed products and gives an excellent stability for the skin creams and emulsions. Xanthan has the advantage that is resistant to the action of common enzymes, the properties of its solution are not affected by pH change in the range of 1–13 and it presents excellent thermal stability. These properties along with its capacity to increase the consistency and the stability of the systems even at low concentration (up to 0.5%) allow the xanthan application in the formulation of various cosmetics (e.g., skin creams, whitening cosmetics, shampoo, toothpaste, etc.) (Knowlton and Pearce, 1993).

The addition of xanthan into pharmaceutical formulations in the form of emulsion or suspension prevents the separation of the components insoluble in water (Katzbauer, 1998). For applications as drug delivery, the xanthan is added as excipient in tablets or it is used as supporting hydrogels (Andreopoulos and Tarantili, 2001; Petri, 2015). Frequently, the xanthan is used for drug delivery systems in combination with other polymers to get the desired properties, for example, chitosan (Kulkarni et al., 2015) or guar gum (Patel and Patel, 2007). The development of the new composites, which are mineralized like the bone, represents a hot area of research due to the possibility to obtain the materials for bone engineering applications. Such materials were obtained by combining the xanthan with chitosan and/or gellan gum. The investigations of such gels evidenced that the intracellular controlled delivery of dexamethasone disodium phosphate through xanthan/gellan hydrogel containing chitosan nanoparticles represents a promising approach for inducing osteo-differentiation (Sehgal et al., 2017). The presence of chitosan nanoparticles of 297 ± 61 nm diameter in the gellan-xanthan gel caused an increase of the cell proliferation and differentiation as a result of the sustained release of growth factors (Dyondi et al., 2013).

Due to its viscoelastic properties (pseudoplasticity), compatibility with salt and thermal stability, xanthan is frequently used in petroleum industry as an additive for drilling fluids or in pipeline cleaning, completion/work-over operations, etc. (BeMiller, 1998). The addition of xanthan in formulations used in agriculture industry (fungicides, herbicides, and insecticides) leads to the uniformity of the solid component in the suspension system improving their flow-ability (Rosalam and England, 2006).

Xanthan is added into water-based paints in order to improve their rheological properties and to stabilize the pigment in system. Moreover, by addition of the xanthan the paint diffusion and its spraying are more efficiently controlled.

KEYWORDS

- xanthan
- conformational transition
- degradation
- ultrasound
- radiation
- hydrogels

REFERENCES

Ahlgren, J. A. Purification and Characterization of a Pyruvated-Mannose-Specific Xanthan Lyase from Heat-stable, Salt-Tolerant Bacteria. *Appl. Environ. Microbiol.* **1991,** *57* (9), 2523–2528.

Al-Akayleh, F., et al. Using Chitosan and Xanthan Gum Mixtures as Excipients in Controlled Release Formulations of Ambroxol HCl—In Vitro Drug Release and Swelling Behavior. *J. Excip. Food Chem.* **2014,** *5* (2), 140–148.

Al-Assaf, S., et al. The Enhanced Stability of the Cross-linked Hylan Structure to Hydroxyl (OH) Radicals Compared with the Uncross-linked Hyaluronan. *Rad. Phys. Chem.* **1995,** *46* (2), 207–217.

Alupei, I. C., et al. Superabsorbant Hydrogels Based on Xanthan and Poly(Vinyl Alcohol) 1. The Study of the Swelling Properties. *Eur. Polym. J.* **2002,** *38* (11), 2313–2320.

Annable, P.; Williams, P. A.; Nishinari, K. Interaction in Xanthan-Glucomannan Mixtures and the Influence of Electrolyte. *Macromolecules* **1994,** *27* (15), 4204–4211.

Andreopoulos, A. G.; Tarantili, P. A. Xanthan Gum as a Carrier for Controlled Release of Drugs. *J. Biomater. Appl.* **2001,** *16* (1), 34–46.

Aslanzadeh, M., et al. Rheological Properties of Low Fat Mayonnaise with Different Levels of Modified Wheat Bran. *J. Food Biosci. Technol.* **2012,** *2,* 27–34.

Behari, K., et al. Graft Copolymerization of Acrylamide onto Xanthan Gum. *Carbohydr. Polym.* **2001,** *46* (2), 185–189.

BeMiller, J. N.; Gums. In *Concise Polymeric Materials Encyclopedia*; Salamone, J. C., Ed.; CRC Press: Boca Raton, FL, 1998; p 597.

Bilanovic, D.; Starosvetsky, J.; Armon, R. H. Cross-Linking Xanthan and other Compounds with Glycerol. *Food Hydrocoll.* **2015,** *44,* 129–135.

Bilanovic, D.; Starosvetsky, J.; Armon, R. H. Preparation of Biodegradable Xanthan–Glycerol Hydrogel, Foam, Film, Aerogel and Xerogel at Room Temperature. *Carbohydr. Polym.* **2016,** *148,* 243–250.

Binh, N. V., et al. Low Molecular Weight Xanthan Prepared by Gamma Irradiation and Its Effects on Development of Seedlings. *RAD Conference Proc.* **2016,** *1,* 95–98.

Borges, C. D., et al. The Influence of Thermal Treatment and Operational Conditions on Xanthan Produced by *X. arboricola pv pruni strain 106*. *Carbohydr. Polym.* **2009,** *75* (2), 262–268.
Borges, C. D., et al. Influence of Agitation and Aeration in Xanthan Production by *Xanthomonas Campestris pv Pruni* Strain 101. *Rev. Argent. Microbiol.* **2008,** *40* (2), 81–85.
Brunchi, C. E.; Ghimici, L. Xanthan: Physicochemical Properties and Applications in Separation Processes. In *Polymer Materials with Smart Properties*; Bercea, M., Ed.; Nova Science Publishers Inc.: New York, NY, 2013; pp 151–187.
Brunchi, C. E.; Morariu, S.; Bercea, M. Intrinsic Viscosity and Conformational Parameters of Xanthan in Aqueous Solutions: Salt Addition Effect. *Colloids Surf. B* **2014,** *122,* 512–519.
Brunchi, C. E., et al. Some Properties of Xanthan Gum in Aqueous Solutions: Effect of Temperature and pH. *J. Polym. Res.* **2016a,** *23* (7), 123.
Brunchi, C. E., et al. Investigations on the Interactions Between Xanthan Gum and Poly (Vinyl Alcohol) in Solid State and Aqueous Solutions. *Eur. Polym. J.* **2016b,** *84,* 161–172.
Bueno, V. B., et al. Synthesis and Swelling Behavior of Xanthan-based Hydrogels. *Carbohydr. Polym.* **2013,** *92,* 1091–1099.
Bueno, V. B.; Petri. D. F. S. Xanthan Hydrogel Films: Molecular Conformation, Charge Density and Protein Carriers. *Carbohydr. Polym.* **2014,** *101,* 897–904.
Bueno, V. B., et al. Biocompatible Xanthan/Polypyrrole Scaffolds for Tissue Engineering. *Mater. Sci. Eng. C Mater. Biol. Appl.* **2015,** *52,* 121–128.
Cadmus, M. C., et al. Colonial Variation in *Xanthomonas Campestris* NRRL B–1459 and Characterization of the Polysaccharide from a Variant Strain. *Can. J. Microbiol.* **1976,** *22,* 942–948.
Cadmus, M. C., et al. Synthetic Media for Production of Quality Xanthan Gum in 20 Liter Fermentors. *Biotechnol. Bioeng.* **1978,** *20* (7), 1003–1014.
Cai, J., et al. Physicochemical Properties of Hydrothermally Treated Glutinous Rice Flour and Xanthan Gum Mixture and Its Application in Gluten-Free Noodles. *J. Food Eng.* **2016,** *186,* 1–9.
Callet, F.; Milas, M.; Rinaudo, M. On the Role of Thermal Treatments on the Properties of Xanthan Solutions. *Carbohydr. Polym.* **1989,** *11* (2), 127–137.
Camesano, T. A.; Wilkinson, K. J. Single Molecule Study of Xanthan Conformation Using Atomic Force Microscopy. *Biomacromolecules* **2001,** *2* (4), 1184–119.
Capron, I.; Brigandt, G.; Muller, G. About the Native and Renatured Conformation of Xanthan Exopolysaccharide. *Polymer* **1997,** *38* (21), 5289–5295.
Capron, I.; Brigand, G.; Muller, G. Thermal Denaturation and Renaturation of a Fermentation Broth of Xanthan: Rheological Consequences. *Int. J. Biol. Macromol.* **1998,** *23* (3), 215–225.
Casas, J. A.; Santos, V. E.; García-Ochoa, F. Xanthan Gum Production under Several Operational Conditions: Molecular Structure and Rheological Properties. *Enzyme Microbiol. Technol.* **2000,** *26* (2–4), 282–291.
Charlesby, A. Use of High Energy Radiation for Crosslinking and Degradation. *Rad. Phys. Chem.* **1977,** *9* (1–3), 17–29.
Chen, L., et al. Effect of Xanthan-Chitosan-Xanthan Double Layer Encapsulation on Survival of Bifidobacterium BB01 in Simulated Gastrointestinal Conditions, Bile Salt Solution and Yogurt. *LWT Food Sci. Technol.* **2017,** *81,* 274–280.

100   Smart Materials

Christensen, B. E.; Smidsrød, O. Hydrolysis of Xanthan in Dilute Acid: Effects on Chemical Composition, Conformation, and Intrinsic Viscosity. *Carbohydr. Polym.* **1991**, *214* (1), 55–69.

Christensen, B. E., et al. Depolymerization of Double-Stranded Xanthan by Acid Hydrolysis: Characterization of Partially Degraded Double Stranded and Single-stranded Oligomers Released from the Ordered Structures. *Macromolecules* **1993**, *26* (22), 6111–6120.

Christensen, B. E.; Myhr, M. H.; Smidsrød, O. Degradation of Double-Stranded Xanthan by Hydrogen Peroxide in the Presence of Ferrous Ions: Comparison to Acid Hydrolysis. *Carbohydr. Res.* **1996**, *280* (1), 85–99.

Ciurzynska, A., et al. The Effect of Composition and Aeration on Selected Physical and Sensory Properties of Freeze-Dried Hydrocolloid Gels. *Food Hydrocoll.* **2017**, *67*, 94–103.

Clough, R. L. High-energy Radiation and Polymers: A Review of Commercial Processes and Emerging Applications. *Nucl. Instrum. Methods Phys. Res. Sect. B Beam Interact. Mater. Atoms* **2001**, *185* (1–4), 8–33.

Covielo, T., et al. Solution Properties of Xanthan. 1. Dynamic and Static Light Scattering from Native and Modified Xanthans in Dilute Solutions. *Macromolecules* **1986**, *19* (11), 2826–2831.

Czechowska-Biskup, R., et al. Degradation of Chitosan and Starch by 360-kHz Ultrasound. *Carbohydr. Polym.* **2005**, *60* (2), 175–184.

Dea, I. C. M., et al. Associations of Like and Unlike Polysaccharides: Mechanism and Specificity in Galactomannans, Interacting Bacterial Polysaccharides, and Related Systems. *Carbohydr. Res.* **1977**, *57*, 249–272.

Dentini, M.; Crescenzi, V.; Blasi, D. Conformational Properties of Xanthan Derivatives in Dilute Aqueous Solution. *Int. J. Biol. Macromol.* **1984**, *6* (2), 93–98.

Dintzis, F. R.; Babcock, G. E.; Tobin, R. Studies on Dilute Solutions and Dispersion of the Polysaccharide from *Xanthomonas Campestris* NRRL B-1459. *Carbohydr. Res.* **1970**, *13* (2), 257–267.

Dyondi, D.; Webster, T. J.; Banerjee, R. A Nanoparticulate Injectable Hydrogel as a Tissue Engineering Scaffold for Multiple Growth Factor Delivery for Bone Regeneration. *Int. J. Nanomed.* **2013**, *8* (1), 47–59.

Dwek, R. A. Glycobiology: Toward Understanding the Function of Sugars. *Chem. Rev.* **1996**, *96* (2), 683–720.

El-Mohdy, H. L. A. Radiation-induced Degradation of Sodium Alginate and Its Plant Growth Promotion Effect. *Arab. J. Chem.* **2017**, *10* (1), S431–S438.

Faria, S., et al. Characterization of Xanthan Gum Produced from Sugar Cane Broth. *Carbohydr. Polym.* **2011**, *86* (2), 469–476.

Fitzpatrick, P., et al. Control of the Properties of Xanthan/Glucomannan Mixed Gels by Varying Xanthan Fine Structure. *Carbohydr. Polym.* **2013**, *92* (2), 1018–1025.

Foss, P.; Stokke, B. T.; Smidsrød, O. Thermal Stability and Chain Conformational Studies of Xanthan at Different Ionic Strengths. *Carbohydr. Polym.* **1987**, *7* (6), 421–433.

Fujiwara, J., et al. Structural Change of Xanthan Gum Association in Aqueous Solutions. *Thermochim. Acta.* **2000a**, *352–353*, 241–246.

Fujiwara, J., et al. Gelation of Hyaluronic Acid through Annealing. *Polym. Int.* **2000b**, *49*, 1604–1608.

Gamini, A.; de Bleijser, J.; Leyte, J. C. Physico-chemical Properties of Aqueous Solutions of Xanthan: An NMR Study. *Carbohydr. Res.* **1991**, *220*, 33–47.

García-Ochoa, F., et al. Xanthan Gum: Production, Recovery and Properties. *Biotechnol. Adv.* **2000**, *18* (7), 549–579.

Ghali, Y., et al. Modification of Corn Starch and Fine Flour by Acid and Gamma Irradiation. Part 1. Chemical Investigation of the Modified Product. *Starch Stärke.* **1979**, *31* (10), 325–358.

Gilbert, L., et al. Stretching Properties of Xanthan, Carob, Modified Guar and Celluloses in Cosmetic Emulsions. *Carbohydr. Polym.* **2013**, *93* (2), 644–650.

Gils, P. S.; Ray, D.; Sahoo, P. K. Characteristics of Xanthan Gum-based Biodegradable Superporous Hydrogel. *Int. J. Biol. Macromol.* **2009**, *45* (4), 364–371.

Goodwin, D. J., et al. Ultrasonic Degradation for Molecular Weight Reduction of Pharmaceutical Cellulose Ethers. *Carbohydr. Polym.* **2011**, *83* (2), 843–851.

Gulrez, S. K. H., et al. Revisiting the Conformation of Xanthan and the Effect of Industrially Relevant Treatments. *Carbohydr. Polym.* **2012**, *90* (3), 1235–1243.

Habibi, H.; Khosravi-Darani, K. Effective Variables on Production and Structure of Xanthan Gum and Its Food Applications: A Review. *Biocatal. Agric. Biotechnol.* **2017**, *10*, 130–140.

Han, G., et al. Preparation of Xanthan Gum Injection and Its Protective Effect on Articular Cartilage in the Development of Osteoarthritis. *Carbohydr. Polym.* **2012**, *87* (2), 1837–1842.

Hashimoto, W., et al. Xanthan Lyase of *Bacillus* sp. Strain GL1 Liberates Pyruvylated Mannose from Xanthan Side Chains. *Appl. Environ. Microbiol.* **1998**, *64* (10), 3765–3768.

Hassler, R. A.; Doherty. D. H. Genetic Engineering of Polysaccharide Structure: Production of Variants of Xanthan Gum in *Xanthomonas Campestris*. *Biotechnol. Prog.* **1990**, *6* (3), 182–187.

Hayashi, T.; Aoki, S. Effect of Irradiation on Carbohydrate Metabolism Responsible for Sucrose Accumulation in Potatoes. *J. Agric. Food Chem.* **1985**, *33* (1), 14–17.

Hamcerencu, M., et al. New Unsaturated Derivatives of Xanthan Gum: Synthesis and Characterization. *Polymer* **2007**, *48* (7), 1921–1929.

Holzwarth, G.; Prestridge, E. B. Multistranded Helix in Xanthan Polysaccharide. *Science* **1977**, *197* (4305), 757–759.

Holzwarth, G.; Ogletree, J. Pyruvate Free Xanthan. *Carbohydr. Res.* **1979**, *76* (1), 277–280.

Horvat, G., et al. Novel Ethanol-Induced Pectin–Xanthan Aerogel Coatings for Orthopedic Applications. *Carbohydr. Polym.* **2017**, *166*, 365–376.

Iseki, T., et al. Viscoelastic Properties of Xanthan Gum Hydrogels Annealed in the Sol State. *Food Hydrocoll.* **2001**, *15* (4–6), 503–506.

Jalali, M. A.; Dadvand Koohi, A.; Sheykhan, M. Experimental Study of the Removal of Copper Ions Using Hydrogels of Xanthan, 2-Acrylamido-2-Methyl-1-Propane Sulfonic Acid, Montmorillonite: Kinetic and Equilibrium Study. *Carbohydr. Polym.* **2016**, *142*, 124–132.

Jampala, S. N., et al. Plasma-enhanced Modification of Xanthan Gum and Its Effect on Rheological Properties. *J. Agric. Food Chem.* **2005**, *53* (9), 3618–3625.

Jamshidian, M., et al. Stretching Properties of Xanthan and Hydroxypropyl Guar in Aqueous Solutions and in Cosmetic Emulsions. *Carbohydr. Polym.* **2014**, *112*, 334–341.

Katzbauer, B. Properties and Applications of Xanthan Gum. *Polym. Degard. Stab.* **1998,** *59* (1–3), 81–84.

Kennedy, J. F.; Bradshaw, I. J. Production, Properties and Applications of Xanthan. *Prog. Ind. Microbiol.* **1984,** *19,* 319–371.

Khouryieh, H. A.; et al.; Influence of Deacetylation on the Rheological Properties of Xanthan–Guar Interactions in Dilute Aqueous Solutions. *J. Food Sci.* **2007,** *72* (3), 173–181.

Kierulf, C. The Thermal Stability of Xanthan. Ph.D. Thesis, Heriot-Watt University, 1988.

Kim, J., et al. Engineered Chitosan–Xanthan Gum Biopolymers Effectively Adhere to Cells and Readily Release Incorporated Antiseptic Molecules in a Sustained Manner. *J. Ind. Eng. Chem.* **2017,** *46,* 68–79.

Knowlton, J. L.; Pearce, S. E. M. *Handbook of Cosmetic Science & Technology,* 1st ed.; Elsevier Advanced Technology: Amsterdam, Netherlands, 1993; p 20.

Koda, S.; Taguchi, K.; Futamura, K. Effects of Frequency and a Radical Scavenger on Ultrasonic Degradation of Water-Soluble Polymers. *Ultrason. Sonochem.* **2011,** *18* (1), 276–281.

Kool, M. M., et al. Comparison of Xanthans by the Relative Abundance of Its Six Constituent Repeating Units. *Carbohydr. Polym.* **2013,** *98* (1), 914–921.

Kool, M. M., et al. The Influence of the Six Constituent Xanthan Repeating Units on the Order-disorder Transition of Xanthan. *Carbohydr. Polym.* **2014,** *104,* 94–100.

Kulicke, W. M.; van Eikern, A. Determination of the Microstructure of the Fermentation Polymer Xanthan. *Makromol. Chem. Macromol. Symp.* **1992,** *61* (1), 75–93.

Kulicke, W. M., et al. Characterization of Aqueous Carboxymethylcellulose Solutions in Terms of Their Molecular Structure and Its Influence on Rheological Behaviour. *Polymer.* **1996,** *37* (13), 2723–2731.

Kulkarni, N.; Wakte, P.; Naik, J. Development of Floating Chitosan–Xanthan Beads for Oral Controlled Release of Glipizide. *Int. J. Pharm. Investig.* **2015,** *5* (2), 73–80.

Kumar, R.; Srivastava, A.; Behari, K. Graft Copolymerization of Methacrylic Acid onto Xanthan Gum by Fe^{2+}/H_2O_2 Redox Initiator. *J. Appl. Polym. Sci.* **2007,** *105* (4), 1922–1929.

Kumar, R.; Srivastava, A.; Behari, K. Synthesis and Characterization of Polysaccharide Based Graft Copolymer by Using Potassium Peroxymonosulphate/Ascorbic Acid as an Efficient Redox Initiator in Inert Atmosphere. *J. Appl. Polym. Sci.* **2009a,** *112* (3), 1407–1415.

Kumar, A.; Singh, K.; Ahuja, M. Xanthan-g-Poly(acrylamide): Microwave-Assisted Synthesis, Characterization and in Vitro Release Behavior. *Carbohydr. Polym.* **2009b,** *76* (2), 261–267.

Kuo, S. M., et al. Evaluation of the Ability of Xanthan Gum/Gellan Gum/Hyaluronan Hydrogel Membranes to Prevent the Adhesion of Postrepaired Tendons. *Carbohydr. Polym.* **2014,** *114,* 230–237.

Lagoueyte, N.; Paquin, P. Effects of Microfluidization on the Functional Properties of Xanthan Gum. *Food Hydrocoll.* **1998,** *12* (3), 365–371.

Lambert, F.; Rinaudo, M. On the Thermal Stability of Xanthan Gum. *Polymer* **1985,** *26* (10), 1549–1553.

Li, R.; Feke, D. L. Rheological and Kinetic Study of the Ultrasonic Degradation of Xanthan Gum in Aqueous Solutions. *Food Chem.* **2015a,** *172,* 808–813.

Li, R.; Feke, D. L. Rheological and Kinetic Study of the Ultrasonic Degradation of Xanthan Gum in Aqueous Solution: Effects of Pyruvate Group. *Carbohydr. Polym.* **2015b,** *124,* 216–221.

Li, S., et al. Molecular Modification of Polysaccharides and Resulting Bioactivities. *Compr. Rev. Food Sci. Food Saf.* **2016,** *15* (2), 237–250.

Li, Y. J., et al. Effect of Irradiation on Molecular Weight and Antioxidant Activity of Xanthan Gum. *J. Nucl. Agric. Sci.* **2010,** *24* (6), 1208–1213.

Li, Y. J., et al. Effect of Irradiation on the Molecular Weight, Structure and Apparent Viscosity of Xanthan Gum in Aqueous Solution. *Adv. Mat. Res.* **2011,** *239–242,* 2632–2637.

Liu, Z.; Yao, P. Injectable Thermo-responsive Hydrogel Composed of Xanthan Gum and Methylcellulose Double Networks with Shear-thinning Property. *Carbohydr. Polym.* **2015,** *132,* 490–498.

Liu, et al. Thermally Induced Conformational Change of Xanthan in 0.01M Aqueous Sodium Chloride. *Carbohydr. Res.* **1987,** *160,* 267–281.

Luan, L. Q., et al. Biological Effect of Radiation-degraded Alginate on Flower Plants in Tissue Culture. *Appl. Biochem. Biotechnol.* **2003,** *38* (3), 283–288.

Lui, W.; Norisuye, T. Thermally Induced Conformational Change of Xanthan: Interpretation of Viscosity Behaviour in 0.01 M Aqueous Sodium Chloride. *Int. J. Biol. Macromol.* **1988,** *10* (1),44–50.

Mark, G., et al. OH-radical Formation by Ultrasound in Aqueous Solution—Part II: Terephthalate and Fricke Dosimetry and the Influence of Various Conditions on the Sonolytic Yield. *Ultrason. Sonochem.* **1998,** *5* (2), 41–52.

Malhotra, S. L. Ultrasonic Degradation of Hydroxypropyl Cellulose Solutions in Water, Ethanol, and Tetrahydrofuran. *J. Macromol. Sci. Chem.* **1982,** *17* (4), 601–636.

Matsuda, Y.; Biyajima, Y.; Sato, T. Thermal Denaturation, Renaturation, and Aggregation of a Double-helical Polysaccharide Xanthan in Aqueous Solution. *Polym. J.* **2009,** *41* (7), 526–532.

Milas, M.; Rinaudo, M. Conformational Investigation on the Bacterial Polysaccharide Xanthan. *Carbohydr. Res.* **1979,** *76* (1), 189–196.

Milas, M.; Rinaudo, M. On the Existence of Two Different Secondary Structures for the Xanthan in Aqueous Solutions. *Polym. Bull.* **1984,** *12* (6), 507–514.

Milas, M.; Rinaudo, M. Properties of Xanthan Gum in Aqueous Solutions: Role of the Conformational Transition. *Carbohydr. Res.* **1986,** *158,* 191–204.

Milas, M.; Reed, W. F.; Printz, S. Conformations and Flexibility of Native and Re-natured Xanthan in Aqueous Solutions. *Int. J. Biol. Macromol.* **1996,** *18* (3), 211–221.

Milas, M.; Rinaudo, M.; Tinland, B. Comparative Depolymerization of Xanthan Gum by Ultrasonic and Enzymic Treatments. Rheological and Structural Properties. *Carbohydr. Polym.* **1986,** *6* (2), 95–107.

Moffat, J., et al. Visualisation of Xanthan Conformation by Atomic Force Microscopy. *Carbohydr. Polym.* **2016,** *148,* 380–389.

Moreira, A. S., et al. Screening Among 18 Novel Strains of *Xanthomonas Campestris pv Pruni. Food Hydrocoll.* **2001,** *15* (4–6), 469–474.

Mundargi, R. C.; Patil, S. A.; Aminabhavi, T. M. Evaluation of Acrylamide-grafted-Xanthan Gum Copolymer Matrix Tablets for Oral Controlled Delivery of Antihypertensive Drugs. *Carbohydr. Polym.* **2007,** *69* (1), 130–141.

Nishinari, K., et al. Characterization and Properties of Gellan-k-carrageenan Mixed Gels. *Food Hydrocoll.* **1996,** *10* (3), 277–283.

Norton, I. T., et al. Mechanism and Dynamics of Conformational Ordering in Xanthan Polysaccharide. *J. Mol. Biol.* **1984,** *175* (3), 371–394.

Orentas, D. G.; Sloneker, J. H.; Jeanes, A. Pyruvic Acid Content and Constituent Sugars of Exocellular Polysaccharides from Different Species of the Genus Xanthomonas. *Can. J. Microbiol.* **1963,** *9* (3), 427–430.

Oviatt, H. V.; Brant, D. A. Viscoelastic Behavior of Thermally Treated Aqueous Xanthan Solutions in the Semidilute Concentration Regime. *Macromolecules* **1994,** *27* (9), 2402–2408.

Palaniraj, A.; Jayaraman, V. Production, Recovery and Applications of Xanthan Gum by *Xanthomonas campestris.* *J. Food. Eng.* **2011,** *106* (1), 1–12.

Pandey, P. K., et al. Graft Copolymerization of Acrylic Acid onto Xanthan Gum Using a Potassium Monopersulfate/Fe²⁺ Redox Pair. *J. Appl. Polym. Sci.* **2003,** *89* (5), 1341–1346.

Pandey, S.; Mishra, S. B. Graft Copolymerization of Ethylacrylate onto Xanthan Gum, Using Potassium Peroxydisulfate as an Initiator. *Int. J. Biol. Macromol.* **2011,** *49* (4), 527–535.

Papagianni, M., et al. Xanthan Production by *Xanthomonas Campestris* in Batch Cultures. *Proc. Biochem.* **2001,** *37* (1), 73–80.

Patel, V. F.; Patel, N. M. Statistical Evaluation of Influence of Xanthan Gum and Guar Gum Blends on Dipyridamole Release from Floating Matrix Tablets. *Drug Dev. Ind. Pharm.* **2007,** *33* (3),327–334.

Petri, D. F. S. Xanthan Gum: A Versatile Biopolymer from Biomedical and Technological Applications. *J. Appl. Polym. Sci.* **2015,** *132* (23), 42035–42048.

Pfefferkorn, P., et al. Determination of the Molar Mass and the Radius of Gyration, Together with Their Distributions for Methylhydroxyethylcelluloses. *Cellulose* **2003,** *10* (1), 27–36.

Pinto, E. P.; Furlan, L.; Vendruscolo, C. T. Chemical Deacetylation Natural Xanthan (Jungbunzlauer^R). *Polímeros* **2011,** *21* (1), 47–52.

Price, G. J.; Smith, P. F. Ultrasonic Degradation of Polymer Solutions. 1. Polystyrene Revisited. *Polym. Int.* **1991,** *24* (3), 159–164.

Price, G. J.; Smith, P. F. Ultrasonic Degradation of Polymer Solutions: 2. The Effect of Temperature, Ultrasound Intensity and Dissolved Gases on Polystyrene in Toluene. *Polymer* **1993a,** *34* (19), 411–424.

Price, G. J.; Smith, P. F. Ultrasonic Degradation of Polymer Solutions—III. The Effect of Changing Solvent and Solution Concentration. *Eur. Polym. J.* **1993b,** *29* (2–3), 419–424.

Quinn, F. X.; Hatakeyama, T. The Effect of Annealing on the Conformational Properties of Xanthan Hydrogels. *Polymer.* **1994,** *35* (6), 148–152.

Raschip, I. E., et al. In Vitro Evaluation of the Mixed Xanthan/Lignin Hydrogels as Vanillin Carriers. *J. Mol. Struct.* **2011,** *1003* (1–3), 67–74.

Raffi, J. J., et al. Study of γ–Irradiated Starches Derived from Different Food Stuffs: A Way for Extrapolating Wholesomeness Data. *J. Agric. Food. Chem.* **1981,** *29* (6), 1227–1232.

Rather, S. A., et al. Application of Guar–Xanthan Gum Mixture as a Partial Fat Replacer in Meat Emulsions. *J. Food Sci. Technol.* **2016,** *53* (6), 2876–2886.

Rinaudo, M.; Milas, M. Enzymic Hydrolysis of the Bacterial Polysaccharide Xanthan by Cellulase. *Int. J. Biol. Macromol.* **1980,** *2* (1), 45–48.

Rinaudo, M., et al. Physical Properties of Xanthan, Galactomannan and Their Mixtures in Aqueous Solutions. *Macromol. Symp.* **1999**, *140* (1), 115–124.

Rocherfort, W. E.; Middleman, S. Rheology of Xanthan Gum: Salt, Temperature, and Strain Effects in Oscillatory and Steady Shear Experiments. *J. Rheol.* **1987**, *31*, 337–370.

Rosalam, S.; England, R. Review of Xanthan Gum Production from Unmodified Starches by *Xanthomonas Campestris* sp. *Enzyme Microb. Technol.* **2006**, *39* (2), 197–207.

Rosiak, J. M., et al. Radiation Formation of Hydrogels for Biomedical Purposes. Some Remarks and Comments. *Rad. Phys. Chem.* **1995**, *46* (2), 161–168.

Roy, A., et al. Hydrophobically Modified Xanthan: An Amphiphilic but Not Associative Polymer. *Biomacromolecules* **2014**, *15* (4), 1160–1170.

Saha, D.; Bhattacharya, S. Hydrocolloids as Thickening and Gelling Agents in Food: A Critical Review. *J. Food Sci. Technol.* **2010**, *47* (6), 587–597.

Sakeer, K., et al. Use of Xanthan and Its Binary Blends with Synthetic Polymers to Design Controlled Release Formulations of Buccoadhesive Nystatin Tablets. *Pharm. Dev. Technol.* **2010**, *15* (4), 360–368.

Sand, A.; Yadav, M.; Behari, K. Graft Copolymerization of 2-Acrylamidoglycolic Acid on to Xanthan Gum and Study of Its Physiochemical Properties. *Carbohydr. Polym.* **2010**, *81* (3), 626–632.

Sandford, P. A., et al. Variation in *Xanthomonas Campestris* NRRL B-1459: Characterization of Xanthan Products of Differing Pyruvic Acid Content. In *Extracellular Microbial Polysaccharides*; ACS Symp. Ser. 45. ACS: Washington, DC, 1977; pp 192–210.

Sathish, K. R., et al. Effective Removal of Humic Acid Using Xanthan Gum Incorporated Polyethersulfone Membranes. *Ecotoxicol. Environ. Saf.* **2015**, *121*, 223–228.

Schittenhelm, N.; Kulicke, W. M. Producing Homologous Series of Molar Masses for Establishing Structure-property Relationships with the Aid of Ultrasonic Degradation. *Macromol. Chem. Phys.* **2000**, *201* (15), 1976–1984.

Schorsch, C.; Garnier, C.; Doublier, J. L. Viscoelastic Properties of Xanthan Galactomannan Mixtures: Comparison of Guar Gum with Locust Bean Gum. *Carbohydr. Polym.* **1997**, *34* (3), 165–175.

Sehgal, R. R., et al. Nanostructured Gellan and Xanthan Hydrogel Depot Integrated within a Baghdadite Scaffold Augments Bone Regeneration. *J. Tissue Eng. Regen. Med.* **2017**, *11* (4), 1195–1211.

Shatwell, K. P.; Sutherland, I. W.; Ross-Murphy, S. B. Influence of Acetyl and Pyruvate Substitutions on the Solution Properties of Xanthan Polysaccharide. *Int. J. Biol. Macromol.* **1990**, *12* (2), 71–78.

Shiroodi, S. G.; Lo, Y. M. The Effect of pH on the Rheology of Mixed Gels Containing Whey Protein Isolate and Xanthan-Curdlan Hydrogel. *J. Dairy Res.* **2015**, *82* (4), 506–512.

Smidsrød, O.; Haug, A. Estimation of the Relative Stiffness of the Molecular Chain in Polyelectrolytes from Measurements of Viscosity at Different Ionic Strengths. *Biopolymers* **1971**, *10* (7), 1213–1227.

Smith, I. H., et al. Influence of the Pyruvate Content of Xanthan on Macromolecular Association in Solution. *Int. J. Biol. Macromol.* **1981**, *3* (2), 129–134.

Southwick, J. G., et al. Self-association of Xanthan in Aqueous Solvent-Systems. *Carbohydr. Res.* **1980**, *84* (2), 287–295.

Srivastava, A.; Behari, K. Modification of Natural Polymer via Free Radical Graft Copolymerization of 2-Acrylamido-2-Methyl-1-Propane Sulfonic Acid in Aqueous Media and

Study of Swelling and Metal Ion Sorption Behavior. *J. Appl. Polym. Sci.* **2009,** *114* (3), 1426–1434.

Sutherland. I. W. *Xanthomonas* Polysaccharides Improved Methods for Their Comparison. *Carbohydr. Polym.* **1981,** *1* (2), 107–115.

Sutherland, I. W. An Enzyme System Hydrolyzing the Polysaccharides of *Xanthomonas* Species. *J. Appl. Microbiol.* **1982,** *53* (3), 385–393.

Sutherland, I. W. Xanthan Lyases-novel Enzymes Found in Various Bacterial Species. *J. Gen. Microbiol.* **1987,** *133* (11), 3129–3134.

Şen, M., et al. Radiation Induced Degradation of Xanthan Gum in the Solid State. *Rad. Phys. Chem.* **2016,** *124,* 225–229.

Tako, M.; Nakamura, S. Rheological Properties of Deacetylated Xanthan in Aqueous Media. *Agric. Biol. Chem.* **1984,** *48,* 2987–2993.

Tako, M.; Nakamura, S. Evidence for Intramolecular Associations in Xanthan Molecules in Aqueous Media. *Agric. Biol. Chem.* **1989,** *53* (7), 1941–1946.

Tako, M. Synergistic Interaction Between Deacylated Xanthan and Galactomannan. *J. Carbohydr. Chem.* **1991,** *10* (4), 619–633.

Thonart, Ph., et al. Xanthan Production by *Xanthomonas Campestris* NRRL B-1459 and Interfacial Approach by Zeta Potential Measurement. *Enzyme Microbiol. Technol.* **1985,** *7* (5), 235–238.

Tinland, B.; Rinaudo, M. Dependence of the Stiffness of the Xanthan Chain on the External Salt Concentration. *Macromolecules* **1989,** *22* (4), 1863–1865.

Viebke, C. Order-Disorder Conformational Transition of Xanthan Gum. In *Polysaccharides: Structural Diversity and Functional Versatility*; Dumitriu, S., Ed.; Marcel Dekker: New York, NY, 2005; pp 459–474.

Wang, T., et al. Rheological, Textural and Flavour Properties of Yellow Mustard Sauce as Affected by Modified Starch, Xanthan and Guar Gum. *Food Bioproc. Technol.* **2016,** *9* (5), 849–858.

Yan-Jie L.; et al.; Effect of Irradiation on Molecular Weight and Antioxidant Activity of Xanthan Gum. *J. Nucl. Agricul. Sci.* **2010,** *24* (6), 1208–1213.

Yang, L.; Zhang, Li-M. Chemical Structural and Chain Conformational Characterization of Some Bioactive Polysaccharides Isolated from Natural Sources. *Carbohydr. Polym.* **2009,** *76* (3), 349–361.

Zirnsak, M. A.; Boger, D. V.; Tirtaatmadja, V. Steady Shear and Dynamic Rheological Properties of Xanthan Gum Solutions in Viscous Solvents. *J. Rheol.* **1999,** *43* (3), 627–650.

CHAPTER 4

AMPHIPHILIC POLYMERS DESIGNED FOR BIOMEDICAL APPLICATIONS

MARIETA NICHIFOR*, GEORGETA MOCANU, and CRISTINA M. STANCIU

Department of Natural Polymers, Bioactive and Biocompatible Materials, "Petru Poni" Institute of Macromolecular Chemistry, Aleea Grigore Ghica Voda 41A, 700487 Iasi, Romania

*Corresponding author. E-mail: nichifor@icmpp.ro

CONTENTS

ABSTRACT

A large variety of amphiphilic polymers were obtained by chemical modifications of biocompatible polysaccharides such as dextran or pullulan. Attachment of hydrophobic segments (long alkyl chains and bile acid moieties) either along polysaccharide main chain or at its chain end provided several classes of amphiphilic polymers with ability to form self-organized structures by intra- or/and intermolecular hydrophobic interactions, with more or less well defined hydrophobic and hydrophilic microdomains. The self-organized structure properties such as size, morphology, polarity, and compactness of the microdomains can be tuned by an appropriate choice of the segments (blocks) chemical composition. The synthesized amphiphilic polysaccharides can have intrinsic biological activity with potential use as antibacterial agents or hypolipemic drugs, can encapsulate hydrophobic drugs and act as controlled drug delivery systems, or can interact with liposomes and improve their stability in biological environment.

4.1   INTRODUCTION

Amphiphiles are compounds possessing both hydrophilic and hydrophobic (lipophilic) structural segments. They can self-assembly by a spontaneous process based on specific inter-segments interactions, which determine a transition from an unordered structure to a well-organized one (Raffa et al., 2015). The main (non-covalent) forces acting in amphiphile self-assembly are hydrophobic associations accompanied by a decrease in the free energy of the system due to the removal of hydrophobic segments from the aqueous phase, with the formation of hydrophobic clusters stabilized by hydrophilic segments exposed to water. The self-assembling is a discontinuous process, as it starts only above a certain amphiphile concentration, called *critical micelle concentration* (*CMC*). The formed aggregates are in a thermodynamic equilibrium with nonaggregated molecules. The best known amphiphiles are surfactants, the name of which stems from their significant surface activity, and they can form spherical, cylindrical, lamellar, or vesicular aggregates. *CMC* value and aggregate morphology are specific to each amphiphile and depend on its chemical structure and medium properties (solvent, ionic strength, pH, and presence of other molecules) (Calandra et al., 2015; Raffa et al., 2015).

In comparison with low molecular surfactants, macromolecular amphiphiles, also called polymeric surfactants, amphiphilic polymers, micellar polymers, hydrophobically modified water-soluble polymers, associative polymers, exhibit a higher aggregate stability due to a lower aggregation concentration, a slower exchange rate between hydrophobes in free or associated form, a lower mobility of the hydrophobic core. Besides, the structural complexity of polymeric amphiphiles, given by the relative number, length and distribution of hydrophilic, and hydrophobic segments along the polymer chain, can generate different aggregate size, shape, morphologies, and colloidal solution properties (Mai and Eisenberg, 2012; Raffa et al., 2015), requiring a sustained scientific effort for their understanding and elucidation. The amphiphilic polymers have proved a great potential in application as water-borne paints and coating fluids, emulsion and dispersion agents, cosmetics, personal care goods, drug delivery systems, tissue regeneration, reservoirs for chemical or enzymatic reactions, pseudostationary phases in electrokinetic chromatography, and membrane protein solubilizers (Alexandridis and Lindman, 2000; Cotanda and O'Reilly, 2012; Durand and Marie, 2009; Tang et al., 2016). Moreover, the resemblance of amphiphilic polymers to biological macromolecules like RNA, the major macromolecule of the living systems, or serum albumin able to solubilize and transport different hydrophobes, can help to improve the knowledge of biological matter behavior. This resemblance also prompted the recent achievements in building bioinspired materials, such as cellinspired biomimetic materials or macromolecular assemblies mimicking the functions of biopolymers (Zhao et al., 2013a).

In the following, we will focus on polysaccharide (dextran and pullulan) based amphiphilic polymeric systems with potential biomedical applications. The chemical structures of several classes of amphiphilic polymers, their self-assembling properties, as well as strategies for application in biomedical field will be described. The applications that have been foreseen for these amphiphilic polymers include antimicrobial agents, hypolipemic drugs, drug delivery system, and liposome stabilizers.

4.2 CHEMICAL STRUCTURES OF AMPHIPHILIC POLYMERS BASED ON POLYSACCHARIDES

Amphiphilic polymers have various chemical structures derived from the synthetic methods and the relative positions and length of hydrophilic and hydrophobic components. If they contain ionic charges, the amphiphilic polyelectrolytes so obtained exhibit solution properties resulting from both the hydrophobic attractions and the electrostatic repulsions. Using different synthetic procedures, amphiphilic polymers can be designed with a large variety of structural architectures: random or alternated copolymers from hydrophobic and hydrophilic monomers, homo- or copolymers of monomers containing both ionic and hydrophobic groups (surfactant monomers or surfmers), hydrophobically modified water-soluble polymers, where hydrophobic moieties (without or with charged groups) are located on the side chains, graft or star polymers, block copolymers prepared from hydrophilic and hydrophobic blocks or hydrophobically end-capped water-soluble polymers. The use of polysaccharides as water-soluble polymers in the design of amphiphilic polymers takes the advantage of polysaccharides' low toxicity, good biocompatibility and biodegradability, relatively chemical stability and immunogenicity (depending of the individual structure), availability of different functionalities with selective reactivity (Tang and Huang, 2016; Yang et al., 2015). Presence of numerous functional groups (hydroxyl, carboxyl, and amine) in the repeating carbohydrate units provides reactive sites for attachment of different molecules by oxidation, esterification, etherification, and amidation for the design of various side chain amphiphilic polymers (Hassani et al., 2012). Moreover, each polysaccharide owes one reductive terminal group, which can be the subject of a reductive amination with an amine group-containing derivative. The proper derivatization of the reductive end groups was successfully used for the design of amphiphilic block-copolymers or hydrophobically end-capped polymers where a polysaccharide was the hydrophilic block. Many of these amphiphilic polysaccharides were designed for biomedical applications (Hassani et al., 2012; Liu et al., 2016; Mizrahyab and Peer, 2012).

The chemical structures of several amphiphilic polymer classes based on polysaccharides as water-soluble polymers (blocks) will be presented in the following. The polysaccharides used were dextran and pullulan. Dextran is a microbial polysaccharide composed of D-glucopyranose units linked by α-(1,6) glycosidic bonds with a low percentage of (1,3)-linked side chains (Raemdonck et al., 2013). Pullulan is a polysaccharide polymer consisting of repetitive maltotriose units connected each other via an α-1 and 6-glycosidic bond (Singh and Saini, 2012). Both polysaccharides and their derivatives attracted increasing attention in pharmaceutical applications (Ishak et al., 2016; Singh and Saini, 2012; Varshosaz, 2012).

4.2.1   SIDE-CHAIN AMPHIPHILIC POLYMERS

The synthesis of the polymers with side chains can be obtained by chemical modification of a hydrophilic polymer for attachment of hydrophobic or amphiphilic pendent segments. Pendants can be short or long (grafted polymers), and hydrophobes can belong to known classes of lipophilic compounds, such as long alkyls, steroids, or can become hydrophobic by changing external conditions, for example, poloxamers which become hydrophobic by raising temperature. Polymers with charges (polyelectrolytes) can have charges and hydrophobes located on different side chains, or located on the same side chains. In the latter case, polymers are called polysoaps or polysurfmers, because the side chains are actually surfactants bound to the polymer backbone through their charged groups (head-attached) or hydrophobes (tail-attached).

In the following, the amphiphilic polysaccharide structures will be classified as a function of the chemical structure of the attached side groups: (1) bile acid and (2) quaternary ammonium groups. Dextran was chosen for exemplification of the chemical structures of the synthesized amphiphilic polymers.

4.2.1.1   AMPHIPHILIC POLYMERS WITH BILE ACID SIDE GROUPS

Bile acids are natural amphiphilic compounds with an important contribution to biological processes, such as emulsification and membrane transport of cholesterol (CH), vitamins, retinol, β-carotene (Small, 1971). Rigid steroid skeleton and facial amphiphilicity (Fig. 4.1) lead to formation of micelles and other supramolecular structures with specific properties. Their hydroxyl and carboxylic acid groups can be easy chemically modified; therefore, bile acids have been often used as building materials for various low molecular or polymeric structures (Cunningham and Zhu, 2016; Durand, 2007; Zhu and Nichifor, 2002). Bile acids were covalently bound to many polysaccharides, such as heparin (Bae et al., 2013; Hwang and Lee, 2016), chitosan (Cadete et al., 2012; Park et al., 2014), carboxymethyl curdlan (Yan et al., 2015), starch (Yang et al., 2014), or hyaluronic acid (Wei et al., 2015), and the resulting amphiphilic derivatives found different biomedical application, especially as micellar carriers for hydrophobic drug delivery (Cadete et al., 2012; Park et al., 2014; Yan et al., 2015).

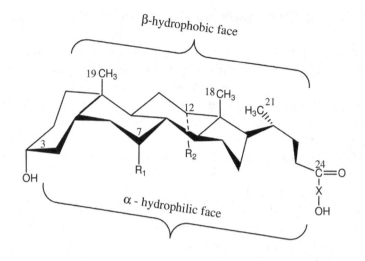

Bile acid	(BA)	R_1	R_2	X
Cholic acid	(CA)	OH	OH	-
Deoxycholic acid	(DCA)	H	OH	-
Lithocholic acid	(LCA)	H	H	-
Glycocholic acid	(GCA)	OH	OH	NH-CH$_2$-CO

FIGURE 4.1   Chemical structure of bile acids.

In the most of the abovementioned reports, the bile acids were attached to a polysaccharide through their COOH groups. Nichifor and Carpov (1999) prepared two classes of polysaccharide derivatives with the bile acid connected to polysaccharide OH groups either at C^{24} carboxylic group (DM-BA(m)) or at C^3 OH group (DM-CACOONa(m)) (Nichifor et al., 2004a) (Fig. 4.2). The direct attachment of a bile acid by esterification between polysaccharide OH groups and bile acid carboxylic groups was easily achieved in the presence of N,N-dicyclohexylcarbodiimide (DCC) as a coupling agent and N,N-dimethylaminopyridine (NMAP) as a catalyst, in a mixture N-methylformamide/DMF/dichloromethane 10/9/1 v/v/v. Several natural bile acids were used (BA = CA, DCA, and GCA), as well as a cholic acid derivative, N-(3α,7α,12α–trihydroxy-5-β-cholan-24-carboxy)-ε-aminocaproic acid) (in this case X in Figs. 4.1 and 4.2 is NH—(CH$_2$)$_5$—CO). The degree of substitution (DS) = m ≤6 mol%, as at higher bile acid content the water solubility is lost. A similar synthetic procedure was used for the preparation of polymers DM-CACOONa(m), by dextran reaction with a cholic acid derivative, 3-(succinoyloxy)cholic acid trichloroethyl ester, in the presence of DCC and NMAP. The attachment of this derivative was followed by a selective elimination of the trichloroethyl groups, which afforded amphiphilic polymers with bile acid side groups connected to polymer backbone at their C^3 positions and carrying free carboxylic groups at C^{24} positions. The presence of anionic carboxylic groups at the end of the side chains endowed these polymers with a polyelectrolyte character and a better water solubility than that of non-charged DM-BA polymers, which was preserved until a DS ≤ 25 mol%.

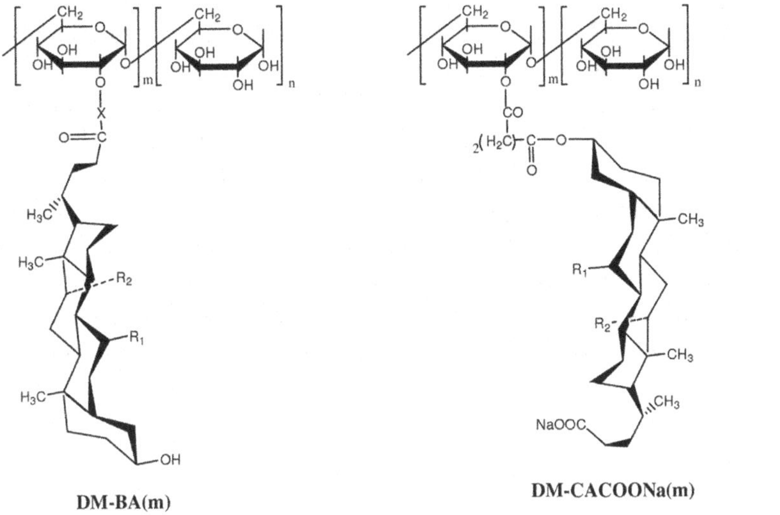

DM-BA(m)

DM-CACOONa(m)

FIGURE 4.2   Chemical structures of amphiphilic dextrans with bile acid side groups. D means dextran, M is dextran molar mass, in kDa, and m is substitution degree, in mol%. m + n = 100.

4.2.1.2 AMPHIPHILIC POLYMERS WITH QUATERNARY AMMONIUM SIDE GROUPS

Hydrophilic and biocompatible polysaccharides with quaternary ammonium groups have found numerous pharmaceutical, industrial, and environmental applications as additives for paper, textile, food, cosmetics, flocculants, or antibacterial agents (Belalia et al., 2008). Their amphiphilic analogs display similar properties and applications, but also additional features due to the surface activity and ability to self-organize in micelle-like aggregates. The simultaneous presence of ammonium and hydrophobic groups enhances polymer biological activity; for example, stronger interactions with some components of skin and hair provide an improved delivery of drugs from their topical formulations (Ballarin et al., 2011).

In order to prepare new amphiphilic polysaccharides with quaternary ammonium pendent groups and variable charge and hydrophobe content, Nichifor et al. (2010) developed a synthetic procedure, which avoids the intermediate synthesis of the quaternization reagent. According to this procedure, a neutral polysaccharide (dextran, pullulan, and hydroxyethyl cellulose) was treated with an equimolar mixture of epichlorohydrin (ECH) and a tertiary amine ($R_1R_2R_3N$), where R_1, R_2, R_3, are identical or different, and at least one of the R substituents is an alkyl chain with 2–16 carbons, or a benzyl (Bz) group. The procedure can be applied to various water-soluble or water-swellable polysaccharides in linear or cross-linked forms. The procedure provided several classes of cationic amphiphilic polymers, the self-assembling properties, and applications of which could be tuned by variation of DS (5–90 mol%), the length or structure of one R substituent, presence of a cross-linker and its nature. The general chemical structure of various classes of dextran carrying pendent N-(2-hydroxypropyl)-N,N-dimethyl-N-(R_1/R_2)-ammonium chloride groups is depicted in Figure 4.3 and details about the chemical composition of each class is presented in Table 4.1.

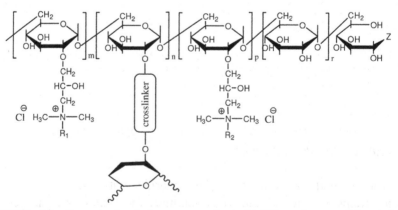

FIGURE 4.3 General chemical structure of amphiphilic polymers with pendent quaternary ammonium groups. D is dextran, M is dextran molar mass, in kDa, QR refers to the quaternary ammonium group with an R_1/R_2 substituent, $m + n + p + r = 100$.

TABLE 4.1 Chemical Composition of Cationic Dextran Derivatives with General Chemical Structure Presented in Figure 4.3.

R_1	R_2	Cross-linker structure	Z	Substitution degree, mol%		
				m	n	p
Linear monopolar cationic dextran DM-QR(m)						
$R_1 = R_2 = C_2, C_4, C_8, C_{12}, C_{16}$ or Bz		—	CHO	5–90	0	0
Monopolar cationic cross-linked dextran DM-QR(m)-ECH						
$R_1 = R_2 = C_2, C_4, C_8, C_{12}, C_{16}$ or Bz		$CH_2-CH(OH)-CH_2$	CHO	25–30	10–20	0
Bipolar cationic cross-linked dextran DM-QR$_1$(m)R$_2$(p)-ECH						
C_2	C_{12}, C_{16}	$CH_2-CH(OH)-CH_2$	CHO	35–50	10–20	10–25
End-modified cationic dextran Z-DM-QR (m)						
$R_1 = R_2 = C_2, C_8$, Benzyl		—	C_{12}, C_{18}	≤30	—	—
End-modified cross-linked cationic dextran Z-DM-QR (m)-DVS						
$R_1 = R_2 = Bz$		$(CH_2)_2-O{=}S{=}O-(CH_2)_2$	C_{18}	≤20	10–20	—

Polymers DM-QR(m) are water soluble with DS up to 90 mol% and variable hydrophobicity given by the length of R substituent (C_2 –C_{16}). The water solubility of polymers with $R = C_2$–C_8 or benzyl could be preserved until DS = 70–90 mol%, but was limited to DS = 30 mol% when $R = C_{12}$–C_{16}.

Quaternization reaction performed on polysaccharide particles previously cross-linked with epichlorohydrin (ECH) afforded cationic amphiphilic hydrogels. The hydrophilic/lipophilic balance (HLB) of these hydrogels was varied in two ways: (1) by changing the hydrophobicity of one substituent R in a single step quaternized gels, that is, monopolar gels DM-QR(m)-ECH, where $R = C_2$–C_{16} or Bz (Nichifor et al., 2001a; Nichifor et al., 2001b); (2) by changing the ratio between DS_1 and DS_2 in two step quaternized gels, designed as bipolar gels DM-QR_1(m)R_2(p)-ECH, where $R_1 = C_2$ and $R_2 = C_{12}$ or C_{16} (Mocanu and Nichifor, 2014). The synthesis of bipolar gels offered the opportunity to combine an appropriate HLB with a high cationic group content required by some potential applications, for example, as bile acid sequestrants. The swelling capacity of the hydrogels, the water amount retained (WR) by the gel at equilibrium, was influenced by the content in amino groups, hydrophobicity of the R_1/R_2 substituents, and the amount of ECH used in crosslinking step.

4.2.2 BLOCK COPOLYMERS

Poly(ethylene glycol) was the first choice as the hydrophilic segment for the synthesis of amphiphilic block copolymers or hydrophobically end-modified polymers, but polysaccharides have been recently taken into account as alternative hydrophilic blocks (Schatz and Lecommandoux, 2010; Tizzotti et al., 2010). Amphiphilic block copolymers of dextran with polystyrene (Chen et al., 2014; Houga et al., 2009; Lepoittevin et al., 2011), poly(ε-caprolactone) (Li et al., 2013; Sun et al., 2010), poly(LD-lactide) (Verma et al., 2012; Zhao et al., 2013b), or poly(γ-benzyl L-glutamate) (Schatz et al., 2009) were reported, as well as hydrophobically end-modified dextrans (Hirsjärvi et al., 2013; Zhang and Marchant, 1996). The starting point for these block and block-like copolymer synthesis is the selective reductive amination of polysaccharide reductive end, which allows the preparation of linear polymers without affecting the OH groups of the polysaccharide backbone.

4.2.2.1 BLOCK COPOLYMERS DEXTRAN-BILE ACID OLIGOESTERS

In the attempt to obtain new block copolymers with improved biocompatibility, required by biotechnological and pharmaceutical potential applications, Stanciu and Nichifor (2015) synthesized new amphiphilic diblock copolymers from dextran as the hydrophilic block and a bile acid oligoester as the hydrophobic block. The coupling of a semi-flexible dextran chain and a rigid bile acid oligoester chain can be a premise for new and interesting self-aggregate shape and size. The copolymers with the chemical structure presented in Figure 4.4 could be prepared by end-to-end reaction between end-aminated dextran (obtained by dextran end reductive amination with ethylenediamine) and the carboxylic end group of a bile acid oligoester. Bile acid oligoesters with molecular weight 2–10 kDa resulted from the polycondensation between a bile acid derivative, 3α-(succinoyloxy)-hydroxy-5β-cholan-24-oic acid, and diethylene glycol (DEG) (Stanciu and Nichifor, 2013; Stanciu et al., 2009). One block copolymer D6-b-(LCA-DEG)$_{15}$, containing 15 wt% oligoester, was prepared from dextran with MW 6 kDa and a LCA-DEG oligoester, and was further used for self-assembling studies.

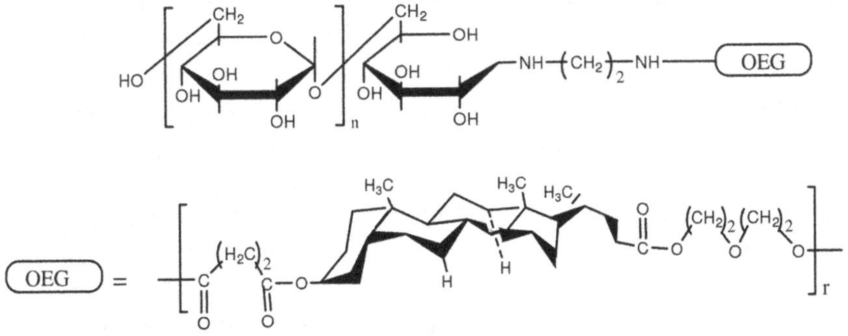

FIGURE 4.4   Chemical structure of the diblock copolymer D6-b-(LCA-DEG) based on dextran and an oligoester of LCA with DEG.

4.2.2.2 HYDROPHOBICALLY END-MODIFIED DEXTRAN

Attachment of a hydrophobic molecule at one polymer chain end provides hydrophobically end-modified polymers (or α-modified and semi-telechelic), which are similar to amphiphilic diblock copolymer, but have shorter hydrophobic blocks. In comparison to block copolymers, semi-telechelic polymers can be prepared by easier, more reproducible and less expensive methods, and gained attention due to their applications for the steric stabilization of liposomes (Pignatello et al., 2013; Torchilin et al., 2001), nanoscaled colloidal drug carriers (Kuskov et al., 2010) or nanoreactors (Patterson et al., 2013).

Nichifor et al. (2014) prepared a new series of semitelechelic amphiphilic polymers by end attachment of hydrophobic segments (alkyl, dialkyl, and bile acids) to dextran of different molar masses. Reductive end amination of dextran was carried out in the presence of a large excess of an alkyl amine (C_{12}, C_{16}, C_{18}) or a (2'-aminoethylene) 5β-cholane amide and $NaBH_3CN$ as reducing agent, with high conversion degree (95–98%) of dextran chain end. The resulting polymers, Z-DM, have a structure similar to that presented in Figure 4.3, without pendent amino groups or crosslinks, and $Z = C_n$ or BA. Chemical modifications of the dextran main chain of polymers Z-DM were also performed in the attempt to improve the stability of aggregates formed by these polymers or to extend their applicability. Crosslinking of the outer shell or of the core of micelles formed by block or block-like polymers can lead to well-defined nanosized particles with stable size and shape when diluted, for example, in the bloodstream, where micelles systemically administrated as drug carriers are prone to extensive dilution (Wooley, 2000). Polymers Z-D10 were crosslinked with divinyl sulfone (DVS), in an aqueous solution of pH 12, obtaining derivatives Z-D10-DVS, with a DVS content up to 20 wt%. Polymers Z-DM and Z-DM-DVS were further modified by quaternization in order to introduce cationic charges, which could improve their applicability as antibacterial agents or as multifunctional drug delivery carriers. Quaternization reaction was performed under similar conditions described before and the chemical structure and composition of the resulting Z-DM-QR(m) and Z-DM-QR(m)-DVS are presented in Figure 4.3 and Table 4.1 (Mocanu et al., 2015; Nichifor et al., 2014).

4.3 AMPHIPHILIC POLYMERS SELF ASSEMBLY

Amphiphilic polymers self-organize in water medium giving rise to self-aggregate with a tremendous variety of possible morphologies (spherical micelles, vesicles, rods, lamellae) and sizes, which ranges from 10 nm to several μm, as a function of polymer architecture. The most important differences can be found between side chain polymers (hydrophobically modified polymers) and block copolymers.

4.3.1 SIDE CHAIN POLYMERS

Side chain polymers have no clear spatial segregation between the hydrophilic and hydrophobic segments, consequently, their aggregates have no well-defined boundaries between hydrophobic and hydrophilic domains, but a hydrophilic-rich surface is always present for aggregate stabilization. The aggregation takes usually place by the simultaneous action of intra- and intermolecular hydrophobic associations, at a concentration called *critical aggregation concentration* (*CAC*) similar to CMC_s of classical surfactants, with $CAC << CMC_s$. Aggregation process is progressive and stretches over a large concentration range, due to the structural heterogeneity and the lack of association cooperativity (Nichifor and Mocanu, 2005; Petit-Agnely et al., 2000). The predominance of intra- or inter-molecular hydrophobic association depends mainly on the content in hydrophobes and polymer concentration. The evolution of aggregate types with increasing content in pendent hydrophobic segments is illustrated in Figure 4.5. At low DS, intermolecular associations are predominant and lead to very viscous aqueous solutions and eventually to a physically cross-linked gel at high polymer concentrations. These polymers are called associative and are applied as thickening agents in different food, cosmetics, or care-good formulations. Polymers with a moderate DS (10–30 mol%) have the most heterogeneous self-assembly behavior, with simultaneous presence of intramolecular hydrophobic microdomains, nonaggregated hydrophobic side chains, and some intermolecular associations (neck-lace model, Dobrynin and Rubinstein, 2000). Presence of a high amount of side chain hydrophobes (DS > 25–30 mol%) results only in formation of intramolecular hydrophobic microdomains, which give rise to very small colloidal particles with an inner hydrophobic cluster covered by an outer hydrophilic layer (Yusa et al., 2002). These polymers are called polysoaps and the resemblance of their aggregates with the core–shell structure of micelles makes them very suitable for encapsulation of hydrophobic compounds, for example, drugs.

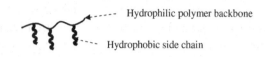

Hydrophilic polymer backbone

Hydrophobic side chain

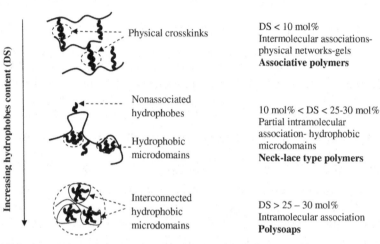

Increasing hydrophobes content (DS)

Physical crosskinks

DS < 10 mol%
Intermolecular associations-
physical networks-gels
Associative polymers

Nonassociated
hydrophobes

Hydrophobic
microdomains

10 mol% < DS < 25-30 mol%
Partial intramolecular
association- hydrophobic
microdomains
Neck-lace type polymers

Interconnected
hydrophobic
microdomains

DS > 25 – 30 mol%
Intramolecular association
Polysoaps

FIGURE 4.5   Schematic representation of hydrophobically modified polymers self-assembly, as a function of hydrophobic side-chain content (DS).

Besides the content in hydrophobic side chains, increasing polymer concentration (C_p) can change the type of association from intramolecular in diluted solutions to intermolecular in more concentrated solutions, and increasing polymer molecular mass can enhance intermolecular associations. The predominance of one or another type of association can be highlighted by polymer viscosimetric behavior in aqueous solutions, as Figure 4.6 shows for polymers DM-QR(m). With increasing m values from 5 to 30 mol% in polymers D40-QC$_{16}$(m), variation of the specific viscosity with polymer concentration changes from an exponential profile (associative polymers), to a linear one (polysoaps), with an intermediate behavior for $m = 10$ and 15 mol% (Fig. 4.6A). Associative effect of polymers DM-QC$_{12}$(10) is clearly enhanced by increasing polymer molecular mass from 40 to 200 kDa (Fig. 4.6B) (Nichifor et al., 2011).

The DS values indicated in Figure 4.5 for the transition between different types of aggregates are valid for polymers carrying long alkyl side chain, but they can be different for other hydrophobic pendent groups, for example, polymers D40-BA(m) which form compact intramolecular aggregates at DS as low as 4–5 mol% (Nichifor et al., 1999). Presence of charges will shift the transition to higher DS values.

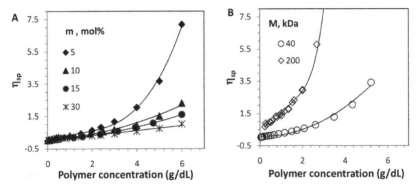

FIGURE 4.6 Variation of specific viscosity with concentration in aqueous solutions of polymers DM-QC$_n$(m), as a function of side chain content (A) and dextran molar mass M (B). Modified from Nichifor et al. (2011).

The chemical composition of the side chains can also influence the polarity and compactness (viscosity) of hydrophobic microdomains. These characteristics can be evaluated by fluorescence measurements in the presence of a fluorescence probe, for example, 1-(N-phenyl)-naphthylamine (NPN), which is very sensitive to the polarity and viscosity of the environment (Brito and Vaz, 1986). With increasing polymer concentration above *CAC*, a progressive partition of NPN into microdomains takes place, accompanied by a progressive blue shift of the maximum (λ_{max}) of the NPN emission spectrum and an increase in the intensity of this maximum (both proving a decrease of environment polarity), and an increase of anisotropy (due to the increase of medium viscosity). Figure 4.7 shows the variation of these parameters with polymer concentration of two classes of polymers, containing very rigid bile acid side chains or more flexible long alkyl side chains. The polarity of D40-DCA(4) microdomains is very low, similar to that of pure dioxan, but that of D40-QC$_{16}$(30) microdomains is similar to a mixture 1/1 water/dioxan (Fig. 4.7A). A similar difference could be observed for the microdomain viscosity, as anisotropy r = 0.13 for D40-DCA(4) and 0.1 for D40-QC$_{16}$(30) (r is 0 in water and 0.17 in rigid media) (Fig. 4.7B) (Nichifor et al., 1999; Nichifor et al., 2004b). These results indicate a much higher tendency of bulky steroid nucleus to form very compact and nonpolar intramolecular hydrophobic

122   Smart Materials

microdomains, despite the much lower DS of polymers with bile acid side chains than that of polymers with alkyl side chains. The presence of charged groups in D40-QC$_{16}$(30) can also be responsible for the less compact hydrophobic microdomains. However, the comparison of the polarity and viscosity, determined from λ_{max} and intensity of NPN emission spectra, respectively, for the hydrophobic microdomains formed by charged polymers D40-CACOONa(25) and D40-QC$_{16}$(25) (Fig. 4.7C,D), reveals that at similar hydrophobe and charge content ($DS\approx25$ mol%), the microdomains formed by association of steroid side chains are again much more compact and hydrophobic than those formed by C$_{16}$ alkyl chains (Nichifor et al., 2004a, 2004b).

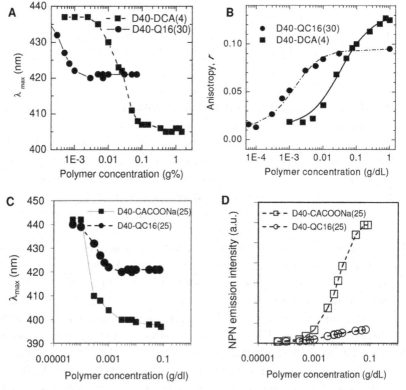

FIGURE 4.7 Influence of side chains' chemical composition on polarity (given by λ_{max} of NPN emission spectra) and compactness (given by NPN anisotropy or fluorescence emission intensity) of aggregates formed by polymers carrying alkyl (D40-QC$_{16}$(m)) or bile acid ((D40-DCA(4) and D40-CACOOH(25)) pendent groups. Adapted from Nichifor et al. (1999), Nichifor et al. (2004b), and Nichifor et al. (2004a).

The size of aggregates falls in nanometric range (10–200 nm) and depends on chemical structure and concentration. Polymers DM-BA(m) form large and loose aggregates at low concentration (200 nm at $C_p \leq 0.02$ g/dL) and very small and compact aggregates at high concentrations (15–20 nm at $C_p \geq 0.2$ g/dL) (Nichifor et al., 1999). The number of hydrophobic microdomains/polymer chain (N_{ch}) decreases and the number of hydrophobes included in one aggregate (aggregation number [N_{agg}]) increases with increasing concentration of polymers DM-QC$_n$(25–30). For example, aggregates formed by D40-QC$_{12}$(30) in aqueous solution have N_{ch} = 40 and N_{agg} = 2 at C_p = 0.01 g/dL, but N_{ch} = 2–3 and N_{agg} = 40 at C_p = 1 g/dL (Nichifor et al., 2004b).

4.3.2   BLOCK COPOLYMERS

Block copolymers and hydrophobically end-modified polymers with clear spatial separation of hydrophilic and hydrophobic segments give exclusively intermolecular aggregation in water solutions, even in conditions of extreme dilution and give rise to aggregates with well-defined hydrophobic and hydrophilic domains, for example, micelles with an inner hydrophobic core and an outer hydrophilic shell (Raffa et al., 2015; Riess, 2003, Mai and Eisenberg, 2012). Their *CAC* is very low (several mg/L), depends on the length of the hydrophobic block and sometimes it cannot be measured experimentally, as no break-point in a certain property (fluorescence probe intensity, surface activity, and viscosity) variation with concentration could be noticed, indicating that the copolymers are only in aggregated form, irrespective of concentration, and only the number of aggregates increases with increasing C_p.

The copolymer D10-*b*-(LCA-DEG)$_{15}$ self-organized in aqueous solutions (*CAC* = 28 mg/L) with formation of spherical micelles, obeying the empirical general rule proposed by Discher and Eisenberg (2002) according to which the block copolymers with a hydrophilic weight fraction >45% will form spherical micelles. The hydrophobicity of the micelle inner core was high (the pyrene polarity parameter $I_1/I_3 \sim 1$, compared with 1.3 for classical surfactants), indicating the ability of oligoester chains to self-associate. The micelle size distribution was unimodal and the average size was 150 nm (Stanciu and Nichifor, 2015). The bigger size in comparison with other block-copolymers obtained from more flexible hydrophobic blocks, such as poly(ε-caprolactone) was probably due to the rigid conformation of LCA oligoester chain and to the less compact packing characteristic to the facially amphiphilic bile acids (Wu et al., 2011).

Hydrophobically end-modified polymers Z-DM formed also spherical micelles with sizes in the range 20–150 nm, with uni- or bimodal distribution. CAC values were in the range 0.1–0.8 mg/mL and the polarity parameter was about 0.8–1.0. The size, N_{agg}, and core hydrophobicity increased and CAC values decreased with increasing the length of alkyl chain end Z and the decrease in dextran molar mass. Cross-linking of the dextran outer shell (Z-DM-DVS) improved the micelle stability to dilution, as shown by fluorescence measurements in the presence of pyrene, according to which hydrophobicity (given by the ratio I_1/I_3) of the micelle inner core was kept to high values even at very low concentrations, even below CAC of uncrosslinked polymer. This finding shows that cross-linked derivatives Z-DM-DVS can be used as carriers for hydrophobic drug delivery. Chemically modified derivatives Z-DM-R(m) preserved self-assembling ability when R was C_8 or Bz and m < 30 mol%. The size of the aggregates formed by these cationic derivatives was higher than that of neutral analogs (200–350 nm) due to the extension of charged dextran chains as a consequence of electrostatic repulsion between charged sites (Nichifor et al., 2014).

4.4 BIOMEDICAL APPLICATIONS OF AMPHIPHILIC POLYMERS

Amphiphilic polymers have their own (intrinsic) biological activity, for example, they can act as antimicrobial or hypolipemiant agents, or they can be a component of a formulation designed for an improved bioavailability and release of biologically active compounds (drugs, peptides, and vaccines).

4.4.1 POLYMERS WITH INTRINSIC BIOACTIVITY

4.4.1.1 ANTIMICROBIAL AGENTS

Low-molecular quaternary ammonium compounds have excellent antibacterial activity and are frequently applied as disinfectants or biocides in medical, industrial, or household areas (Merianos, 1991). Their macromolecular analogs, that is, amphiphilic cationic polymers, display similar, or superior efficacy, accompanied by a long-term activity and lower toxicity to humans and environment due to lack of volatility and skin permeation ability (Wang et al., 2015; Waschinski et al., 2008) Antimicrobial activity of both types of cationic derivatives is the result of the interaction of cationic sites and hydrophobic components with negatively charged microbial cell surface, which leads to an irreversible damage of the cell membrane and eventually to the cell death (Jennings et al., 2015; Munoz-Bonilla and Fernandez-García, 2012; Timofeeva and Kleshcheva, 2011). This specific action mechanism makes cationic compounds less susceptible to develop bacterial resistance, a drawback often observed during the treatment with traditional antibiotics.

Many cationic polymers with different chemical composition and relative position of cationic and hydrophobic groups have been synthesized and evaluated for their antimicrobial activity (Kenawy et al., 2007; Deka et al., 2015; Nichifor and Stanciu, 2011; Carmona-Ribeiro and Carrasco, 2013). Few polysaccharides were chosen for the design of antimicrobial agents based on natural biocompatible polymers, for example, chitosan and its derivatives (Hosseinnejada and Jafari, 2016; Kong et al., 2010), Konjac glucomannan (Yu et al., 2007), and cashew gum (Quelemes et al., 2017).

Cationic dextran derivatives with the structure D40-QR(m) with different hydrophobicity of substituent R (C_2, C_4, C_8, C_{12}, C_{16}, Bz) and different m values (30 and 70 mol%) were tested for their activity against several bacterial and fungi strains (Nichifor et al., 2010). The results showed that only polymers with R = Bz or C_8 displayed antibacterial activity against all tested strains, except for *Aspergillus niger* ATCC 16888 (Table 4.2). However, minimal inhibitory concentration (*MIC*) was relatively high (3–5 mg/mL) for these class of polymers, therefore, new cationic amphiphilic polymers were designed to combine the pendent type structure with that placing the hydrophobic group at the end of the polymer chain (Nichifor et al., 2014), in the attempt to obtain chemical structures where cationic and hydrophobic components are spatially separated, as some polymers obtained from poly(oxazoline) with the components placed at different polymer chain ends proved to be very active (Krumm et al., 2014; Waschinski et al., 2008).

Polymers Z-DM-QR(m), where Z was C_{12} or C_{18}, M was 6 or 10 kDa, R was Bz or C_8, and m ≤ 30 mol% were prepared and their antimicrobial activity was evaluated against several reference bacterial and fungal strains, as well as against clinical pathogenic *S. aureus* species (Tuchilus et al., 2017). All tested polymers had a significant activity against all strains tested, except for *Pseudomonas aeruginosa* ATCC 27853 and methicillin-resistant *S. aureus* 68 (isolated from surgical wounds). The best activity was found against Gram-positive bacteria (*Staphylococcus aureus* ATCC 25923, *Sarcina lutea ATCC* 9341) and pathogenic yeasts (*Candida albicans* ATCC 90028, *Candida glabrata* ATCC MYA 2950, and *Candida parapsilosis* ATCC 22019). Gram-negative bacteria (*Escherichia coli* ATCC 25922) and methicillin-resistant *S. aureus*100 displayed a moderate response to treatment.

TABLE 4.2 Inhibition of Microbial Growth by Cationic Amphiphilic Polymers D40-QR(m), After Addition of 10 μL Aqueous Polymer Solution (5 mg/mL) to Microbial Culture Medium.

Polymer	Microbial growth				
	Staphylo-coccus aureus ATCC 6538P	*Escherichia coli* ATCC 8739	*Pseudomonas aeruginosa* ATCC 9027	*Candida albicans* ATCC 10231	*Aspergillus niger* ATCC 16888
D40-QC$_2$(30)	++	++	+	+	++
D40-QBz(30)	+	+	−	−	++
D40-QBz(70)	−	−	−	−	+
D40-QC$_8$(30)	−	−	−	−	++
D40-QC$_8$(70)	−	−	−	+	++
D40-QC$_{12}$(30)	++	++	++	++	+++

(−) Microbial growth was missing (<100 CFU/mL); (+) low (10^2 FU/mL); (++) moderate (10^3 CFU/mL); and (+++) significant (10^4 CFU/mL).

Polymers based on shorter chain dextran ($M = 6$ kDa) were more efficient, irrespective of the length of Z alkyl attached to the chain end, showing that an appropriate balance between hydrophobic and hydrophilic groups of polymers is required for a good antimicrobial activity. MIC values determined for the systems C$_{12}$-D10-QBz(30) or C$_{18}$-D6-QBz(30)/*S. aureus* (60 μg/mL) were similar to those reported for other cationic polymers (Strassburg et al., 2015, Wang et al., 2015, Quelemes et al., 2017) and recommend these polymers as broad-spectrum external biocides.

4.4.1.2 HYPOCHOLESTEREMIC DRUGS

Cholesterol, a vital component of cell membranes, is transported in blood-stream by lipoproteins, is metabolized in liver to bile acids and is excreted in bile inside bile acid micelles or phospholipid vesicles. Excess of plasma total cholesterol and low-density lipoprotein cholesterol (LDL-CH) leads to accumulation of cholesterol in blood vessels (atherosclerotic plaque), followed by the obstruction of blood vessels (atherosclerosis), which increases the risk for the development of coronary heart disease (CHD) (Grundy, 1998). Clinical therapies developed for plasma CH level reduction use different medication, such as statins, niacin, and bile acid seques-trants (BAS). Statins are the first-line agents for the treatment of primary hypercholesterolemia and fenofibrate is used in the treatment of mixed hyperlipidemia, but they are associated with serious systemic adverse effects (Bays and Goldberg, 2007), therefore, other drugs such as BAS used alone or as adjuvants in combined therapies can improve the results and reduce side effects (Robinson and Keating, 2007; Watts et al., 2016).

4.4.1.2.1 Bile Acid Sequestrants

BAS are orally administered insoluble cationic polymeric resins able to selectively bind negatively charged bile acids in the small intestine and eliminate them with feces. In order to re-establish the bile acid pool in the gall-bladder, new CH amounts will be metabolized in liver, reducing, therefore, the serum CH levels (Shepherd et al., 1980; Bays and Goldberg, 2007). Cholestyramine (an anion-exchange resin that contains benzyl trimetyl ammonium chloride groups attached to a styrene–divinylben-zene cross-linked copolymer (Shepherd et al., 1980)), and Colestipol (a cross-linked epichlorohydrin-tetraethylenepentamine resin (Class, 1991)) were for long time the only BAS in clinical use. Some inconvenience in the administration (high doses required, unpleasant taste and smell, side effects, such as nausea, abdominal discomfort, indigestion, and constipa-tion) determined a decline in the use of these drugs, but also prompted the development of more biocompatible and compliant BAS, for example, DMP-504, SK & F 97426-A, and Colesevelam hydrochloride, the last one being now commercially available in USA (Dahl et al., 2005; Alberto et al., 2009). More recent research was focused on the design of BAS with improved selectivity for bile acids, increased binding capacity, and stability (Wang et al., 2016; Polomoscanik et al., 2012; Zhu et al., 2015; Mendonça et al., 2016; Lopez-Jaramillo et al., 2015), mainly by creating new chemical structures with an appropriate balance between cationic and hydrophobic groups, in order to obtain a synergistic effect of electrostatic and hydrophobic interactions between BAS and bile acids. Besides, the use of natural polymers, such as polysaccharides can significantly improve the BAS biocompatibility.

In the attempt to develop new BASs with a superior binding capacity, high selectivity for bile acids over other anions and a better biocompatibility, Nichifor and co-workers focused their research on aminated cross-linked polysaccharides and studied in detail the influence of the nature of polysaccharide support (dextran, pullulan, and microcrystalline cellulose), its swelling porosity and the chemical structure of amino groups (tertiary ammine versus quaternary ammonium group), on the in vitro bile acid binding capacity and affinity (Nichifor et al., 1998, 2000). Further, they studied the effect of the presence of hydrophobic segments in the structure of cationic groups of D40-QR(m)-ECH derivatives. The increase of the length of substituent R from C_2 to C_{12} determined a significant increase in complex BAS/bile acid stability, both in the absence and presence of small electrolytes with competing anions (Nichifor et al., 2001a, 2001b). In order to combine a high charge density with an appropriate ratio hydrophilic/hydrophobic segments, bipolar amphiphilic polymers D40-QR$_1$(m) R$_2$(p)-ECH were prepared and its in vitro evaluation showed that amphiphilic dextran-based BAS had much higher binding affinity for bile acid than Cholestyramine. In vivo administration of a hydrophilic dextran resin (D40-QC2(60)-ECH), an amphiphilic resin (D40-QC$_2$(40)C$_{12}$(20)-ECH) and cholestyramine to normolipemic rats (Trinca et al., 2007) demonstrated a better hypolipemic effect of the amphiphilic resin (Fig. 4.8) in comparison with hydrophilic resin and Cholestyramine. D40-QC$_2$(40) C$_{12}$(20)-ECH determined a significant decrease in total CH, low and very low-density lipoprotein (LDL and VLDL), and triglycerides, without modification of liver, pancreas, or GIT functions. The most important effect of the amphiphilic dextran resin was the increase in the ratio anti-atherogenic high-density lipoprotein (HDL)/atherogenic low-density lipoproteins (LDL + VLDL) from 1.1 (control group) to 2.4. A higher ratio is more efficient in preventing the occurrence of atheromatous lesions than the level of each single lipid fraction.

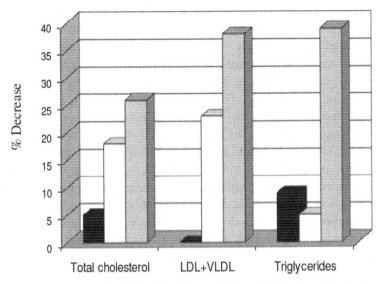

FIGURE 4.8 Decrease of lipid level (% of control) after administration of Cholestyramine (white columns), D40-QC$_2$(40)C$_{12}$(20)-ECH (grey column) and D40-QC$_2$(60)-ECH (black columns) to normolipemic rats (21 days and 1.2 g/kg/day).

4.4.1.2.2 Cholesterol Solubilizers

Besides its contribution to heart diseases, excess of CH is also the main risk factor for gallstone occurrence and growth due to CH crystallization. Under normal conditions, the process is inhibited by CH dissolution in bile salt micelles, therefore, the intake of bile salts was perceived as a basis for nonsurgical therapy (Bhat et al., 2006). Despite good results obtained with chenodeoxycholic acid and ursodeoxycholic acid, clinical use of such a treatment is limited due to bile acid poor efficacy, toxic side effects, and long time required (months to years) (Konikoff, 2003; Hofmann and Hagey, 2014; Abendan and Swift, 2013), being recommended for prevention or in special clinical situation when surgery is not possible (Konikkof, 2003).

By analogy to unbound bile acid micelles, hydrophobic microdomains formed in aqueous solutions of dextran derivatives DM-BA(m) by association of their BA pendent moieties led as to supposition that these aggregates could trap CH molecules inside their hydrophobic microdomains and increase thus CH solubility. The results obtained with different bile acid modified dextrans (Fig. 4.9) proved this hypothesis. As expected, the amount of solubilized CH increased with polymer concentration, but the variation profile depended on the bound BA. The amount of solubilized CH increased almost linearly with DCA containing polymers concentration, but an exponential increase was observed in the presence of DM-CA polymers. A similar behavior was observed for solubilization of CH by corresponding bile acid salts (Nagadome et al., 1995) and was related to the variation of micelle number and size with each bile acid concentration. This similarity between the behavior of the BA-modified polymer and the free BA might indicate a similar arrangement of BA aggregates, irrespective of their bound or unbound state. The increase in the dextran molar mass enhanced CH solubilization, what can be explained by the presence of a higher number of bile acid bound per polymer chain, resulting in a higher number, or size of hydrophobic microdomains/chain. Water solubility of CH is very low (3×10^{-8} g/mL, 0.007 mM) (Saad and Higuchi, 1965), and DM-BA polymers were able to increase this solubility up to 60 times. A direct comparison between free and bound bile acid ability to solubilize CH is difficult, but the molar ratio BA:solubilized CH is much lower in case of polymer. For example, NaCDCA:CH = 20:1 (in 2.5 mM bile salt), NaCA:CH = 63:1 (in 50 mM bile salt solution) (Mukhopadhyay

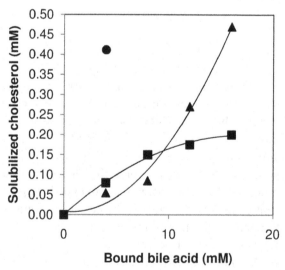

FIGURE 4.9 Variation of the amount of solubilized cholesterol with polymer concentration (expressed as bound bile acid concentration). Polymers are DM-BA(3.6), where M = 40 kDa (squares and triangles) or 200 kDa (circle) and BA is CA (triangles and circle) or DCA (squares).

and Maitra, 2004), and CA:CH = 10 (in D200-CA (3.6) solution containing 4 mM bound CA).

The capacity of polymers DM-BA to solubilize CH can find application not only in gallstone dissolution, but also in different domains, such as food or cosmetic industries.

4.4.2 DRUG DELIVERY

An amphiphilic polymer self-assembly consists in an inner hydrophobic cluster where a hydrophobic drug can be solubilized, and an outer hydrophilic shell which can protect the drug from the degradation under the action of biological fluid components, for example, plasma proteins. The properties of polymers can be tuned to create a hydrophobic core with an appropriate size and affinity for a given drug. Additionally, the outer shell can be chemically modified by crosslinking for aggregate stabilization to dilution, or by attachment of specific ligands which facilitate the active targeting of the drug to its site of action (Guo et al., 2016; Dong et al., 2015; Simoes et al., 2014; Tan et al., 2013). Attachment of charged groups to chains forming outer corona can enhance the interaction polymer-drug by electrostatic attraction, or can allow a simultaneous encapsulation of hydrophobic and hydrophilic drugs (Martin et al., 2016).

Mocanu and Nichifor (2014) have studied the capacity of amphiphilic polymers D40-QR$_1$(m)R$_2$(p)-ECH with different chemical composition for retention and release of several dyes (as drug model) and drugs with variable hydrophobicities and molecular masses. The order of the dye retention (Table 4.3), that is, Rose Bengal (RB) > Brilliant Blue (BB) > vitamin B12 (VB12) is in good agreement with the order of dye hydrophobicity, what can be considered as a proof that the retention driving forces are hydrophobic interactions between solute and gel hydrophobic clusters. The gels D40-QC$_{12}$(25)-ECH and D40-QC$_2$(35)C$_{12}$(25)-ECH with the highest amount of hydrophobic R substituent have the highest loading capacity. Some electrostatic interactions can also occur between the cationic gels and anionic dyes RB and BB. The retention of two protein with high molar mass, bovine serum albumin (BSA, MW~60 kDa), and tetanus anatoxin (TA, MW~150 kDa), was strongly influenced by their amphiphilic character and isoelectric point. Despite its higher molar mass, TA retention was much better than that of BSA (Table 4.3), perhaps due to a stronger electrostatic interaction with cationic gels and a more compact structure.

TABLE 4.3   Loading Capacity of Amphiphilic Gels D40-QR$_1$(m)R$_2$(p)-ECH for Different Dyes and Drugs. Retentions Were Performed from Aqueous Solution of pH = 5.5, Except for BSA, Where pH = 6.9.

Polymer characteristics			Loading capacity, mg/g					
R$_1$(m)	R$_2$(p)	WR, g/g	Dye				Drug	
			RB	BB	VB12	BSA	TA	α-TF
C$_2$(50)	–	3.97	90	60	45	26	110	140
–	C$_{12}$(25)	3.96	105	55	42	18	152	120
C$_2$(35)	C$_{12}$(25)	4.58	107	47	46	15	125	180
C$_2$(47)	C$_{12}$(10)	9.30	88	47	47	51	153	140
C$_2$(47)	C$_{16}$(10)	6.70	105	40	40	42	154	–
Bz(28)	–	3.07	75	38	39	35	125	130
C$_8$(22)	–	3.30	90	60	41	750	180	160
C$_8$(20)	–	60.00	120	75	40	800	190	280

Source: Adapted from Mocanu and Nichifor (2014).

This hypothesis is supported by a lower influence of gel chemical composition and its swelling capacity on the loading capacity. The influence of the HLB of gel chemical structure is more evident in case of α-tocopherol (α-TF), a low molecular uncharged hydrophobic drug (Fig. 4.10). In this case, the higher loading capacity was found for the gel D40-QC$_2$(35)C$_{12}$(25)-ECH with higher content in hydrophobic groups. It is also worth mentioning that the best retention results for all studied drugs were obtained with gels D40-QC$_8$(20)-ECH with a moderate hydrophobicity and high swelling capacity. The release of all drugs from amphiphilic gels was very slow, what suggests the presence of a strong interaction gel-drug and shows the potential of these polymers as carriers for controlled and sustained drug delivery.

Mocanu et al. (2015, 2017) have also examined the potential use of polymers Z-DM-DVS and Z-DM-QR(m)-DVS as carriers for drug delivery. Several drugs with different hydrophobicities and chemical structures were used for these experiments (Fig. 4.10). The retention of some of these drugs on amphiphilic polymer C$_{18}$-D10-DVS and its cationic derivative C$_{18}$-D10-QBz(10)-DVS is presented in Figure 4.11 and the results allow us to highlight a few aspects: (1) the loading capacity of the neutral polymer C$_{18}$-D10-DVS increases with increasing drug hydrophobicity (decreasing water solubility) in the order: rifampicin (RIF, 2.77 mg/mL)

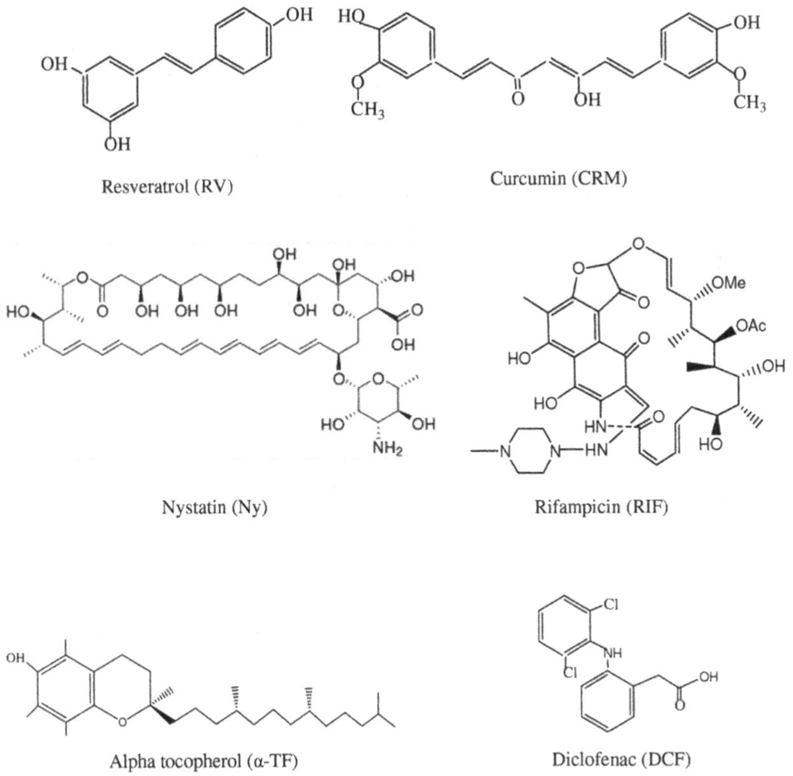

Resveratrol (RV)

Curcumin (CRM)

Nystatin (Ny)

Rifampicin (RIF)

Alpha tocopherol (α-TF)

Diclofenac (DCF)

FIGURE 4.10 Chemical structures of drugs used for loading/release studies.

< nystatin (Ny, 0.36 mg/mL) < resveratrol (RV, 0.03 mg/mL) < curcumin (CRM, 11 ng/mL) < α-TF (below 1 µg/mL). This behavior clearly indicates the preferential solubilization of more hydrophobic drugs inside the hydrophobic core of polymeric micelles; (2) the higher loading capacity for α-TF than for CRM, which have similar water solubility and molecular weights, can be assigned to their different chemical structure. CRM has a more bulky structure, with two phenyl cycles at each end of an unsaturated alkene chain, but α-TF has a long alkyl tail, which probably fits better inside the micelle core formed also from alkyl chains. This suggests that a structural affinity between drug and polymer hydrophobic segments is required for an enhanced drug retention; and (3) comparison between the loading capacity of the polymer C_{18}-D10-DVS and its cationic derivative shows an overall improvement of the retention on cationic gel, probably

due to some electrostatic interactions between cationic groups of the gel and ionizable OH groups present in most of the drugs' structures (see Fig. 4.11). The lowest improvement was observed for α-TF, with only one OH group, and the highest one for diclofenac (DCF) carrying a carboxylic group in its molecule. DCF loading was 6 times higher on C_{18}-D10-Q(10)-DVS than on the neutral analog, due to the stronger electrostatic attraction between gel cationic groups and drug carboxylic groups. This finding might represent an opportunity to use amphiphilic cationic gels for simultaneous loading of very hydrophobic uncharged drugs and more hydrophilic negatively charged drugs.

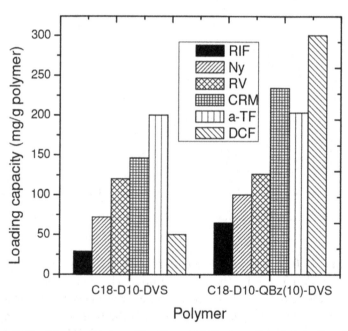

FIGURE 4.11 Drug-loading capacity of amphiphilic cross-linked derivatives C_{18}-D10-DVS and C_{18}-D10-QBz(10)-DVS. Adapted from Mocanu et al. (2015) and Mocanu et al. (2017).

The rate of drug release from drug-loaded polymers was always slower than the free drug recovery rate, and decreased with increasing drug hydrophobicity. Thus, the cumulative drug release (% from the initial amount bound to polymer) from the gel C_{18}-D10-DVS decreased in the order RIF(95) > Ny (57) > RV (40) >> CRM $\approx$ α-TF (5–10), and was

1–5% lower for release of each drug from C_{18}-D10-QBz(10)-DVS cationic gel. It seems that both hydrophobic and electrostatic forces between drugs and polymers are strong enough to slow down the drug release and to provide a prolonged and sustained drug bioavailability.

The gradual release of CRM and α-TF from the drug-loaded polymers could be also proved by following the drug antioxidant activity, using specific analytic procedures (Mocanu et al., 2015). Antioxidant activity of each drug loaded in polymer was lower than that of the same amount of free drug. In case of CRM, a similar time evolution of the drug release and its antioxidant activity in release medium was found. Based on these experimental results, we can assume that the loaded drugs preserve their antioxidant activity, but this activity is expressed only after drug release.

4.5   CONCLUDING REMARKS

Chemical modifications of polysaccharides, by attachment of hydrophobic segments either along polysaccharide main chain or at its chain end provide a large variety of amphiphilic polymers with ability to form organized structures by intra- or/and intermolecular hydrophobic interactions, with more or less defined hydrophobic and hydrophilic microdomains. The self-organized structure properties, such as size, morphology, polarity, and compactness of the microdomains can be tuned by an appropriate choice of the segments (blocks) chemical composition. The synthesized amphiphilic polysaccharides have great potential for biomedical applications as antimicrobial agents, hypolipemiants, or as carriers for drug delivery.

KEYWORDS

- polysaccharides
- amphiphiles
- self-organization
- antimicrobial
- hypolipemiants
- drug delivery

REFERENCES

Abendan, R. S.; Swift, J. A. Cholesterol Monohydrate Dissolution in the Presence of Bile Acid Salts. *Cryst. Growth Des.* **2013**, *13* (8), 3596–3602.

Alberto, C.; Eberhard, W.; Michel, F. Colesevelam Hydrochloride: Usefulness of a Specifically Engineered Bile Acid Sequestrant for Lowering LDL-cholesterol. *Eur. J. Cardiovasc. Prevent. Rehabil.* 2009, *16* (1), 1–9.

Alexandridis, P.; Lindman, B. *Amphiphilic Block Copolymers: Self-assembly and Applications*; Elsevier: Amsterdam, Netherlands, 2000; p 448.

Bae, S. M., et al. An Apoptosis-homing Peptide-conjugated Low Molecular Weight Heparin-taurocholate Conjugate with Antitumor Properties. *Biomaterials* 2013, *34* (8), 2077–2086.

Ballarin, B., et al. Effect of Cationic Charge and Hydrophobic Index of Cellulose-based Polymers on the Semipermanent Dyestuff Process for Hair. *Int. J. Cosmetic Sci.* **2011**, *33* (3), 228–233.

Bays, H. E.; Goldberg, R. B. The "Forgotten" Bile Acid Sequestrants: Is Now a Good Time to Remember? *Am. J. Ther.* **2007**, *14* (6), 567–580.

Belalia, R., et al. New Bioactive Biomaterials Based on Quaternized Chitosan. *J. Agric. Food Chem.* **2008**, *56* (5), 1582–1588.

Bhat, S., et al. Use of Novel Cationic Bile Salts in Cholesterol Crystallization and Solubilization In Vitro. *Biochim. Biophys. Acta.* **2006**, *1760* (10), 1489–1496.

Brito, R. M. M.; Vaz, W. L. C. Determination of the Critical Micelle Concentration of Surfactants Using the Fluorescent Probe *N*-phenyl-1-naphthylamine. *Anal. Biochem.* 1986, *152* (2), 250–255.

Cadete, A., et al. Development and Characterization of a New Plasmid Delivery System Based on Chitosan-sodium Deoxycholate Nanoparticles. *Eur. J. Pharm. Sci.* **2012**, *45* (4), 451–458.

Calandra, P., et al. How Self-assembly of Amphiphilic Molecules Can Generate Complexity in the Nanoscale. *Colloids Surf. A Physicochem. Eng. Asp.* **2015**, *484*, 164–183.

Carmona-Ribeiro, A. M.; Carrasco, L. D. M. Cationic Antimicrobial Polymers and Their Assemblies. *Int. J. Mol. Sci.* **2013**, *14* (5), 9906–9946.

Chen, S., et al. Synthesis of Amphiphilic Diblock Copolymers Derived from Renewable Dextran by Nitroxide Mediated Polymerization: Towards Hierarchically Structured Honeycomb Porous Films. *Polym. Chem.* **2014**, *5*, 5310–5319.

Class, S. D. Quaternized Colestipol, an Improved Bile Salt Adsorbent: In Vitro Studies. *J. Pharm. Sci.* 1991, *80* (2), 128–131.

Cotanda, P.; O'Reilly, R. K. Molecular Recognition Driven Catalysis Using Polymeric Nanoreactors. *Chem. Commun.* **2012**, *48* (83), 10280–10282.

Cunningham, A. J.; Zhu, X. X. Polymers Made of Bile Acids: From Soft to Hard Biomaterials. *Can. J. Chem.* **2016**, *94* (8), 659–666.

Deka, S. R.; Sharma, A. K.; Kumar, P. Cationic Polymers and Their Self-assembly for Antibacterial Applications. *Curr. Top. Med. Chem.* **2015**, *15* (13), 1179–1195.

Dhal, P. K.; Huval, C. C.; Holmes-Farley, S. R. Functional Polymers as Human Therapeutic Agents. *Ind. Eng. Chem. Res.* **2005**, *44* (23), 8593–8604.

Discher, D. E.; Eisenberg, A. Polymer Vesicles. *Science* **2002**, *297* (5583), 967–973.

Dobrynin, A. V.; Rubinstein, M. Hydrophobically Modified Polyelectrolytes in Dilute Salts-free Solutions. *Macromolecules* **2000**, *33* (21), 8097–8105.

Dong, R., et al. Functional Supramolecular Polymers for Biomedical Applications. *Adv. Mater.* **2015**, *27* (3), 498–526.

Durand, A. Bile Acids as Building Blocks of Amphiphilic Polymers. Applications and Comparison with Other Systems. *Collect. Czech. Chem. Commun.* 2007, *72* (11), 1553–1578.

Durand, A.; Marie, E. Macromolecular Surfactants for Miniemulsion Polymerization. *Adv. Colloid Interface Sci.* **2009**, *150* (2), 90⁻105.

Grundy, S. M. The Role of Cholesterol Management in Coronary Disease Risk Reduction in Elderly Patients. *Endocrinol. Metab. Clin. North Am.* **1998**, *27* (3), 655–675.

Guo, X., et al. Polymer-based Drug Delivery Systems for Cancer Treatment. *J. Polym. Sci. A Polym. Chem.* **2016**, *54* (22), 3525–3550.

Hassani, L. N.; Hendra, F.; Bouchemal, K. Auto-associative Amphiphilic Polysaccharides as Drug Delivery Systems. *Drug Discov. Today* **2012**, *17* (11–12), 608–614.

Hirsjärvi, S., et al. Surface Modification of Lipid Nanocapsules with Polysaccharides: From Physicochemical Characteristics to In Vivo Aspects. *Acta Biomater.* **2013**, *9* (5), 6686–6693.

Hofmann, A. F.; Hagey, L. R. Key Discoveries in Bile Acid Chemistry and Biology and Their Clinical Applications: History of the Last Eight Decades. *J. Lipid Res.* **2014**, *55* (8), 1553–1596.

Hosseinnejada, M.; Jafari, S. M. Evaluation of Different Factors Affecting Antimicrobial Properties of Chitosan. *Int. J. Biol. Macromol.* **2016**, *85*, 467–475.

Houga, C., et al. Micelles and Polymersomes Obtained by Self-assembly of Dextran and Polystyrene Based Block Copolymers. *Biomacromolecules* **2009**, *10* (1), 32–40.

Hwang, H. H.; Lee, D. Y. Antiangiogenic Actions of Heparin Derivatives for Cancer Therapy. *Macromol. Res.* **2016**, *24* (9), 767–772.

Ishak, R. A. H.; Osman, R.; Awad, G. A. S. Dextran-based Nanocarriers for Delivery of Bioactives. *Curr. Pharm. Design.* **2016**, *22* (22), 3411–3428.

Jennings, M. C.; Minbiole, K. P. C.; Wuest, W. M. Quaternary Ammonium Compounds: An Antimicrobial Mainstay and Platform for Innovation to Address Bacterial Resistance. *ACS Infect. Dis.* **2015**, *1* (8), 288–303.

Kenawy, E. R.; Worley, S. D.; Broughton, R. The Chemistry and Applications of Antimicrobial Polymers: A State-of-the-art Review. *Biomacromolecules* **2007**, *8* (5), 1359–1384.

Kong, M., et al. Antimicrobial Properties of Chitosan and Mode of Action: A State of the Art Review. *Int. J. Food Microbiol.* **2010**, *144* (1), 51–63.

Konikoff, F. M. Gallstones—Approach to Medical Management. *Med. Gen. Med.* **2003**, *5* (1), 1–9.

Krumm, C., et al. Antimicrobial Poly(2-methyloxazoline)s with Bioswitchable Activity through Satellite Group Modification. *Angew. Chem. Int. Ed.* **2014**, *53* (15), 3830–3834.

Kuskov, A. N., et al. Preparation and Characterization of Amphiphilic Poly-N-vinylpyrrolidone Nanoparticles Containing Indomethacin. *J. Mat. Sci. Mat. Med.* **2010**, *21* (5), 1521–1530.

Lepoittevin, B., et al. Easy Access to Amphiphilic Glycosylated-functionalized Polystyrenes. *Carbohydr. Polym.* **2011**, *83* (3), 1174–1179.

Li, B., et al. Preparation, Drug Release and Cellular Uptake of Doxorubicin-loaded Dextran-b-poly(ε-caprolactone) Nanoparticles. *Carbohydr. Polym.* **2013**, *93* (2), 430–437.

Liu, K.; Jiang, X.; Hunziker, P. Carbohydrate-based Amphiphilic Nano Delivery System for Cancer Therapy. *Nanoscale* **2016**, *8* (36), 16091–16156.

Lopez-Jaramillo, F. J., et al. In Vitro and In Vivo Evaluation of Novel Cross-linked Saccharide Based Polymers as Bile Acid Sequestrants. *Molecules* **2015**, *20* (3), 3716–3729.

Mai, Y.; Eisenberg, A. Self-assembly of Block Copolymers. *Chem. Soc. Rev.* **2012**, *41* (18), 5969–5985.

Martin, C., et al. Recent Advances in Amphiphilic Polymers for Simultaneous Delivery of Hydrophobic and Hydrophilic Drugs. *Ther. Deliv.* **2016**, *7* (1), 15–31.

Mendonça, P. V., et al. Synthesis of Tailor-made Bile Acid Sequestrants by Supplemental Activator and Reducing Agent Atom Transfer Radical Polymerization. *RSC Adv.* **2016**, *6* (57), 52143–52153.

Merianos, J. J. Quaternary Ammonium Antimicrobial Compounds. In *Disinfection, Sterilization, and Preservation*, 4th ed.; Block, S. S., Ed.; Lea & Febiger: Philadelphia, PA, 1991; pp 225–255.

Mizrahyab, S.; Peer, D. Polysaccharides as Building Blocks for Nanotherapeutics. *Chem. Soc. Rev.* **2012**, *41* (7), 2623–2640.

Mocanu, G.; Nichifor, M. Cationic Amphiphilic Dextran Hydrogels with Potential Biomedical Applications. *Carbohydr. Polym.* **2014**, *99*, 235–241.

Mocanu, G.; Nichifor, M.; Stanciu, M. C. New Shell Crosslinked Micelles from Dextran with Hydrophobic End Groups and Their Interaction with Bioactive Molecules. *Carbohydr. Polym.* **2015**, *119*, 228–235.

Mocanu, G.; Nichifor, M.; Sacarescu, L. Dextran Based Polymeric Micelles as Carriers for Delivery of Hydrophobic Drugs. *Curr. Drug Deliv.* **2017**, *14* (3), 406–415.

Mukhopadhyay, S.; Maitra, U. Facile Synthesis, Aggregation Behavior, and Cholesterol Solubilization Ability of Avicholic Acid. *Org. Lett.* **2004**, *6* (1), 31–34.RGANIC6

Munoz-Bonilla, A.; Fernandez-García, M.; Polymeric Materials with Antimicrobial Activity. *Progr. Polym. Sci.* **2012**, *37* (2), 281–339.

Nagadome, S., et al. Solubilization and Precipitation of Cholesterol in Aqueous Solutions of Bile Salts and Their Mixtures. *Colloid Polym. Sci.* **1995**, *273* (7), 675–689.

Nichifor, M., et al. Aminated Polysaccharides as Bile Acid Sorbents: In Vitro Study. *J. Biomater. Sci. Polym. Ed.* **1998**, *9* (6), 519–534.

Nichifor, M.; Carpov, A. Bile Acids Covalently Bound to Polysaccharides 1. Esters of Bile Acids with Dextran. *Eur. Polym. J.* **1999**, *35* (12), 2125–2129.

Nichifor, M., et al. Aggregation in Water of Dextran Hydrophobically Modified with Bile Acids. *Macromolecules* **1999**, *32* (21), 7078–7085.

Nichifor, M.; Cristea, D.; Carpov, A. Sodium Cholate Sorption on Cationic Dextran Hydrogel Microspheres. 1. Influence of the Chemical Structure of Functional Groups. *Int. J. Biol. Macromol.* **2000**, *28* (1), 15–21.

Nichifor, M., et al. Bile Acid Sequestrants Based on Cationic Dextran Hydrogel Microspheres. 2. Influence of the Length of Alkyl Substituents at the Amino Groups of the Sorbents on the Sorption of Bile Salts. *J. Pharm. Sci.* **2001a**, *90* (6), 681–689.

Nichifor, M., et al. Interaction of Hydrophobically Modified Cationic Dextran Hydrogels with Biological Surfactants. *J. Phys. Chem. B.* **2001b**, *105* (12), 2314–2321.

Nichifor, M.; Stanciu, M. C.; Zhu, X. X. Bile Acids Covalently Bound to Polysaccharides. 2. Dextran with Pendent Cholic Acid Groups. *React. Funct. Polym.* **2004a**, *59* (2), 141–148.

Nichifor, M., et al. Self-aggregation of Amphiphilic Cationic Polyelectrolytes Based on Polysaccharides. *J. Phys. Chem. B* **2004b**, *108* (42), 16463–16472.

Nichifor, M.; Mocanu, G. Self Aggregation of Amphiphilic Polyelectrolytes. In *Focus on Ionic Polymers*; Dragan, E. S., Ed.; Research Signpost: Thiruvananthapuram, India, 2005; pp 153–201.

Nichifor, M.; Stanciu, M. C.; Simionescu, B. C. New Cationic Hydrophilic and Amphiphilic Polysaccharides Synthesized by One Pot Procedure. *Carbohydr. Polym.* **2010**, *82* (3), 965–975.

Nichifor, M.; Stanciu, C. Cationic Polymers as Therapeutic Agents. In *Medical Applications of Polymers*; Popa, M., Ottenbritte, R. M., Uglea, C. V., Eds.; American Scientific Publishers: Valencia, CA, 2011; Vol. 1, pp 179–219.

Nichifor, M., et al. Hydrodynamic Properties of Some Cationic Amphiphilic Polysaccharides in Dilute and Semi-dilute Aqueous Solutions. *Carbohydr. Polym.* **2011**, *83* (4), 1887–1894.

Nichifor, M.; Mocanu, G.; Stanciu, M. C. Micelle-like Association of Polysaccharides with Hydrophobic End Groups. *Carbohydr. Polym.* **2014**, *110*, 209–218.

Park, J. K., et al. Bile Acid Conjugated Chitosan Oligosaccharide Nanoparticles for Paclitaxel Carrier. *Macromol. Res.* **2014**, *22* (3), 310–317.

Patterson, J. P., et al. Catalytic Y-tailed Amphiphilic Homopolymers—Aqueous Nanoreactors for High Activity, Low Loading SCS Pincer Catalysts. *Polym. Chem.* **2013**, *4* (6), 2033–2039.

Petit-Agnely, F.; Iliopoulos, I.; Zana, R. Hydrophobically Modified Sodium Polyacrylates in Aqueous Solutions: Association Mechanism and Characterization of the Aggregates by Fluorescence Probing. *Langmuir* **2000**, *16* (25), 9921–9927.

Pignatello, R., et al. New Amphiphilic Conjugates of Amino–poly(Ethylene Glycols) with Lipoamino Acids as Surface Modifiers of Colloidal Drug Carriers. *Macromol. Chem. Phys.* **2013**, *214* (1), 46−55.

Polomoscanik, S. C., et al. Hydrophobically Modified Poly(allylamine) Hydrogels Containing Internal Quaternary Ammonium Groups as Cholesterol Lowering Agents: Synthesis, Characterization, and Biological Studies. *J. Macromol. Sci. Pure Appl. Chem.* **2012**, *49* (12), 1011–1021.

Quelemes, P. V., et al. Quaternized Cashew Gum: An Anti-staphylococcal and Biocompatible Cationic Polymer for Biotechnological Applications. *Carbohydr. Polym.* **2017**, *157*, 567–575.

Raemdonck, K., et al. Polysaccharide-based Systems in Drug and Gene Delivery. *Adv. Drug Deliv. Rev.* 2013, *65* (9), 1123–1147.

Raffa, P., et al. Polymeric Surfactants: Synthesis, Properties, and Links to Applications. *Chem. Rev.* **2015**, *115* (16), 8504–8563.

Riess, G. Micellization of Block Copolymers. *Prog. Polym. Sci.* **2003**, *28* (7), 1107−1170.

Robinson, D. M.; Keating, G. M. Colesevelam: A Review of Its Use in Hypercholesterolemia. *Am. J. Cardiovasc. Drugs* **2007**, *7* (6), 453–465.

Saad, H. Y.; Higuchi, W. I. Water Solubility of Cholesterol. *J. Pharm. Sci.* **1965**, *54* (8), 1205−1206.

Schatz, C., et al. Polysaccharide-block-polypeptide Copolymer Vesicles: Towards Synthetic Viral Capsids. *Angew. Chem. Int. Ed.* **2009**, *48* (14), 2572–2575.

Schatz, C.; Lecommandoux, S. Polysaccharide-containing Block Copolymers: Synthesis, Properties and Applications of an Emerging Family of Glycoconjugates. *Macromol. Rapid Commun.* **2010**, *31* (19), 1664–1684.

Shepherd, J., et al. Cholestyramine Promotes Receptor-mediated Low-density-lipoprotein Catabolism. *New Engl. J. Med.* **1980**, *30* (22), 1219–1222.

Simoes, S. M. N., et al. Polymeric Micelles for Oral Drug Administration Enabling Locoregional and Systemic Treatments. *Expert Opin. Drug Deliv.* **2014**, *12* (2) 1–22.

Singh, R. S.; Saini, G. K. Biosynthesis of Pullulan and Its Applications in Food and Pharmaceutical Industry. In *Microorganisms in Sustainable Agriculture and Biotechnology*; Satyanarayana, T., Johri, B. N., Eds.; Springer Science + Business Media B. V.: Netherlands, 2012; pp 509–553.

Small, D. M. The Physical Chemistry of Cholanic Acids. In *The Bile Acids. Chemistry, Physiology and Metabolism*; Nair, P. P., Kritchevsky, D., Eds.; Plenum Press: New York, NY, 1971; Vol. 1, pp 249–354.

Stanciu, M. C.; Nichifor, M.; Simionescu, B. C. New Biocompatible Polyesters Based on Bile Acids. *Rev. Roum. Chim.* **2009**, *54* (11–12), 951–955.

Stanciu, M. C.; Nichifor, M. New Degradable Polyesters from Deoxycholic Acid and Oligo(ethylene glycol)s. *Polym. Int.* **2013**, *62* (8), 1236–1242.

Stanciu, M. C.; Nichifor, M. New Biocompatible Amphiphilic Diblock Copolymer Based on Dextran. *Eur. Polym. J.* **2015**, *71*, 352–363.

Strassburg, A., et al. Nontoxic, Hydrophilic Cationic Polymers—Identified as Class of Antimicrobial Polymers. *Macromol. Biosci.* **2015**, *15* (12), 1710–1723.

Sun, H., et al. Shell-sheddable Micelles Based on Dextran-SS-poly(ε-caprolactone) Diblock Copolymer for Efficient Intracellular Release of Doxorubicin. *Biomacromolecules* **2010**, *11* (4), 848–854.

Tan, C.; Wang, Y.; Fan, W. Exploring Polymeric Micelles for Improved Delivery of Anticancer Agents: Recent Developments in Preclinical Studies. *Pharmaceutics* **2013**, *5* (1), 201–219.

Tang, Q.; Huang, G. Progress in Polysaccharide Derivatization and Properties. *Mini-Rev. Med. Chem.* **2016**, *16* (15), 1244–1257.

Tang, Z., et al. Polymeric Nanostructured Materials for Biomedical Applications. *Progr. Polym. Sci.* **2016**, *60*, 86–128.

Timofeeva, L.; Kleshcheva, N. Antimicrobial Polymers: Mechanism of Action, Factors of Activity, and Applications. *Appl. Microbiol. Biotechnol.* **2011**, *89* (3), 475–492.

Tizzotti, M., et al. Modification of Polysaccharides through Controlled/living Radical Polymerization Grafting—Towards the Generation of High Performance Hybrids. *Macromol. Rapid. Commun.* **2010**, *31* (20), 1751–1772.

Torchilin, V. P. Amphiphilic Poly-N-vinyl Pyrrolidones: Synthesis, Properties and Liposome Surface Modification. *Biomaterials* **2001**, *22* (22), 3035–3044.

Trinca, L. C., et al. In Vitro and In Vivo Study of the Hypolipemic Effect of Some Aminated Polysaccharides Polymers. *Sci. Papers Univ. Agric. Sci. Vet. Med. Series* **2007**, *50* (9), 106–113.

Tuchilus, C. G., et al. Antimicrobial Activity of Chemically Modified Dextran Derivatives. *Carbohydr. Polym.* **2017**, *161*, 181–186.

Varshosaz, J. Dextran Conjugates in Drug Delivery. *Expert. Opin. Drug Deliv.* **2012**, *9* (5), 509–523.

Verma, M. S., et al. Size-tunable Nanoparticles Composed of Dextran-b-poly(D,L-lactide) for Drug Delivery Applications. *Nano. Res.* **2012**, *5* (1), 49–61.

Wang, H. X., et al. Antibacterial Activity of Geminized Amphiphilic Cationic Homopolymers. *Langmuir* **2015**, *31* (50), 13469–13477.

Wang, X., et al. Novel Bile Acid Sequestrant: A Biodegradable Hydrogel Based on Amphiphilic allylamine Copolymer. *Chem. Eng. J.* **2016**, *304*, 493–502.

Waschinski, C. J., et al. Insights in the Antibacterial Action of Poly(methyloxazoline)s with a Biocidal End Group and Varying Satellite Groups. *Biomacromolecules* **2008**, *9* (7), 1764–1771.

Watts, G. F., et al. Angiographic Progression of Coronary Atherosclerosis in Patients with Familial Hypercholesterolaemia Treated with Non-statin Therapy: Impact of a Fat-modified Diet and a Resin. *Atherosclerosis* **2016**, *252*, 82–87.

Wei, W. H.; Dong, X. M.; Liu, C. G. In Vitro Investigation of Self-assembled Nanoparticles Based on Hyaluronic Acid-deoxycholic Acid Conjugates for Controlled Release Doxorubicin: Effect of Degree of Substitution of Deoxycholic Acid. *Int. J. Mol. Sci.* **2015**, *16* (4), 7195–7209.

Wooley, K. L. Shell Crosslinked Polymer Assemblies: Nanoscale Constructs Inspired from Biological Systems. *J. Polym. Sci. A Polym. Chem.* **2000**, *38* (9), 1397–1407.

Wu, J.; Pan, X.; Zhao, Y. Time-dependent Shrinkage of Polymeric Micelles of Amphiphilic Block Copolymers Containing Semirigid Oligocholate Hydrophobes. *J. Coll. Interface Sci.* **2011**, *353* (2), 420–425.

Yan, J. K., et al. Self-aggregated Nanoparticles of Carboxylic Curdlan-deoxycholic Acid Conjugates as a Carrier of Doxorubicin. *Int. J. Biol. Macromol.* **2015**, *72*, 333–340.

Yang, J., et al. Physicochemical Characterization of Amphiphilic Nanoparticles Based on the Novel Starch–Deoxycholic Acid Conjugates and Self-aggregates. *Carbohydr. Polym.* **2014**, *102*, 838–845.

Yang, J., et al. Preparation and Application of Micro/nanoparticles Based on Natural Polysaccharides. *Carbohydr. Polym.* **2015**, *123*, 53–66.

Yu, H., et al. Preparation and Characterization of Quaternary Ammonium Derivative of Konjac Glucomannan. *Carbohydr. Polym.* **2007**, *69* (1), 29–40.

Yusa, S., et al. Fluorescence Studies of pH-responsive Unimolecular Micelles Formed from Amphiphilic Polysulfonates Possessing Long-chain Alkyl Carboxyl Pendents. *Macromolecules* **2002**, *35* (27), 10182–10188.

Zhang, T.; Marchant, R. E. Novel Polysaccharide Surfactants: The Effect of Hydrophobic and Hydrophilic Chain Length on Surface Active Properties. *J. Colloid Interface Sci.* **1996**, *177* (2), 419–426.

Zhao, Y., et al. Progressive Macromolecular Self-assembly: From Biomimetic Chemistry to Bio-inspired Materials. *Adv. Mater.* **2013a**, *25* (37), 5215–5256.

Zhao, Z., et al. Biodegradable Stereocomplex Micelles Based on Dextran-block-polylactide as Efficient Drug Deliveries. *Langmuir* **2013b**, *29* (42), 13072–13080.

Zhu, X. X.; Nichifor, M. Polymeric Materials Containing Bile Acids. *Acc. Chem. Res.* 2002, *35* (7), 539–546.

Zhu, X., et al. Cationic Amphiphilic Microfibrillated Cellulose (MFC) for Potential Use for Bile Acid Sorption. *Carbohydr. Polym.* **2015**, *132*, 598–605.

CHAPTER 5

SELF-ASSEMBLY IN SOLUTION OF METAL COMPLEXES BASED ON ORGANO-SILOXANE LIGANDS

MIRELA-FERNANDA ZALTARIOV*, MARIA CAZACU, and CARMEN RACLES

Inorganic Polymers Department, "Petru Poni" Institute of Macromolecular Chemistry, Grigore Ghica-Voda Alley 41A, 700487 Iasi, Romania

*Corresponding author. E-mail: zaltariov.mirela@icmpp.ro

CONTENTS

ABSTRACT

The self-assembly of amphiphilic molecules or block copolymers is widely investigated, providing a fascinating route to functional nanomaterials with potential for a variety of applications. However, the self-assembly of metal complexes has been relatively less explored compared to the purely organic systems.

This chapter presents some general aspects regarding this phenomenon in siloxane-based structures, focusing on metal complexes. In particular, we investigated the ability of Cu^{II}, Co^{II}, and Zn^{II} complexes of a poly-azomethine ligand derived from bis (formyl-p-phenoxymethyl) tetra-methyldisiloxane and 2,5-bis (p-aminophenyl)-1,3,4-oxadiazole, to form aggregates in selective solvents. The self-organization of metal complexes was evaluated in DMF solution by UV–Vis and dynamic light scattering techniques while the preservation or evolution of the assemblies after solvent evaporation was evidenced by atomic force microscopy and transmission electron microscopy.

5.1 INTRODUCTION

Self-assembly of macromolecules represents a current topic of nanotechnology focusing on the design of novel materials with improved structural and functional properties (Lehn, 2002). Recent development in the field of metal-containing polymers has highlighted among other interesting properties, their ability to generate assemblies with exciting functions useful for future development of new materials. A particular attention is given to the approaches to the well-defined structures having controlled size and shape and to the bioinspired and biomimetic materials with a special arrangement at nanometric level of the components (Ruiz-Hitzky et al., 2008).

The self-organization of polymers/copolymers in ordered morphologies at nanoscale is due to the thermodynamic incompatibility between their assembled blocks: the hydrophilic or polar and lipophilic or nonpolar components. In the most common situation, as a result of composition differences, self-assembly of amphiphilic compounds occurs in water due to specific interaction (affinity) with the environment: the hydrophilic parts of the molecules interact with water, while the hydrophobic parts tend to migrate away from it, thus forming a layer at the interface (in particular at the surface). This behavior is the key of surface activity in the case of surfactants. Micellization is the prototype for their self-assembly based upon the structural separation depending on the hydrophilic-hydrophobic balance of a molecule. The self-assembly can be designed into various geometries: rods, disk, ordered liquid crystalline phases, bilayer vesicles, reverse micelles, etc. by changing the control parameters (molecular structure, solvent, concentration, temperature, or pH). In a similar way, microphase separation between components of a copolymer, blend, or amphiphilic compound occurs in selective solvents and/or in bulk.

Currently, a major interest in the field of polymer science is given to the design of metal-containing ordered structures by proper choice of the building blocks. The selectivity toward a predefined morphology of the hybrid material is the result of the choice of appropriate organic and inorganic building units and the suitable processing conditions (Brus et al., 2004). In the same time, the supramolecular interactions, such as ionic, hydrogen and coordination bonds or hydrophobic ones have a key role in the intrinsic mobility leading to ordered nanostructures (Busseron et al., 2013).

This chapter presents some aspects regarding the ability of self-assembly in solution of metallic complexes containing dimethylsiloxane spacers in general with focus on those based on polyazomethine type ligand derived from bis(formyl-p-phenoxymethyl) tetramethyldisiloxane and 2,5-bis(p-aminophenyl)-1,3,4-oxadiazole with metals such as Cu^{II}, Co^{II}, and Zn^{II}. The self-organization of metal complexes was evaluated in DMF solution by UV–Vis and dynamic light scattering (DLS) techniques while the preservation or evolution of the assemblies after solvent evaporation was evidenced by atomic force microscopy (AFM) and transmission electron microscopy (TEM).

5.2 DIMETHYLSILOXANE-BASED COMPOUNDS AND THEIR PARTICULARITIES

Although the siloxane bond has a polar character conferred by the electronegativity difference between oxygen and silicon, the dimethylsiloxane sequences, either in very short variant (e.g., in disiloxane) or as oligomers and polymers, show an extreme hydrophobic character. This character is conferred by the methyl groups, which are one of the most hydrophobic organic groups. The high and flexible bond angle around the O atom in the siloxane backbone, ranging between 135° and 180° confers high rotational and oscillatory freedom of the methyl side groups around the backbone axis and thus enhances the role of methyl groups in shielding the polar character of Si—O bond. As a result, the intermolecular interactions of the dimethylsiloxane are extremely low. Another distinctive aspect of the diorganosiloxane, in general, is the irregular cross section of the backbone, which is large at the substituted silicon atom and small at the un-substituted oxygen atom (Cazacu, 2008). These structural particularities have great effects on the properties of silicone materials being responsible of, among others, good solubility in nonpolar solvents, low bulk viscosity, high volatility at low molecular weight, high permeability to gases, etc. Another key feature of polysiloxanes, namely, their low surface energy, or surface tension (21–23 mJ m^{-2} for PDMS) is a result of closely packed methyl groups at the surface (Hillborg et al., 2000). This has led to the development of the silicone surfactants class (Randall, 1999) consisting in shorter or longer siloxane moieties having attached polar or ionic groups. In fact, amphiphilic silicones or surfactants were first introduced in the 1950s for the development of industrial-grade polyurethane foam. They are surface-active agents that lower the interfacial surface tension and allow spreading. They are known for their unique properties to maintain surface activity in both water and nonaqueous systems, such as oil, and polyols and are able to significantly lower the surface tension (Hasan et al., 2016). Additionally, silicone surfactants have special aggregation properties in comparison with those of hydrocarbon surfactants, forming bilayer phases, and vesicles rather than micelles and gel phases (Lin and Alexandridis, 2002). Due to their increased hydrophobicity, they generally have very low critical micellization concentrations (CMC) and thus are effective in very low amounts (Racles et al., 2010). The most known types of siloxane surfactants are based on poly(ethylene oxide) (PEO) hydrophilic block attached in different architectures to the polysiloxane block, while more recently other classes of surfactants have been proposed (Racles et al., 2010). A fascinating aspect is the versatility of siloxane-based compounds: their modification can be achieved in a variety of ways to get a wide range of molecular architectures, such as comb-like siloxanes, functionalized siloxanes, branched siloxanes (Soni et al., 2002), amphiphilic block-copolymers and modified cyclosiloxanes (Racles and Hamaide, 2005; Racles et al., 2006), or telechelic amphiphilic polysiloxanes (Racles et al., 2014a, 2014b, 2014c). Organo-alkoxysilanes nanohybrids were also reported as a new family of biomimetic materials that mimic liposomes, derived from the self-assembly of surfactants covalently bonded to a silica-based network (Ruiz-Hitzky et al., 2008). The low toxicity of the siloxane segments, in addition to properly chosen hydrophilic groups attached have issued numerous biocompatible silicone surfactants, which found a wide range of applications.

Silicone surfactants have been widely employed as stabilizers for nanoparticles (NPs) either polymer NPs (Racles and Hamaide, 2005), metal (e.g., silver) NPs (Racles et al., 2011) or polymer-hydrophobic drug mixtures (Racles, 2013). They may act as solubility enhancers for hydrophobic drugs (Racles et al., 2014a, 2014b, 2014c) and their encapsulation for target delivery systems. Other recent developments in the field of siloxane surfactants are directed toward high-performance applications as molecular transport, nanoreactors, microemulsions, and reaction in supercritical CO_2 (Racles et al., 2010).

In addition to their surface activity, the use of these amphiphilic compounds as ligands for metals is very important in decontamination or in micellar solubilization of nonsoluble drugs very useful in biomedical applications (Racles et al., 2014a, 2014b, 2014c).

Examples of siloxane surfactants developed in our group and their tested applicabilities are presented in Table 5.1.

5.3 POLAR–NONPOLAR COMPOUNDS AS TEMPLATES FOR METAL-CONTAINING NANOSTRUCTURES

Currently, the main focus is on exploit of polar–nonpolar compounds, such as organosiloxanes as templates for developing supramolecular composites with metal complexes designed to function as flexible nanostructural materials. Hydrogen-bonding and metal – ligand interactions are among the most powerful directing forces for the self-assembly processes, which allow the construction of large suprastructures. In particular, such interactions led to the development of discrete metal-containing nanostructures, which are promising candidates in fields such as molecular recognition, photophysics, and catalysis (Company et al., 2006).

Amphiphilic supermolecules can be designed by employing noncovalent interactions, electrostatic, and complementary hydrogen bonding, toward hierarchically self-assemblies in different solvent media. To date,

TABLE 5.1   Several Siloxane Amphiphiles and Their Applicability.

Structure of various siloxane amphiphiles	Applicability	References
	• Biocompatible nonionic polymer surfactant	Racles and Hamaide (2005)
	• Colloidal stabilization of nanoparticles	Racles and Cozan (2012)
	• Nonionic and ionic surfactants	Racles et al. (2006)
	• Polymer nanoparticles formulations	Racles et al. (2009)
	• Biocompatible nonionic surfactant	Racles et al. (2014a, 2014b, 2014c)
	• Polymer nanoparticles formulations	
	• Stabilization of nanoparticles	
	• Aqueous dispersion of metal oxide nanoparticle	

TABLE 5.1 *(Continued)*

Structure of various siloxane amphiphiles	Applicability	References
	• Biocompatible compounds	Racles (2010)
	• Stabilization of nanoparticles	Racles et al. (2011)
	• Ligand for *4f* metals	Racles et al. (2014a, 2014b, 2014c)
	• Aqueous dispersion of metal oxide nanoparticle	
	• Micellar solubilization of nonsoluble drugs	

metal–ligand coordination bonds have been largely used for the synthesis of discrete molecules and extended networks, coordination polymers, or bulk organogels (Liu et al., 2016).

Metal-directed self-assembly is one of the most useful approaches for generation of complex and highly elaborated molecular architectures. In this case, the advantage of metal-directed self-assembly is conferred by the control over different geometries and bond strengths depending on the metallic centers used in the system. The rational concepts regarding the building blocks: the number of ligands, their relative orientation in space, and the geometry of metal ions are key factors which predict what species are able to self-assemble in solution. A broad variety of nano-objects such as nanotubes, micelles, or vesicles has been obtained from various molecules: synthetic amphiphilic block copolymers, amphiphilic perylene bisimides, porphyrins, terpyridines, inclusion complexes of β-cyclodextrin, or metal clusters (Baytekin et al., 2009).

The template method toward metallosupramolecular structures consists in using preorganized building blocks to predict and build up molecular structures via hierarchical self-assembly (Menozzi et al., 2006).

Metalloproteins and enzymes can be easily designed starting from complex amino acids and metal centers found in biological systems. Diblock copolypeptide amphiphiles are representative synthetic materials, which can be used in this purpose due to their ability to form double-walled vesicles or fibrillar nanostructures based on their self-assembly capacity, which recommend them as templates for inorganic compounds or for their transformation via dynamic tuning of the electronic state of the materials (Lutolf et al., 2005).

For the construction of such structures, Schiff bases have been often used especially for their directionality, thermodynamic stability, and coordination ability of nitrogen atom (Nakamura et al., 2016). Pyridyl-imine based ligands have been used to design a wide variety of self-assembled structures: from macrocycles, helicates, and polymers to metal–organic capsules as well as interlocked structures (Castilla et al., 2014).

Polyazomethines, known as Schiff base polymers or polyimines represent one of the most known classes of macromolecular ligands, and are synthesized from a wide variety of carbonyl and amine precursors (Calligaris et al., 1972; Garnovskii et al., 1993). It is well known that the properties of the polymers can be radically modified by minor changes in chemical structure, the possibility to use metal ions for the preparation of polymeric structures with interesting and useful properties being very attractive. The most known strategy for the design of such compounds is the use of polydentate linkers containing donor atoms or groups of atoms with ability to coordinate metal ions to form polymeric structures with different architectures. The presence of the lone electron pair in a sp^2 hybridized orbitals of the nitrogen atom of the imine (CH=N) group (Kianfara et al., 2010) in the structure of polyazomethines gives them a remarkable ability to coordinate various metals ions from s, p, d, or f blocks with different oxidation state (Andruh et al., 2009; Sui et al., 2010; Hazra et al., 2009; Dolai et al., 2013).

Design of molecular structures with targeted properties (optical, magnetic, or electrical) based on these types of compounds having different number of coordination sites is an innovative direction regarding the development of new metal complexes that have proven their applicability in various fields, such as solar cells, electrochromic devices, organic field-effect transistors (OFET) (Hindson et al., 2010; Petrus et al., 2015; Isik et al., 2012; Sicard et al., 2013) analytical chemistry (optical, electrochemical, and chromatographic sensors) (Jungreis and Thabet, 1969; Ibrahim and Sharif, 2007), and material science (emitting diode OLED and PLED) (Katsuki, 1995). Their valuable properties recommend polyazomethines as thermostable polymers with good hydrolytic stability and enhanced mechanical properties (Saugusa et al., 1992; Cooper et al., 2009).

However, one of the highest limitations of this class of compounds is their insolubility and infusibility due to the rigidity of the macromolecular chains. These limitations can be overcome by few approaches: the introduction of the alkyl or alkoxy groups in ortho position of the aromatic ring, the presence of some irregularities in the polymeric chain, inserting flexible groups in the main chain or pendant in order to disturb the well-organized packing of the polymer chains, or by inclusion complexes of rotaxanes (Samal et al., 1999; Shulpin, 2002; Venegas-Yazigi et al., 2010). One alternative to these modifications, toward improved solubility and transition temperatures, is the introduction of siloxane segments in the structure of polyazomethines (Racles, 2008; Racles et al., 2007).

Given the various possibilities of obtaining such structures, our goal was directed toward the preparation of metal complexes derived from a polyazomethine having flexible siloxane sequences in the structure, by using metal ions with different coordination geometries, namely copper(II), cobalt(II), and zinc(II). The use of flexible ligands having siloxane sequence as building blocks for the coordination compounds is an attractive direction for obtaining new metal complexes with interesting properties (Vasiliu et al., 2005; Cazacu et al., 2003).

The right choice of polar-nonpolar diblock composition, the flexibility of the backbone chains, and experimental parameters allowed obtaining various micellar morphologies, polymeric hollow structures (Li et al., 2015), nanoporous thin films (Wang et al., 2008), NPs, nanocrystals, nanotubes, nanowires, etc. (Liu et al., 2013; Perez-Pagea et al., 2016).

5.4 METAL COMPLEXES CONTAINING SILOXANE LIGANDS ABLE TO SELF-ASSEMBLE IN SELECTIVE SOLVENTS

The self-assembly of block copolymers in either bulk state or solution provides a fascinating route to functional nanomaterials with the potential for a variety of applications: porous membranes, lithographic templates, and photonic band-gap materials. On the other hand, micelles formed from block copolymers in solution have been used as nanoreactors, drug delivery carriers, and templates for the fabrication of one-dimensional nanostructures (Wang et al., 2007). Block copolymers containing metal centers and related segmented architectures show a considerable interest as a class of materials with properties and functions complementary with the organic analogs. These materials present remarkable flexibility in terms of design and show phase-separation into nanoscopic, metal-rich domains, either in bulk or in thin films, and are able to form core–shell NPs in selective solvents (Zhou et al., 2014).

The versatility in the coordination modes of metal ions, the proper choice of the ligands together with a careful control of their solvophobicity and solvophilicity, as well as subtle changes in the structures of complexes can lead to interesting supramolecular self-assembled structures and topologies and/or self-organization or aggregation behavior (Yam et al., 2015). In particular, molecular assemblies based on d^8 and d^{10} transition metal complex systems represent important classes of luminescent self-assemblies, while their aggregation ability in solution depending on the solvent nature could be the basis for catalytic activity in homogeneous systems (Soroceanu et al., 2015). However, the self-assemblies of metal complexes have been relatively less explored than the organic systems.

Research in this area began in 1955 with the synthesis of poly(vinyl ferrocene), which exhibits a side-chain location of the metal centers. For instance, polycondensation, ring-opening polymerization, electropolymerization, and living anionic, as well as controlled radical polymerization, have been used to form metallopolymers with either side- or main-chain metal centers (Zhou et al., 2014). Subsequently, an important milestone in the development of ferrocenyl-containing polymers was the preparation of linear, branched or cross-linked polyorganosiloxanes by varying the chemical nature of the substituent attached to the silicon atom. The self-assembly of these amphiphilic block copolymers in solution was a powerful tool to obtain ordered structures in both the nano- and micro-domains. In selective solvents, a self-organization of these structures in micelles with a wide range of morphologies was observed. In an earlier work, we have prepared a ferrocenyl-siloxane polyamide, poly{1,10-ferrocene-diamide-[1,3-bis(propylene)tetramethyldisiloxane]} (Fig. 5.1) by polycondensation in solution (methylene chloride) of 1,10-di(chlorocarbonyl)ferrocene with 1,3-bis(aminopropyl)tetramethyldisiloxane (Cazacu et al., 2006).

Alternating along the chain of nonpolar bis(propyl)tetramethyldisiloxane sequence with more polar ferrocenyl-diamide group creates the premise for aggregation in selective solvents. The self-assembling ability of this compound was experimentally proved both in solution and in film (Cazacu et al., 2009; Cazacu et al., 2010), the pattern differing in dependence on the solvent polarity. A package of techniques has been used to study this behavior: tensiometry, viscometry, UV–Vis spectrophotometry, DLS, scanning electron microscopy, and AFM but also ^{1}H-NMR spectroscopy less approached in this purpose. It has been found that the signals corresponding to protons in the polar segments of macromolecule (ferrocene diamide) are better solved in DMSO-d6, whereas in $CDCl_3$ the signal corresponding to the protons in the dimethylsiloxane units is better observed. The temperature and concentration of the solution also influence the ratio of the signals.

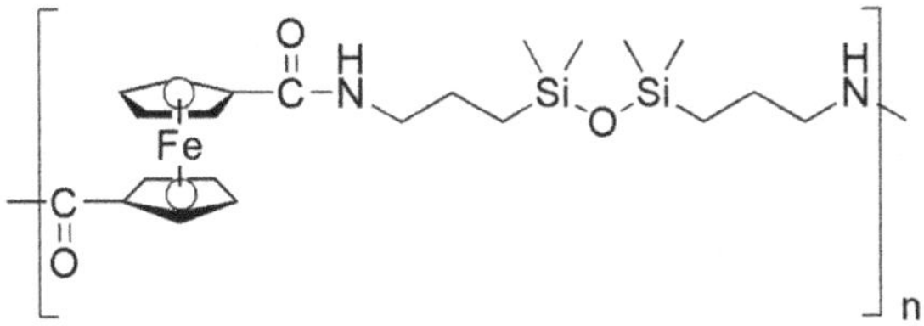

FIGURE 5.1   Poly{1,10-ferrocene-diamide-[1,3-bis(propylene)tetramethyldisiloxane]}.

The employment of such metallo-block copolymers able to self-organize in solution was particularly attractive due to the additional electronic, magnetic, and optical properties that these have provided (Massey et al., 1998a, 1998b; Resendes et al., 2000; Cazacu et al., 2006). More recently, different approaches to build up supramolecular nano architectures are focused on the use of coordination polymers (Li et al., 2014) with different molecular architectures including main-chain, side chain, and star-like or branched polymers.

Self-assembled structures of discrete metal complexes have also attracted considerable attention due to their exceptional optical, photophysical, electronic, and magnetic properties and their great potential in various research fields, such as biomedicine and optoelectronics (Herkert et al., 2016).

As has been already shown, most siloxane-containing copolymers are able to self-assemble in solution especially due to the unique properties of siloxanes which lead to differences in polarity, solubility, flexibility, etc., and incompatibility with any organic polar building blocks (Racles et al., 2008). The presence of highly flexible and hydrophobic dimethylsiloxane moieties in coordination compounds, either as small molecules or polymers has proved to further favor the structuring process (self-assembling in different objects) in certain solvents. Lately, many metal complex structures have been reported that contain this fragment, and often-such behavior has been highlighted (Table 5.2).

TABLE 5.2 Several Examples of Siloxane-based Structures (Small Molecules, Polymers, and Copolymers) Capable of Self-assembling in Solution.

Reported self-assembled siloxane structures in solution	Type of aggregates	References
	Rod-like micellar structures	Massey et al. (1998)
	Micelles	Cazacu et al. (2009); Cazacu et al. (2010)
	Micelles	Cazacu et al. (2013)
	Disc-shaped aggregates	Soroceanu et al. (2015)
	Micelles and vesicles	Zaltariov et al. (2016)

5.4.1 PRINCIPLES OF MOLECULAR SELF-ASSEMBLY

A self-assembling system consists of a set of structures, same or different, able to interact with one another. Their *interaction* leads to the formation of more or less ordered states in solution from disordered aggregates to random coils depending on the flexibility and the complementarity in shapes of the building blocks. These interactions are generally weak, self-assembly occurring when molecules interact through equilibrium between attractive and repulsive forces.

The generation of ordered structures is influenced by the *reversibility* of the association between complementary building blocks which allow the components of the system to adjust their position within aggregation process. The interactions of the building blocks with their *environment* are crucial for molecules to self-assemble, either in solution or at an interface (Whitesides and Boncheva, 2002). Achieving these interactions is only possible through the contribution of the mobility of the molecules in solution depending on their flexibility (Wu et al., 2017). In the case of metal complexes-containing highly flexible dimethylsiloxane sequences, the self-assembling process is controlled by the coexistence of hydrophobic and flexible units with polar ones. The polar component of the system could be an organic, ionic, or inorganic (metal ions, clusters, or metal complexes) fragment (a few examples in Table 5.2).

Some time ago our group reported different polyazomethines and their metal complexes by using at least a siloxane-based component (either carbonyl or amine) (Cazacu et al., 2003; Cazacu et al., 2008; Vlad et al., 2008) as well as other coordination polymers (Cazacu et al., 2013; Zaltariov et al., 2016) where the highly flexible tetramethyldisiloxane segments imparted increased solubility, an amphiphilic character due to their coexistence with the polar building blocks.

Many amphiphilic polymers/copolymers are able to generate various morphologies, including micelles, vesicular aggregates, or rods by self-assembling processes. In all cases the prediction of the structure and morphology of aggregated systems is difficult. Compared with surfactants, the self-organization in polar solvents of polymeric structures with conjugated π-electron systems becomes more complex, especially due to the increased intermolecular force equilibrium induced by fluctuation of the delocalized π-electrons (Kirstein, et al., 2000).

The structure-function relationship in conjugated polymer aggregates has been explained in terms of H- or J-aggregation, which is characterized by new optical bands, with bathochromic (J-aggregates) or hypsochromic (H-aggregates) shifts. The self-assembly ability of such systems is determined by the molecular electronic interaction and optical behavior (Zhu, et al., 2015). Balancing out attractive forces, such as dipole interactions, π–π stacking, or solvophobic effects with electrostatic repulsive ones has a high impact on the self-assembly mechanism and the resulting morphology of the nanoscale objects: helical or columnar aggregates. The shifts of the UV bands of aromatic moieties in such assemblies are a strong indication for the formation of H- or J-aggregates (Beseniusa et al., 2010).

Recently, porphyrin J-aggregates formed in synthetic liposomes and perylene bisimide derivatives were reported as self-assembled structures in polar solvents for cancer imaging applications due to their aggregation driven by non-covalent interactions, causing remarkable modifications in their optical and electronic properties (Biswas, et al., 2012; Yang, et al., 2017).

5.4.2 PARTICULAR TECHNIQUES USED FOR ANALYSIS OF SELF-ASSEMBLING PROCESS IN SOLUTION

During the past years numerous methodologies have been considerably developed and opened new opportunities to investigate the self-assembled molecular and biomolecular structures: high-resolution optical microscopy, cryo-TEM, time-resolved scattering, magnetic resonance, and computer simulations, as well as classical methods: surface tension, spectral (FT-IR, UV–Vis, and fluorescence), or viscometry measurements.

Surface tension measurement is a suitable method to study micellization, self-assembly into micelles being an important characteristic of surfactants. This method can be used to obtain information on the thermodynamic parameters of the micellization process, by evaluation of CMC. Above a critical concentration, amphiphilic molecules spontaneously self-assemble in aqueous solutions to form aggregates of various morphologies. The determination of the CMC involves the measurement of a physical quantity that changes upon micellization, such as surface tension or solution enthalpy (Prazers et al., 2012). Fluorescence spectroscopy is a valuable technique used for evaluation of self-organization in solution of molecular systems. The difference in the fluorescence response of aggregates possessing ground and excited electronic properties distinct from that of the individual molecules gives information about the assembly processes in solution (Nguyen et al., 2007). A common method for evaluation of structural changes in aggregates includes FT-IR spectroscopy, while viscometry is an efficient one to evaluate the propensity of individual molecules to self-assemble into aggregates.

Although there is a wide range of analytical techniques available, the most valuable data about the molecular and/or polymer/copolymer assemblies in solution are provided by complementary information obtained from scattering and microscopy techniques. Usually, the analysis of molecular/polymer aggregates requires the answer to the following two issues: what is their dimension? and what is their morphology? The answers to these questions became accessible as that many studies are presented in literature where the use of the appropriate performed analysis can elucidate the nanostructuration in solution (Patterson et al., 2014).

The most common scattering techniques for soft materials in solution are static and dynamic light (SLS and DLS), small angle X-ray (SAXS), and small angle neutron scattering (SANS). The basic principle in each case consists in illuminating the solution with a radiation of known wavelength and detecting the intensity scattered by the sample, at a given angle of observation with respect to the incident radiation. If the data are collected as a function of time, particle dynamics can be analyzed as in the case of DLS.

Microscopy techniques are complementary with scattering data as they directly image individual particles, typically when deposited onto a substrate or a support. TEM is one of the most powerful methods for analyzing nanomaterials. AFM images are formed by dragging or tapping a sharp tip across the surface of the sample and, similar to SEM, can provide information about the particle surface.

These methods also including UV–Vis technique requires small amount of sample and at the same time provide concrete information about the aggregate size and their distribution as well as preservation or evolution of the morphology after the evaporation of the solvent.

5.4.3 CASE STUDY: SELF-ASSEMBLING OF METAL COMPLEXES WITH POLYAZOMETHINE LIGAND CONTAINING RIGID OXADIAZOLE AND FLEXIBLE TETRAMETHYLDISILOXANE UNITS

In general, the synthesis of metal-containing polymers proceeds through several methods:

1. reaction between a polydentate ligand and metal ions;
2. polymerization of functional monomers containing metal;
3. complexation between macromolecular ligands and metal ions.

The last path has been approached for the preparation of samples taken in the study here. Thus, a polyazomethine was prepared as was previously described (Zaltariov et al., 2015) by the reaction of a siloxane dialdehyde, *bis*(formyl-*p*-phenoxymethyl)-tetramethyldisiloxane (Racles et al., 2002; Marin et al., 2009), with 2,5-*bis*(*p*-aminophenyl)-1,3,4-oxadiazole was used as a ligand for Cu^{II}, Co^{II}, and Zn^{II} ions, based on the presence of the electron-donor nitrogen atoms from azomethine group and oxadiazole ring (Fig. 5.2).

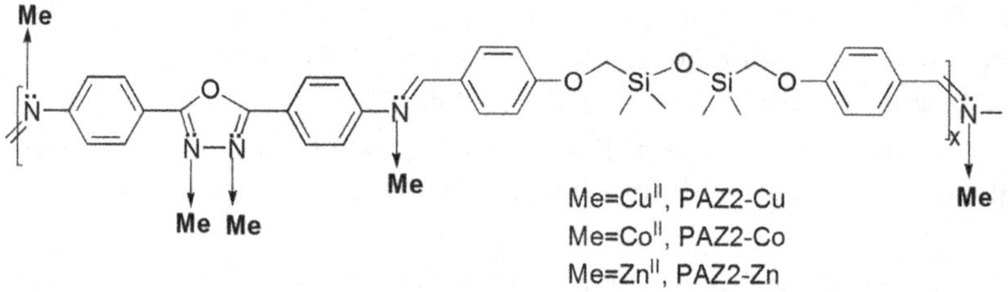

FIGURE 5.2 The general structure of metal complexes derived from polyazomethine containing dimethylsiloxane and oxadiazole units.

The metal complexation was confirmed by FTIR spectroscopy, while the silicon/metal atom ratio was determined by energy-dispersive X-ray fluorescence. The GPC data for the three coordination polymers are provided in Table 5.3. These data are estimated on the basis of the peak corresponding to low molecular fraction because, due to the aggregation in elution solvent, the curves also show peaks that would indicate abnormally high molecular masses for condensation polymer such as polyazomethine.

TABLE 5.3 The Average Numerical Molecular Weight (M_n) and Polydispersity Index (PDI) for Polyazomethine Metal Complexes.

Sample	M_n, g·mol^{-1}	PDI = M_w/M_n
PAZ2-Cu	5350	1.1
PAZ2-Co	3050	1.1
PAZ2-Zn	2800	1.1

The presence of the tetramethyldisiloxane segments, well known for their hydrophobicity, alternating with polar azomethine and phenylene-1,3,4-oxadiazole units within the polyazomethine chain creates the premise for self-assembling in selective solvents. Depending on the solvent polarity, the compound would adopt geometry that to minimize the contact between the solvent and the part of the molecule that possesses opposite character as is suggested in Figure 5.3. In this particular case, DMF is a selective solvent for the polar part of the macromolecular complexes. This is not an appropriate solvent for PDMS, which is reflected in the large difference between the solubility parameters: 7.4 (cal/cm^3)$^{1/2}$ for PDMS and 12.1 (cal/cm^3)$^{1/2}$ for DMF.

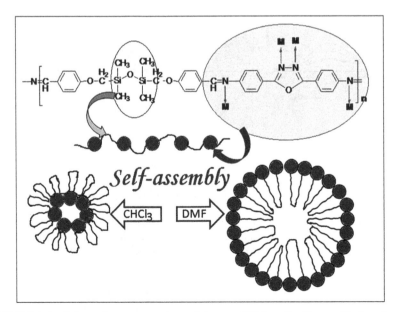

FIGURE 5.3 Schematic representations of the possibilities of self-assembly in solvents with different polarities.

5.4.3.1 SELF-ASSEMBLY OF METAL COMPLEXES IN SOLUTION

5.4.3.1.1 Determination of Critical Aggregation Concentration by UV–Vis Spectroscopy

UV–Vis spectra of metal complexes derived from the siloxane-based poly-azomethine were recorded on a Shimadzu UV-1700 spectrophotometer in DMF solutions using quartz cuvettes of 1 cm, at 24°C and showed two main absorption bands blue shifted by 12 nm as compared with those of the polyazomethine. The first maximum located at 284 nm is assigned to the $S_2 \leftarrow S_0$ (π^*, π) transition in benzene ring (Mitra and Tamai, 1999) and the band at 340 nm is characteristic for $S_1 \leftarrow S_0$ (π^*, π) transition of the azomethine-phenyl-oxadiazole system (Marin et al., 2011) (Fig. 5.4).

In order to follow possible modifications in the UV–Vis spectra, concentration-dependent UV–Vis measurements were performed. Concentrations used have ranged from 3×10^{-6} M to 10^{-1} M.

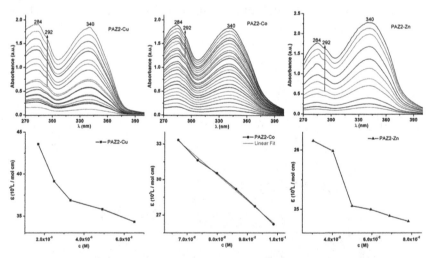

FIGURE 5.4 **(See color insert.)** Concentration-dependent UV–Vis absorption spectra of metal complexes in DMF solution at different concentrations and plots of extinction coefficient of absorbance maximum (ε) at 340 nm vs. molar concentration (c (M)).

UV–Vis spectra show an increase in intensity of the absorption bands gradually with concentration, which does not follow the Lambert–Beer law in the case of PAZ2-Cu and PAZ2-Zn.

This aspect was evidenced by the trend of molar extinction coefficient, which drops severely when the concentration increases. Further increasing the concentration, a weak band around 292 nm was observed, as a result of the aggregation process in solution according to other reports (Berlepsch and Bottcher, 2013). The aggregation concentration values for metal complexes in DMF were obtained from the first inflection in the variation of the extinction coefficient of the absorbance maximum at 340 nm. The found values of the evaluated critical aggregation concentration were 3×10^{-3} M for PAZ2-Cu and 5×10^{-2} M for PAZ2-Zn. In the case of PAZ2-Co, a slight deviation from linearity was observed at about 7×10^{-2}M (Fig. 5.4).

5.4.3.1.2 DLS Measurements

The DLS technique was used to estimate the size and distribution of aggregates in DMF on a Malvern Instruments Autosizer Lo-C 7032 Multi-8 correlator (Malvern Instruments, UK) equipped with a HeNe laser ($\lambda = 632.8$ nm). Before measurement, the solutions were filtered through 0.5 μm filter. In all registrations, the fluctuations in the intensity of the scattered light were analyzed to obtain an autocorrelation function.

The structure of metal complexes based on siloxane-containing polyazomethine consists of hydrophobic tetramethyldisiloxane block alternating with polar moieties formed by the metal coordinated to the azomethine group and 1,3,4-oxadiazole. Their self-organization was tested in the DMF solution at critical aggregation concentration estimated by UV–Vis experiments.

In the case of PAZ2-Cu for the studied concentration, the DLS analysis without sonication revealed populations of large particles (diameter >450 nm). In the case of PAZ2-Zn, populations of at least three particle sizes: small particles (<25 nm), large particles (diameter >100 nm), and very large particles centered at 5.5 μm were identified (Fig. 5.5). The existence of large particles suggests aggregation in solution of the metal complexes.

After sonication, the results evidenced the presence of aggregates whose average diameter is centered around 100 nm for PAZ2-Cu, 120 nm for PAZ2-Co, and 150 nm for PAZ2-Zn. The midrange values of polydispersity index (PDI) were found to be 0.265 for PAZ2-Cu, 0.207 for PAZ2-Co and 0.441 for PAZ2-Zn (Fig. 5.6).

162   Smart Materials

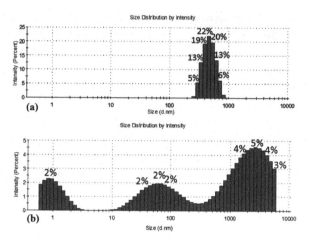

FIGURE 5.5   Size distribution by intensity for aggregates formed in DMF for (a) PAZ2-Cu; and (b) PAZ2-Zn without sonication. The numbers close to the histogram bars indicate the percentage of particles or aggregates of a particular size.

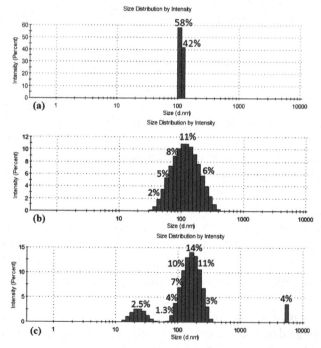

FIGURE 5.6   Size distribution by intensity for aggregates formed in DMF for (a) PAZ2-Cu; (b) PAZ2-Co; and (c) PAZ2-Zn with sonication. The numbers close to the histogram bars indicate the percentage of particles or aggregates of a particular size.

5.4.3.1.3 Microscopical Observation of the Self-Assembled Aggregates Remaining after Solvent Evaporation

The aggregation of the siloxane-containing metal complexes was investigated by TEM on the films born from proper solutions. TEM micrographs were taken using a Hitachi HT7700 microscope. The equipment was operated in high contrast mode at 100 kV accelerating voltage. Carbon-coated copper grids, 300 mesh size, were used to support the samples. For this purpose, solutions of the metal-containing polymer in DMF of the same concentrations as those used for DLS measurements were spread on the grids placed on filter paper and then dried under vacuum overnight.

The three complexes adopt different structuration patterns with entities of different sizes between them: PAZ2-Cu and PAZ2-Zn show spherical aggregates (Fig. 5.7a,c), while PAZ2-Co shows irregularly shaped particles. The dimensions of the aggregates in the case of PAZ2-Cu and PAZ2-Zn are lower than those estimated by DLS suggesting that in solution the aggregates are swollen, strongly interacting with solvent molecules (DMF). The drying process before the TEM analysis leads to shrinking of aggregates to lower dimensions.

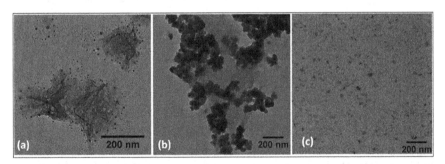

FIGURE 5.7 TEM images of films born from DMF solution: (a) PAZ2-Cu; (b) PAZ2-Co; and (c) PAZ2-Zn.

The aggregation of the metal complexes was also evaluated through AFM. The samples were prepared by spin coating on glass slides from solution in DMF at the same concentration estimated by UV–Vis spectroscopy.

The surface images were obtained with a Solver PRO-M scanning probe microscope (NT-MDT, Russia) in AFM configuration. Rectangular silicon cantilevers NSG10 (NT-MDT, Russia) with tips of high aspect ratio were used. All images were acquired in air, at room temperature (23°C), in tapping mode, and at a scanning frequency of 1.56 Hz. The scan length ranged between 2 and 10 μm.

AFM images revealed the presence of aggregates with similar dimension as observed by TEM. For PAZ2-Co, the quasi-spherical objects have dimensions of <100 nm. In the case of PAZ2-Cu and PAZ2-Zn (Fig. 5.8), the morphology of the samples prepared by spin coating clearly shows aggregates of smaller particles, as evidenced by DLS.

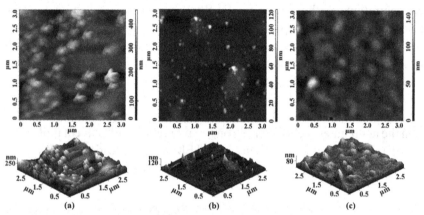

FIGURE 5.8   AFM images on samples remained after solvent evaporation for (a) PAZ2-Cu; (b) PAZ2-Co; and (c) PAZ2-Zn.

5.5   CONCLUDING REMARKS

Coordination polymers of polyazomethine type with complexed metal pending to the main chain and containing tetramethyldisiloxane spacer between azomethine ligand units are able to self-assemble in solution due to the difference between polarities of the two structural units. This ability was emphasized by UV–Vis spectroscopy correlated with DLS analyses. The aggregates are preserved after solvent removal by slow evaporation, being highlighted by TEM and AFM. Thus, with such compounds premises are created to obtain nano-microstructures with specific properties (magnetic, optical, electrical, etc.) induced by the metal presence, of high interest for nanotechnologies. It is supposed that the shape, size, and surface nature of aggregates could be tuned by solvent or solvents mixture used.

ACKNOWLEDGMENTS

This work was supported by a grant of Ministry of Research and Innovation, CNCS–UEFISCDI, project number PN-III-P2-2.1-PED-2016, (Contract 130PED/2017).

KEYWORDS

- metal complexes
- polyazomethine
- siloxane amphiphiles
- self-assembly
- aggregation

REFERENCES

Andruh, M., et al. 3d–4f Combined Chemistry: Synthetic Strategies and Magnetic Properties. *Inorg. Chem.* **2009**, *48*, 3342–3359.

Baytekin, H. T., et al. Metallo-supramolecular Nanospheres via Hierarchical Self-assembly. *Chem. Mater.* **2009**, *21*, 2980–2992.

Berlepsch, H. V.; Bottcher, C. Supramolecular Structure of TTBC J-aggregates in Solution and on Surface. *Langmuir* **2013**, *29* (16), 4948–4958.

Beseniusa, P., et al. Controlling the Growth and Shape of Chiral Supramolecular Polymers in Water. *Proc. Natl. Acad. Sci. USA* **2010**, *107* (42), 17888–17893.

Biswas, S., et al. Two-photon Absorption Enhancement of Polymer-templated Porphyrin-based J-Aggregates. *Langmuir* **2012**, *28* (2), 1515–1522.

Brus, J.; Sÿpırkova, M.; Hlavata, D.; Strachota, A. Self-organization, Structure, Dynamic Properties, and Surface Morphology of Silica/epoxy Films as Seen by Solid-state NMR, SAXS, and AFM. *Macromolecules* **2004**, *37*, 1346–1357.

Busseron, E., et al. Supramolecular Self-assemblies as Functional Nanomaterials. *Nanoscale* **2013**, *5*, 7098–7140.

Calligaris, M.; Nardin, G.; Randaccio, L. Structural Aspects of Metal Complexes with SomeTetradentate Schiff Bases. *Coord. Chem. Rev.* **1972**, *7*, 385–403.

Castilla, A. M.; Ramsay, W. J.; Nitschke, J. R. Stereochemistry in Subcomponent Self-assembly. *Acc. Chem. Res.* **2014**, *47* (7), 2063–2073.

Cazacu, M. Polymers Containing Si, O and Other Elements Within Backbone. In *Advances in Functional Heterochain Polymers*; Cazacu, M., Ed.; Nova Science Publishers: New York, NY, 2008.

Cazacu, M., et al. Chelate Polymers. III. New Polyazomethines of 5,5'-Methylene-bissalicylaldehyde with Siloxane Diamines and Their Divalent Metal Complexes. *J. Pol. Sci. Part A Polym. Chem.* 2003, *41*, 3169–3179.

Cazacu, M., et al. New Organometallic Polymers by Polycondensation of Ferrocene and Siloxane Derivatives. *Macromolecules* **2006**, *39*, 3786–3793.

Cazacu, M., et al. Multifunctional Materials Based on Polyazomethines Derived from 2,5-dihydroxy 1,4-benzoquinone and Siloxane Diamines. *J. Pol. Sci. Polym. Chem.* **2008**, *46* (5), 1862–1872.

Cazacu, M., et al. Association Phenomena of the Ferrocenyl-siloxane Polyamide in Solution. *J. Polym. Sci. Polym. Chem.* **2009**, *47* (21), 5845–5852.

Cazacu, M., et al. New Experimental Insights into Self-organization of Poly(Ferrocenyl Amide-siloxane). *J. Optoelectron. Adv. Mater.* **2010**, *12*, 294–300.

Cazacu, M., et al. Electroactive Composites Based on Polydimethylsiloxane and Some New Metal Complexes. *Smart Mater. Struct.* **2013**, *22* (10), 104008.

Company, A., et al. Molecular Rectangles Resulting from Self-assembly of Dicopper Complexes of Macrocyclic Ligands. *Inorg. Chem.* **2006**, *45*, 2501–2508.

Cooper, L. L.; Samulski, E. T.; Scharrer, E. Towards Room Temperature Biaxial Nematics. *Mol. Cryst. Liq. Cryst.* **2009**, *511*, 203–217.

Dolai, M., et al. Diversity in Supramolecular Self-assembly Through Hydrogen-bonding Interactions of Non-coordinated Aliphatic –OH Group in a Series of Heterodinuclear $Cu^{II}M$ (M = Na^{I}, Zn^{II}, Hg^{II}, Sm^{III}, Bi^{III}, Pb^{II} and Cd^{II}). *Inorg. Chim. Acta.* **2013**, *399*, 95–104.

Garnovskii, A. D.; Nivorozhkin, A. L.; Minkin, V. I. Ligand Environment and the Structure of Schiff Base Adducts and Tetracoordinated Metal-chelates. *Coord. Chem. Rev.* **1993**, *126*, 1–69.

Hasan, S. M., et al. Development of Siloxane-based Amphiphiles as Cell Stabilizers for Porous Shape Memory Polymer Systems. *J. Colloid Interface Sci.* **2016**, *478*, 334–343.

Hazra, S., et al. Cocrystallized Dinuclear–Nuclear–MonoCu^{II}.Na^{I} and Double-decker–Triple-deckerCu^{II}$_5$K^{I}$_3$ Complexes Derived from *N,N'*-Ethylenebis(3-ethoxysalicylaldimine). *Cryst. Growth Des.* **2009**, *9*, 3603–3608.

Herkert, L.; Sampedro, A.; Fernández, G. Cooperative Self-assembly of Discrete Metal Complexes. Cryst. Eng. Comm. 2016, *18*, 8813–8822.

Hillborg, H., et al. Crosslinked Polydimethylsiloxane Exposed to Oxygen Plasma Studied by Neutron Reflectometry and Other Surface Specific Techniques. *Polymer* **2000**, *41*, 6851–6863.

Hindson, J. C., et al. All-aromatic Liquid Crystal Triphenylamine-based Poly(azomethine) s as Hole Transport Materials for Opto-electronic Applications. *J. Mater. Chem.* **2010**, *20* (5), 937–944.

Isık, D., et al. Charge-carrier Transport in Thin Films of π-conjugated Thiopheno-azomethines. *Org. Electron.* **2012**, *13* (12), 3022–3031.

Ibrahim, M. N.; Sharif, S. E. A. Synthesis, Characterization and Use of Schiff Bases as Fluorimetric Analytical Reagents. *J. Chem.* **2007**, *4*, 531–535.

Jungreis, E.; Thabet, S. *Analytical Applications of Schiff Bases*; Marcel Dekker: New York, NY, 1969.

Katsuki, T. Catalytic Asymmetric Oxidations Using Optically Active (Salen) Manganese(III) Complexes as Catalysts. *Coord. Chem. Rev.* **1995**, *140*, 189–214.

Kianfara, A. H.; Zargarib, S.; Khavasic, H. R. Synthesis and Electrochemistry of M(II) N_2O_2 Schiff Base Complexes: X-Ray Structure of {Ni[Bis(3-chloroacetylacetone)Ethylenediimine]}. *J. Iran. Chem. Soc.* **2010**, *7*, 908–916.

Kirstein, S., et al. Chiral J-aggregates Formed by Achiral Cyanine Dyes. *Chemphyschem.* **2000**, *3*, 146–150.

Lehn, J. M. Toward Self-organization and Complex Matter. *Science* **2002**, *295*, 2400–2403.

Li, W., et al. Dynamic Self-assembly of Coordination Polymers in Aqueous Solution. *Soft Matter* **2014**, *10*, 5231–5242.

Li, X., et al. A Facile Approach for the Synthesis of Aromatic Polyazomethine Hollow Structures Employing In Situ Formed Dynamic Imine Crystals as Reactive Templates. *Macromol. Res.* **2015**, *23* (12), 1087–1090.

Lin, Y.; Alexandridis, P. Self-assembly of an Amphiphilic Siloxane Graft Copolymer in Water. *J. Phys. Chem. B.* **2002**, *106*, 10845–10853.

Liu, Y.; Goebla, J.; Yin, Y. Templated Synthesis of Nanostructured Materials. *Chem. Soc. Rev.* 2013, *42*, 2610–2653.

Liu, J., et al. Hierarchical Self-assembly of Luminescent Tartrate-bridged Chiral Binuclear Tb(III) Complexes in Ethanol. *Langmuir* **2016**, *32* (41), 10597–10603.

Lutolf, M. P.; Hubbell, J. A. Synthetic Biomaterials as Instructive Extracellular Microenvironments for Morphogenesis in Tissue Engineering. *Nat. Biotechnol.* **2005**, *23*, 47–55.

Marin, L.; Damaceanu, M. D.; Timpu, D. New Thermotropc Liquid Crystalline Polyazomethines Containing Luminescent Mesogens. *Soft Matter.* **2009**, *7*, 1–20.

Marin, L.; Perju, E.; Damaceanu, M. D. Designing Thermotropic Liquid Crystalline Polyazomethines Based on Fluorene and/or Oxadiazole Chromophores. *Eur. Polym. J.* **2011**, *47*, 1284–1299.

Massey, J., et al. Organometallic Nanostructures: Self-assembly of Poly(ferrocene) Block Copolymers. *Adv. Mater.* **1998a**, *10*, 1559–1562.

Massey, J., et al. Self-assembly of a Novel Organometallic–Inorganic Block Copolymer in Solution and the Solid State: Nonintrusive Observation of Novel Wormlike Poly(ferrocenyldimethylsilane)-*b*-poly(Dimethylsiloxane) Micelles. *J. Am. Chem. Soc.* **1998b**, *120*, 9533–9540.

Menozzi, E., et al. Metal-directed Self-assembly of Cavitand Frameworks. *J. Org. Chem.* **2006**, *71*, 2617–2624.

Mitra, S.; Tamai, N. A Combined Experimental and Theoretical Study on the Photochromism of Aromatic Anils. *Chem. Phys.* **1999**, *246*, 463–475.

Nakamura, T., et al. A Hierarchical Self-assembly System Built Up from Preorganized Tripodal Helical Metal Complexes. *J. Am. Chem. Soc.* **2016**, *138* (3), 794–797.

Patterson, J. P., et al. The Analysis of Solution Self-assembled Polymeric Nanomaterials. *Chem. Soc. Rev.* 2014, *43*, 2412–2425.

Nguyen, T. Q., et al. Self-assembly of 1-D Organic Semiconductor Nanostructures. *Phys. Chem. Chem. Phys.* **2007**, *9*, 1515–1532.

Perez-Pagea, M., et al. Template-based Syntheses for Shape Controlled Nanostructures. *Adv. Colloid Interface Sci.* **2016**, *234*, 51–79.

Petrus, M. L., et al. A Low Cost Azomethine-based Hole Transporting Material for Perovskite Photovoltaics. *J. Mater Chem. A.* **2015**, *3* (23), 12159–12162.

Prazeres, T. J. V., et al. Determination of the Critical Micelle Concentration of Surfactants and Amphiphilic Block Copolymers Using Coumarin 153. *Inorg. Chim. Acta.* **2012,** *381,* 181–187.

Racles, C. Siloxane-containing Liquid Crystalline Polymers. In *Advances in Functional Heterochain Polymers*; Cazacu, M., Ed.; Nova Science Publishers: Hauppauge, NY, 2008.

Racles, C. Siloxane-based Surfactants Containing Tromethamol Units. *Soft Mater.* **2010,** *8* (3), 263–273.

Racles, C. Polydimethylsiloxane–Indomethacin Blends and Nanoparticles. *AAPS Pharm. Sci. Tech.* **2013,** *14* (3), 968–976.

Racles, C.; Cozan, V. Synthesis of Glucose-modified Siloxanes by a Simplified Procedure. *Soft Mater.* **2012,** *10* (4), 413–425.

Racles, C.; Hamaide, T. Synthesis and Characterization of Water Soluble Saccharide Functionalized Polysiloxanes and Their Use as Polymer Surfactants for the Stabilization of Polycaprolactone Nanoparticles. *Macromol. Chem. Phys.* **2005,** *206,* 1757–1768.

Racles, C., et al. Poly(azomethine-ester-siloxane)s: Synthesis and Thermal Behaviour. *High Perform. Polym.* **2002,** *14* (4), 397–413.

Racles, C.; Hamaide, T.; Ioanid, A. Siloxane Surfactants in Polymer Nanoparticles Formulation. *Appl. Organometal. Chem.* **2006,** *20,* 235–245.

Racles, C.; Cozan, V.; Sajo, I. Influence of Chemical Structure on Processing and Thermotropic Properties of Poly(Siloxane-azomethine)s. *High Perform. Polym.* **2007,** *19* (5), 541–552.

Racles, C., et al. Micellization of a Siloxane-based Segmented Copolymer in Organic Solvents and Its Use as a Tool for Metal Complex Nanoparticles. *Macromol. Rapid Commun.* **2008,** *29,* 1527–1531.

Racles, C., et al. On the Feasibility of Chemical Reactions in the Presence of Siloxane-based Surfactants. *Colloid Polym. Sci.* **2009,** *287,* 461–470.

Racles, C.; Hamaide, T.; Fleury, E. Siloxane-containing Compounds as Polymer Stabilizers. In *Polymer Research Developments: Amphiphillic Block Copolymers, Polymer Aging, Block Copolymers/Polymer Stabilizers*; Segewicz, L., Petrowsky, M., Eds.; Nova Science Publishers: New York, NY, 2010.

Racles, C., et al. A Simple Method for the Preparation of Colloidal Polymer-supported Silver Nanoparticles. *J. Nanopart. Res.* **2011,** *13,* 6971–6980.

Racles, C., et al. Aqueous Dispersion of Metal Oxide Nanoparticles, Using Siloxane Surfactants. *Colloids Surf. A Physicochem. Eng. Aspects* **2014a,** *448,* 160–168.

Racles, C.; Mares, M.; Sacarescu, L. A Polysiloxane Surfactant Dissolves a Poorly Soluble Drug (Nystatin) in Water. *Colloids Surf. A Physicochem. Eng. Aspects* **2014b,** *443,* 233–239.

Racles, C.; Silion, M.; Iacob, M. Lanthanum Complex of a Multifunctional Water-soluble Siloxane Compound—Synthesis, Surface Activity and Applications for Nanoparticles Stabilization. *Colloids Surf. A Physicochem. Eng. Aspects* **2014c,** *462,* 9–17.

Randall, M. H. Siloxane Surfactants. In *Silicone Surfactants*; Randall, M. H., Ed.; Marcel Dekker: New York, NY, *1999.*

Resendes, R., et al. A Convenient Transition Metal-Catalyzed Route to Water-Soluble Amphiphilic Organometallic Block Copolymers: Synthesis and Aqueous Self-Assembly of Poly(ethylene oxide)-*block*-poly(ferrocenylsilane). *Macromolecules* **2000,** *33* (1), 8–10.

Ruiz-Hitzky, E.; Darder, M.; Aranda, P. An Introduction to Bio-nanohybrid Materials. In *Bio Inorganic Hybrid Nanomaterials*; Ruiz-Hitzky, E., Ariga, K., Lvov, Y. M., Eds.; Wiley-VCH Verlag GmbH & Co. KGaA: Weinheim, Germany, 2008.

Saegusa, Y.; Koshikawa, T.; Nakamura, S. Synthesis and Characterization of 1,3,4-oxadiazole Containing Polyazomethines. *J. Polym. Sci. Polym. Chem.* **1992**, *30*, 1369–1373.

Samal, S., et al. Chelating Resins VII: Studies on Chelating Resins of Formaldehyde and Furfuraldehyde-condensed Phenolic Schiff Base Derived from 4,4'-diaminodiphenylsulphone and *o*-hydroxyacetophenone. *React. Funct. Polym.* **1999**, *42*, 37–52.

Shul'pin, G. B. Metal-catalyzed Hydrocarbon Oxygenations in Solutions: The Dramatic Role of Additives: A Review. *J. Mol. Catal. A Chem.* **2002**, *189*, 39–66.

Sicard, L., et al. On-substrate Preparation of an Electroactive Conjugated Polyazomethine from Solution-processable Monomers and Its Application in Electrochromic Devices. *Adv. Funct. Mater. 2013, 23 (8), 3549–3559.*

Soni, S. S., et al. Micellar Structure of Silicone Surfactants in Water from Surface Activity, SANS and Viscosity Studies. *J. Phys. Chem. B.* **2002**, *106*, 2606–2617.

Soroceanu, A., et al. Supramolecular Aggregation in Organic Solvents of Discrete Copper Complexes Formed with Organo-siloxane Ligands. *Soft Mater.* **2015**, *13*, 93–105.

Sui, Y., et al. Ionic Ferroelectrics Based on Nickel Schiff Base Complexes. *Inorg. Chem.* **2010**, *49*, 1286–1288.

Vasiliu, M., et al. Chelate Polymers. IV. Siloxanes Functionalized with Chelating Groups Derived from Hydroxy-ketones, Their Metal Complexes and Some Polymers. *Appl. Organomet. Chem.* **2005**, *19*, 614–620.

Venegas-Yazigi, D., et al. Structural and Electronic Effects on the Exchange Interactions in Dinuclear Bis(Phenoxo)-bridged Copper(II) Complexes. *Coord. Chem. Rev.* **2010**, *254*, 2086–2095.

Vlad, A., et al. Polyazomethines Derived from Polynuclear Dihydroxyquinones and Siloxane Diamines. *Eur. Polym. J.* **2008**, *44*, 2668–2677.

Wang, H.; Winnik, M. A.; Manners, I. Synthesis and Self-assembly of Poly (ferrocenyl-dimethylsilane-*b*-2-vinylpyridine) Diblock Copolymers. *Macromolecules* **2007**, *40*, 3784–3789.

Wang, Y.; Angelatos, A. S.; Caruso, F. Template Synthesis of Nanostructured Materials via Layer-by-layer Assembly. *Chem. Mater.* **2008**, *20* (3), 848–858.

Whitesides, G. M.; Boncheva, M. Beyond Molecules: Self-assembly of Mesoscopic and Macroscopic Components. *Proc. Natl. Acad. Sci. USA* **2002**, *99* (8), 4769–4774.

Wu, D.; Huang, Y.; Xu, F.; Mai, Y.; Yan, D. Recent Advances in the Solution Self-assembly of Amphiphilic "Rod-coil" Copolymers. *J. Pol. Sci. Polym. Chem.* **2017**, *55* (9), 1459–1471.

Yam, V. W. W.; Au, V. K. M.; Leung, S. Y. L. Light-emitting Self-assembled Materials Based on d^8 and d^{10} Transition Metal Complexes. *Chem. Rev.* **2015**, *115* (15), 7589–7728.

Yang, C., et al. Robust Colloidal Nanoparticles of Pyrrolopyrrole Cyanines J Aggregates with Bright Near Infrared Fluorescence in Aqueous Media: From Spectral Tailoring to Bioimaging Application. *Chem. Eur. J.* **2017**, *23* (18), 4310–4319.

Zaltariov, M. F., et al. Metallopolymers Based on a Polyazomethine Ligand Containing Rigid Oxadiazole and Flexible Tetramethyldisiloxane Units. *J. Appl. Polym. Sci.* **2015**, *132*, 41631.

Zaltariov, M. F., et al. Oxime-bridged Mn_6 Clusters Inserted in One-dimensional Coordination Polymer. *Macromolecules* **2016,** *49* (17), 6163–6172.

Zhou, J.; Whittell, G. R.; Manners, I. Metalloblock Copolymers: New Functional Nanomaterials. *Macromolecules* **2014,** *47* (11), 3529–3543.

Zhu, J., et al. Controlling Molecular Ordering in Solute-state Conjugated Polymers. *Nanoscale* **2015,** *7*, 15134–15141.

CHAPTER 6

POROUS PARTICLES AS SMART MATERIALS FOR BIOMEDICAL APPLICATIONS

SILVIA VASILIU[1*], STEFANIA RACOVITA[1],
CRISTINA DOINA VLAD[1], IONELA GUGOASA[2], and
MARCEL POPA[2,3]

[1]*"Mihai Dima" Functional Polymers Department, "Petru Poni"
Institute of Macromolecular Chemistry, Grigore Ghica Voda Alley
No 41A, 700487 Iasi, Romania*

[2]*Faculty of Chemical Engineering and Environmental Protection,
"Gheorghe Asachi" Technical University of Iasi, Prof. Dr. Docent
Dimitrie Mangeron Street, No. 73, 700050 Iasi, Romania*

[3]*Academy of Romanian Scientists, Spaiul Independentei Str. 54,
050094 Bucuresti, Romania*

Corresponding author. E-mail: msilvia@icmpp.ro

CONTENTS

ABSTRACT

Porous particles that possess external/internal pores or a combination of both have received a great attention in last decades due to their potential applications in healthcare as carriers for biomacromolecules and other substances, bone tissue engineering and regeneration, pulmonary drug delivery, cell therapy, and dental applications. Porous particles can be obtained from a large variety of natural and synthetic polymers having access to different manufacturing processes like suspension, dispersion, precipitation, multistage, membrane/microchannel emulsification, and microfluidic polymerization. The porous particles possess the excellent properties like greater surface area, lower mass density, superior cell attachment and cell proliferation, higher drug absorption, and appropriate drug release kinetics for desired applications. Also, the use of particulate systems in biomedical applications confers a series of facilities: permits the election and formulation of different combinations polymer-active principle, offers a gradual release of the bioactive principle in such manner as to provide the desired results, allows a diversification in the administration routes, such as oral, transdermal, ophthalmic, nasal, anal, and vaginal administrations.

6.1 INTRODUCTION

Porous materials are inorganic, organic, or hybrid frameworks that contain cavities, channels, or interstices (Wu et al., 2012).

In the human evolution on earth, the porous materials have shown a growing interest due to their advantages. If initially, it has gone from natural materials such as animal bones, stone, and wood in the modern age it has been reached in the manufacture of porous synthetic materials such as ceramics, paper, chalk, textiles, metal foam, and mineral wool or expanded polystyrene (see Fig. 6.1).

The performance and the characteristics of porous materials depend on the porosity, pore size distribution, and specific surface area, while the physical properties of these materials depend on the shape, connectivity, and structure of the pores. Based on various criteria (pore size and shape, material nature) the porous materials can be classified as shown in Figures 6.2 and 6.3, respectively.

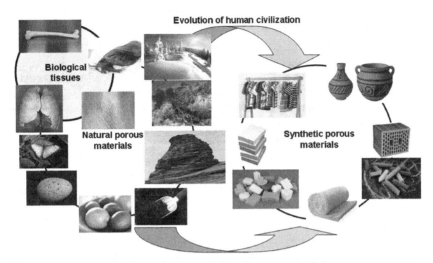

FIGURE 6.1 (See color insert.) The evolution of porous materials.

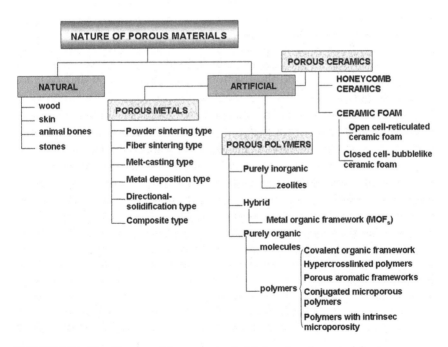

FIGURE 6.2 Classification of the porous materials based on the material nature.

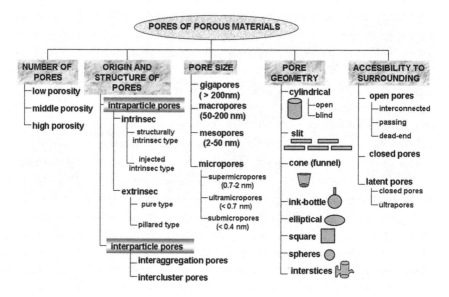

FIGURE 6.3   Classification of the porous materials based on the pore nature.

However, a rigorous classification of these materials is quite difficult. The so-called open porosity refers to the connected cavities while the "closed porosity" refers to the fraction of the total volume that includes the closed pores. When two overlapping pores interact a "dual porosity" having "interconnecting pores" has occurred (Kaneko, 1994). The introduction of open pores in a material leads to a decrease of the material density as well as an increase of specific surface area (Ishizaki et al., 1998).

In order to define the porosity, it has been accepted the use of IUPAC nomenclature dedicated to sorbents and catalysts, because the size of the pores formed in the polymeric network gives a morphology specific to the structure of porous materials (Gokmen and Du Prez, 2012). This morphology determines the destination or area of use of those products. According to this nomenclature, there are three types of porous morphological structures:

- microporous morphological structures with gel-type or swelling porosity that are revealed only in the swollen condition of porous materials, and the size of pores is less than 2 nm;
- mesoporous morphological structures consisting of the so-called transitional pores, when the size of pores is between 2 and 50 nm;

- macroporous morphological structures when the porous materials present pores with sizes higher than 50 nm. The term of macroporosity is usually related to the porosity in dry state, which is characterized by a low density of the network due to the presence of voids in the polymer matrix.

The great variety of porous materials has led to the study of some applications in various fields such as chemistry, environment, engineering, or biomedicine. Thus, in medicine, these materials can be used as implants, controlled drug delivery systems, sensors, membranes, bandages, catheters, valves, prostheses, and agents in medical imaging (Perez et al., 2014; Sene et al., 2008; Yang et al., 2012; Asefa et al., 2009; Zhang et al., 2013a Wally et al., 2015).

In terms of presentation form, the current technologies allow the achievement of polymer biologically active substance systems in different forms such as films, gels, hydrogels, injectable solutions, tablets, and micro- and nanoparticulate systems. Among them, the particulate systems show the highest interest because of the many advantages they present:

- permits the election and formulation of different polymer-drug combinations;
- offers a gradual release of the bioactive principle in such a manner as to provide the desired results;
- allows a diversification in the administration routes, particulate systems being functional in: oral (Salonen et al., 2005), parenteral (Hernan Perez de la Ossa, 2012), transdermal (Badilli et al. 2011), ophthalmic (Wang et al. 2014), nasal (Anton et al., 2012), anal, and vaginal (Yoo et al., 2010) administrations.

Some of these systems are also sensitive to different stimulus such as temperature, pH, light radiation, electric, or magnetic field (Joshi et al., 2013; Sun and Deng, 2005; Gao et al., 2016; Philippova et al., 2011; Meenach et al., 2012). These types of materials are called smart materials.

For this reason, the goal of this chapter is to present the recent development regarding the particulate porous polymer materials and their applications in pharmaceutical and medical fields. The difference between traditional and porous particles is the presence of pores situated on the surface or inside of the particles, leading to a high specific surface and low density of the polymer network.

Also, the pore characteristics (size, amount, and structure of pores) have a great influence on drug adsorption and release processes. For design of new drug delivery systems based on porous polymeric materials, it is very important to control the following parameters: porosity, specific surface area, and pore size. For example, in templating method (polystyrene microspheres as a template) by changing the time and temperature of heat treatment, it adjusts the size of pores situated on the surface of the porous film (Li and Zhang, 2007).

6.2 PREPARATION TECHNIQUES AND BIOAPPLICATIONS OF POROUS POLYMERIC PARTICLES

Porous polymer particles can be obtained from a large variety of natural and synthetic polymers having access to different classical and modern procedures like suspension polymerization (Okay, 2000; Horak et al., 1993; Fang et al., 2007; Wu et al., 2003), precipitation and dispersion polymerization (Cho et al., 2016; Boyere et al., 2012; Moustafa et al., 1999; Kawaguchi and Ito, 2005; Saracoglu et al., 2009; Li and Stover, 2000), multistage heterogeneous polymerization (Okubo et al., 1991; Cai et al., 2013), membrane/microchannel emulsification (Gokmen and Du Prez, 2012; Wu et al., 2015; Sugiura et al., 2001; Yuan et al., 2010), microfluidic polymerization (Watanabe et al., 2014, Wang et al., 2017), spray drying method (Iskandar et al., 2009), and molecular imprinting technology (Lofgreen and Ozin, 2014).

6.2.1 SUSPENSION POLYMERIZATION

The suspension polymerization highlighted for the first time in 1950 (Cui et al., 2008) is the most widely used technique for preparation of the macroporous particles due to some technical and economic advantages:

1. low-cost price compared to the diversity of porous particle properties;
2. excellent heat transfer during the transformation of monomers in polymers;
3. possibility to control the particle size;
4. obtaining a predetermined internal structure;

5. the ease of using and separation of particles resulting from synthesis processes;
6. a low number of components used in the polymerization system compared to the emulsion technique;
7. purification of the final product is made by simple procedures (washing, steam distillation).

There are two types of suspension polymerization reactions, namely, (1) of the organic phase in aqueous phase (Brooks, 2010; Vivaldo-Lima et al., 1997) and (2) of the aqueous phase in organic phase, known in the literature as the inverse suspension polymerization.

In general, the suspension polymerization is a heterogeneous radical polymerization reaction, where initially the monomers containing a radical initiator are dispersed as droplets (discontinuous phase) in the reaction medium (continuous phase), followed by polymerization of monomers and finally leading to the formation of the spherical particles.

In order to have a maximum stability (minimum interfacial energy), these droplets adopt a spherical form, having a medium size controlled by the stirring speed, as well as by the quantity and type of stabilizing agents (Ahmed, 1984; Erbay et al., 1992; Esfandyari-Manesh et al., 2013). The mechanism by which the stabilizer acts depends on its nature. The chemical structure is very important and the balance between strongly polar groups (which ensure the dissolution of the stabilizer in water) and the number of hydrophobic groups (which physically interact with the surface of the monomer droplets) can help to find some effective stabilizers for each suspension polymerization system. The stabilizing system affects the particle size, controls the coalescence of droplets, and reduces the particle agglomeration. A good stabilizer must have some characteristics:

1. does not interfere with polymerization process by inhibiting or retarding effects;
2. must be characterized by an optimal ratio of hydrophilic/hydrophobic groups;
3. ensure both the formation of a mechanically resistant layer around the droplets and the optimal increase of the solution viscosity in order to not hinder the proper dispersion of the monomers.

There is a direct relationship between the stirring speed and the average particle size, for example, a nonuniform stirring cannot generate a monodisperse system. Also, an optimal stirring can ensure a proper coalescence/dispersion ratio.

To establish the stirring conditions, several parameters can be taken into account such as the ratio between the liquid level and the stirrer, type of the stirrer, the ratio between initial viscosity of monomer and the one of aqueous phase, densities of the two phases, fluid flow parameters, and criteria of Reynolds and Weber (Borwankar et al., 1986; Johnson, 1980; Langner et al., 1980).

The temperature of the suspension polymerization reactions is usually in the range of 40–90°C, depending on the specifics of the initiating system as well as the mechanism and kinetics of reactions that occur in the system.

The radical initiators, which are introduced in the reaction system, are organic peroxides (benzoyl peroxide and lauroyl peroxide) or azo compounds (azobisisobutyronitrile) soluble in monomers and insoluble in the aqueous phase. Thus, it was investigated a large number of initiating systems because it has been found that the nature of initiator play an important role in the formation of particles with uniform sizes (Vlad et al., 1978; Conceicao et al., 2011; Qui et al., 2001; Pinto et al., 2001).

The introduction of an inert medium in the polymerization system leads to the formation of permanently heterogeneous structures (containing pores after drying). In order to obtain rigid materials having porosity, the dispersed liquid of the polymer mixture must contain not only the monomers but also an adequate amount of porogenic agent. This porogenic agent should not react during the polymerization but must remain within the newly formed particles. After the completion of the process of the networks preparation, the porogenic agent is removed by the extraction method with a solvent giving to the macromolecular network a micro- or macroporous structure, depending on the affinity of the porogen for the monomers and the copolymer (Okay et al., 1985; Jacobelli et al., 1979; Li et al., 2006).

From the point of view of the molecular weight, the porogenic agents can be either low molecular weight substances or macromolecular compounds. The low molecular weight porogenic agents can be classified into two categories:

• Chemical substances with values of solubility parameter similar to those of monomers and copolymers and represent the class of swelling porogenic agents;

• Chemical substances with values of solubility parameter that are close to those of monomers and differ from those of copolymers, these being part of the category of precipitation porogenic agents (Dragan et al., 1976).

The cross-linked copolymer structure formation depends on the quantity and nature both of monomers and porogenic agents. In order to understand the mechanism of spherical particle preparation, during the suspension polymerization, let us imagine that each droplet is a spherical microreactor, shape that is provided by the stirring speed and interfacial tension. As a result of cross-linking reaction and solubility changes associated with the growth of the macromolecular chain length, the polymer molecules formed in the interior of this "macroreactor" precipitate in the reaction medium (mixture of porogenic agent and monomers).

This phase separation occurs at beginning of polymerization leading to the formation of globular microscopic entities, which begin to grow and are not coalescent due to the cross-linking reaction. Eventually, they come in contact with each other and associate forming clusters composed of interconnected globules and void spaces (pores). In essence, each micro-reactor (droplet) is transformed into a macroporous particle, in which the percentage of void volume correlates well with the percentage of porogenic agent used (Svec and Frechet, 1996). Macroporous structure formation is exclusively determined by the behavior of cross-linked chains during the copolymerization reaction (Rabelo and Couthinho, 1994).

The heterogeneous character of the cross-linking polymerization is not a sufficient condition for the formation of the two phases of the structure. For this reason, it is important that the mechanism of phase separation (Konishi et al., 2003; Schwartz-Linek et al., 2012; Chen, et al., 2003), that occurs close to the limit of thermodynamic stability when it can take place at macroscale (macrosyneresis) or at microscale (microsyneresis). This is typical of the formation of biphasic-dispersed structures including the macroporous networks (Dusek, 1967).

Depending on the stage of copolymerization reaction, the phase separation can take place in three moments of the reaction:

1. Before the gel point, when low-quality macrogranular structures are obtained, due to the fact that one of the phases (polymer or porogenic agent) is in a minority. In this case, the coalescence phenomenon occurs before its structure is fixed by cross-linking.

2. After formation of cross-links (long after the gel point), when the high cross-linking density of the polymer prevents the microseparation of liquid and its exclusion occurs outside of the polymer (macrosyneresis). This creates a fine structure of pores determining the appearance of specific surfaces.

3. The most appropriate moment for starting the phase separation is the gel point when the dispersed structure is fixed by the formation of the new cross-links.

The porous particles obtained by the suspension polymerization technique are used as sorbents in different types of chromatography, in hemoperfusion, in controlled release of pharmaceutical products, immobilization of enzymes, encapsulation of mammary cells, or as immunochemical reagents (Galia, et al., 1994; Senel et al., 2002; Srikanth et al., 2010; Huichao et al., 2014). The mixture of polymerization should be adapted to the required properties of particles, especially for achieving the sphericity, size, and porosity. For example, in order to separate the small molecules or oligomers, particles with small pores and with a very narrow size distribution are used, while to separate the proteins, there are necessary pores with large diameter ensuring a diffusion of the substance dissolved through the polymer. All these applications are possible thanks to the benefits brought by the particles with porous structure, which remain unaffected by the changes occurred during the use. Thus, the macroporous materials are swelling less than microporous gels in all solvents, and the mass transfer inside and outside of the pores is better than the one in the nonswollen homogeneous polymers. The molecules diffuse frequently by pores forming a labyrinth-like some interconnected channels.

Mane et al. (2015) were obtained by suspension polymerization techniques the hyperhydrophilic porous cross-linked microparticles based on acrylic acid and pentaerythritol tetraacrylate that are modified with cobalt and nickel metals in order to obtain a new, efficient, and cost-effective drug carrier. The drug used in this study was metoprolol. In case of cobalt metal supported polymer, the authors were observed a higher adsorption of metoprolol compared to nickel metal supported polymer leading to the conclusion that the first system can be used successfully in drug delivery.

Another possible drug delivery systems are obtained by grafting some polysaccharides [xanthan (XAN), gellan (GLL), and chitosan (CH)] onto porous cross-linked network based on glycidyl methacrylate (GMA) and different cross-linked monomers [ethylene glycol dimethacrylate (EGDMA), diethylene glycol dimethacrylate (DEGDMA) and triethylene glycol dimethacrylate (TEGDMA)] (Lungan et al., 2014; Lungan et al., 2015; Cigu et al., 2016).

The association between natural and synthetic polymers aims to obtain structures that present both the chemical, mechanical, and thermal stability of synthetic polymers, as well as the special properties of natural polymers such as bioadhesion, biodegradability, or biocompatibility. The grafting ways of the polysaccharides onto the polymer network chains are shown in Figure 6.4.

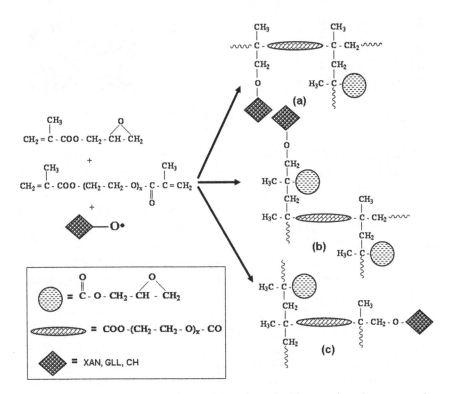

FIGURE 6.4   The grafting possibilities of the polysaccharides onto the polymer networks.

Scanning electron microscopy (SEM) highlights that microparticles are characterized by porous morphologies with defined spherical shape and micrometric sizes (see Fig. 6.5).

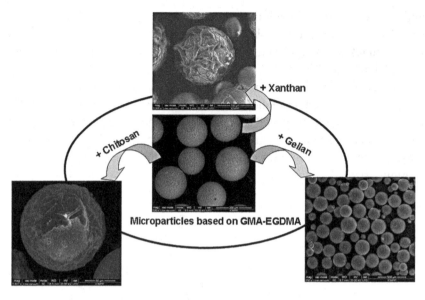

FIGURE 6.5   SEM micrographs of porous microparticles.

It has also been found that there is a difference between the internal and external structure of the microparticles, this nonuniformity being due to the fact that grafting takes place especially on the surface and superficial layers of the microparticles. By changing the reaction parameters (stirring speed, the nature of the cross-linker and of the porogenic agent, and cross-linking degree), particles with different morphologies were obtained. The ability of the porous macromolecular supports to retain and release biologically active principles was evaluated using the following drug models: theophylline (Lungan et al., 2015), chloramphenicol succinate sodium salt (Cigu et al., 2016), and cefuroxime sodium salt (Vasiliu, et al., 2011).

In all cases, a better drug loading was observed on the grafted polysaccharide supports. Taking into account the in vitro drug release mechanism, the model that best fits the release data is the first order kinetic model, indicating that the release rate of drug is mainly controlled by diffusion. In addition, for the microparticles based on GMA-XAN and GMA-CH, the values of the diffusion exponent, n, from the Korsmeyer–Peppas equation were situated between 0.43 and 0.85, respectively, suggesting that the release mechanism is complex, being characteristic of an anomalous diffusion. In this case, the transport of theophylline and chloramphenicol is controlled by diffusion, as well as by the swelling process of the porous macromolecular supports.

Significant progresses have been made in the use of ion exchangers as polymeric carriers for biomedical applications. Thus, core–shell microparticles based on weak base acrylic ion exchangers and two polysaccharides (XAN and GLL) were obtained in order to retain by adsorption two antibiotics, namely, cefotaxime sodium salt and cefuroxime sodium salt. The core of the microparticles is represented by the ion exchanger while the microparticle shell is a layer of polyelectrolyte complex that has been formed through the interaction between the COO^- groups belonging to the GLL or XAN and the NH_3^+ groups of the ion exchanger. The two types of ion exchangers were obtained in two stages as follows:

1. porous copolymers were prepared by using the suspension copolymerization technique of divinylbenzene (DVB), acrylonitrile (AN), and ethyl acrylate (EA);
2. aminolysis reaction with ethylenediamine (ion exchangers 1) or hydrazine hydrate (ion exchangers 2) of the DVB-AN-EA copolymers (Vasiliu et al., 2011; Vasiliu, et al., 2009; Racovita et al., 2016).

The cefotaxime release from the systems based on ion exchangers 1 and the cefuroxime release from the systems based on ion exchangers 2 followed a non-Fickian mechanism in which the release rate is controlled by diffusion and also, by the swelling of the macromolecular carriers. The systems based on XAN/GLL and ion exchangers 1 can be used for oral administration, and the systems based on GLL and ion exchangers 2 can be used in rectal administration and a dermal patch for transdermal administration of drugs.

Materials used for adsorption of bovine serum albumin (BSA) and trypsin proteins may be obtained by modifying the Poly(glycidyl methacrylate-co-divinylbenzene) (poly(GMA-co-DVB)) porous microparticles with poly(ethylene glycol) (PEG). Porous microparticles were obtained by suspension polymerization using as porogen agent a mixture of isooctane and 4-methyl-2-pentonal. To prevent irreversible protein adsorption, the microparticles were modified in the presence of boron trifluoride with PEG through a coupling reaction that takes place between OH groups of the PEG and the epoxy groups belonging to the porous microparticles (Wang et al., 2006).

Recent decades have brought considerable progress in the field of enzyme therapy. Since most enzymes exhibit instability in working conditions (high temperature, pH variation, organic solvents, or other denaturing agents), a number of techniques have been developed in order to try to increase the stability of biocatalysts. Of these, enzyme immobilization represents a physically or chemically binding of the enzyme compounds on insoluble supports in order to increase the stability of the enzyme preparations toward operational conditions (temperature, pH, and organic solvents), without affecting their enzymatic capacity (Brena and Batista-Viera, 2006).

The requirements that an enzyme–substrate system has to meet in order to be effective are (Brodelius and Mosbach, 1987; Buchholz and Klein, 1987):

- to be resistant to detachment, desorption;
- to allow a good accessibility of chemical analysts to active centers;
- to be stable at temperature variations and extreme pH values;
- to show stability over time and a high density of immobilized biomolecules;
- to have high biocatalytic activity;
- to respond quickly in time.

All these considerations should be taken into account before choosing a specific immobilization method (by linking to a support, entrapping, and cross-linking). Regardless of the immobilization method, any enzyme–substrate system has two functions, a noncatalytic function that characterizes the substrate and an enzyme-specific catalytic function. The noncatalytic function of an enzyme complex depends on the nature of the support on which the enzyme is immobilized. In the immobilization process, a support is considered good if it fulfills some properties: permeability, insolubility, resistance to a microbial attack, high rigidity, chemical stability, hydrophilicity, mechanical, and thermal stability and can be easily regenerated. The immobilization of biocatalysts (enzymes, hormones, growth factors, vitamins, microbial cells, or plant and animal cells) on solid, insoluble, organic, or inorganic supports has been of particular interest in pharmaceutical and biomedical applications.

The acrylic copolymers were proven to be ideal for the preparation of solid supports in the form of membranes, gels, films, or micro and nanoparticles for the immobilization of bioactive molecules. This was due to both increased chemical and mechanical stability, as well as to their very good resistance to microbial degradation. In the range of hydrophilic acrylic monomers that can lead to the supports with suitable properties for the immobilization of enzymes, both acrylic and methacrylic acids (MAA) are included. Thus, the copolymerization of acrylic acid (AA) and MAA with TEGDMA by suspension polymerization, dispersion polymerization, or microemulsion in the presence of porogenic agents such as polystyrene microparticles or n-hexane leads to the preparation of the microparticulate porous structures. The particles were used to immobilize the protease either by ionic or by covalent linkages. The ionic binding was accomplished via electrostatic interactions between the acidic groups present on the copolymer chain and the basic groups of the enzyme, while the covalent binding occurred by modifying the carboxyl groups with thionyl chloride to acid chloride, followed by their reaction with the amino groups of the enzyme (Jangde et al., 2008). Also, the porous particles based on GMA and EGDMA were obtained by suspension polymerization in order to immobilize the chymotrypsin. To adjust the hydrophobicity of the copolymer, a third monomer, namely, styrene, was introduced into the system. Toluene was introduced in the organic phase as a porogenic agent, in order to obtain a porous structure. The presence of pores in the copolymer structure has determined a higher surface area, which led to the retention of a large amount of chymotrypsin on the microparticle surface and a better enzymatic activity compared to that of nonporous particles (Pratima et al., 2012). Functionalized and cross-linked acrylic polymers and copolymers are characterized by different chemical structures and various hydrophilicities, which recommend them as supports for the immobilization of bioactive molecules. Thus, macroporous resins based on methyl methacrylate and DVB have been used to immobilize papain. Initially, the macroporous copolymers were obtained by suspension polymerization, in the presence of petroleum ether as a porogenic agent. To create several amine groups, the macroporous copolymers were aminated with hydrazine hydrate (Ding et al., 2003).

The immobilization of papain on macromolecular supports was performed either by cross-linking with glutaraldehyde or by coupling to the diazo groups. Studies on papain immobilization have shown that the system has a high stability to the acid environment and to the working temperature.

6.2.2 EMULSION POLYMERIZATION

The emulsions are dispersed systems obtained by mixing two or more immiscible phases (aqueous phase and oil phase) being stabilized by the presence of an emulsifier (surfactant).

The emulsions can be classified according to several criteria such as (Leal-Calderon et al., 2007; Buszello and Muller, 2000; Garti and Benichou, 2001):

1. Size of dispersed particles:
    *   macroemulsions where the size of dispersed particles is >400 nm, being characterized by pronounced instability and opaque appearance;
    *   microemulsions where the size of particles is between 10 and 80 nm, being characterized by thermodynamic stability;
    *   miniemulsions where the size of particles is between 100 and 400 nm.

2. Type of emulsion:
    *   oil/water emulsion (direct emulsion), where the dispersed phase is a nonpolar liquid, lipophilic (vegetable or mineral oil, organic solvents) and the dispersed phase is hydrophilic (water);
    *   water/oil emulsion (inverse emulsion);
    *   multiple emulsions (water/oil/water or oil/water/oil type emulsions), which are obtained by the emulsification of a direct or inverse emulsion into another oil or aqueous phases.

3. Origin:
    *   natural emulsions milk, latexes, and emulsions of oleaginous seeds (almonds, peanuts) where the emulsification takes place due to the presence of some emulsifiers in the structure of the seeds;
    *   artificial emulsions.

4. Concentration of the dispersed phase:
    *   diluted emulsions where the internal phase is around 0.1% (v/v);
    *   concentrated emulsion, which can be:
    *   monodisperse emulsion where the internal phase is dispersed in the form of regular spheres occupying 74% of the total volume of the phase;

- polydisperse emulsions are the emulsions which show a great globular heterogeneity when the small droplets can slide into the free spaces of the large spherical droplets.
- highly concentrated emulsion, also called viscous semisolid emulsions, where the particles are deformed and become polyhedral;
- solid or dry emulsions;
- adsorbed, dry, pulverulent emulsions which are compressed or introduced into gelatin capsules.

5. Route of administration:
    - oral emulsions for internal use (oil/water type);
    - topical emulsions: dermatological, cosmetic emulsion including creams and ointments;
    - emulsions administered through a mucous membrane: ophthalmic, rhino pharyngeal, otic, rectal, and vaginal;
    - parenteral emulsions: intramuscular (water/oil, oil/water or double emulsions) or intravenous administration (oil/water or water/oil/water).

6. Role and pharmacological action:
    - drug emulsions for oral, transdermal, and parenteral administrations or through mucous membranes;
    - cosmetic emulsions—with emollient action, for skin care, and protection against UV radiation, makeup removal and as a component of body cleansing products, etc.

Two immiscible liquids, upon mixing by vigorous stirring, form temporarily an emulsion. The separation of one of the phases in very small droplets leads to the increase of the total surface of the system and also to the increase of the superficial free energy. The system becomes, therefore, thermodynamically unstable, and for reducing the superficial free energy and return to a thermodynamically stable system, there is a tendency to unite the droplets of the internal phase into larger droplets, producing the coalescence, and phase separation.

On the other hand, for an emulsion to be stable, it must meet the following conditions:

- the dispersed phase should have an advanced degree of dispersion;
- the difference of density between the two phase should be as small as possible;

- the dispersed phase should have a high viscosity;
- the ratio between the two phases should be 50:50 or 40:60;
- the quantity and nature of the emulsifier should be chosen properly.

The emulsifiers are surface-active substances or surfactants which are absorbed in the oil/water interface having the role of reducing the interfacial tension and facilitating the mixing of the two immiscible liquids.

When preparing an emulsion, it should be taken into account of the superficial and electrical properties, as well as the hydrophilic or lipophilic character of the emulsifier.

The classification of the emulsifiers was made according to the following criteria:

1. Mode of production:
   a. natural substances: arabic gum, lecithins, gelatin, and pectin (Malmsten, 2002; Hancock, 1984);
   b. semisynthetic substances: cellulose derivatives (methyl cellulose, carboxymethyl cellulose, and hydroxyethyl cellulose);
   c. synthetic substances: poly(vinyl pyrrolidone), poly(vinyl alcohol) (PVA), esters of fatty acids with sorbitol (SPAN), and polyoxyethylene esters of sorbitan (polysorbates or TWEEN).

2. The dissociation degree:
   a. ionic emulsifiers: anionic (sodium lauryl sulfate), cationic (quaternary ammonium salts), and amphoteric (lecithin);
   b. nonionic emulsifiers: lanolin, liquid paraffin, PEG, PVA, TWEEN, and SPAN.

3. The emulsifying capacity:
   a. primary emulsifiers are substances designed to create alone an emulsion, for example, arabic gum;
   b. secondary emulsifiers (pseudo emulsifiers or cvasi emulsifiers) that are used only in combination with primary emulsifiers.

4. Solubility:
   a. hydrophilic emulsifiers, soluble in water or polar solvents;
   b. lipophilic emulsifiers, soluble in oil phase (nonpolar).

The solubility of emulsifiers is evaluated by means of an empirical measurement unit, called hydrophilic–lipophilic balance (HLB).

HLB expresses the ratio between the hydrophilic and lipophilic character of the emulsifier. Conventionally, this ratio is quantified by numerical values ranging from 1 to 40 (Griffin, 1949; Griffin, 1954). As the HLB value is lower, the emulsifier has a more pronounced hydrophobic nature and vice versa.

An emulsifier must have an affinity for both phases, in order to accumulate at the oil/water interface. If the emulsifier has a too high affinity to one of the phases (strongly hydrophobic or strongly hydrophilic), then its molecules are no longer absorbed at interface, but they are dispersed or dissolved in phase with closed polarity.

The emulsifiers with HLB = 3–6, are soluble in nonpolar solvent and favor the formation of inverse emulsions (water in oil), for example, SPAN 60 (HLB = 4.7). The emulsifiers with HLB = 8–18, are soluble in polar solvent (water) and favor the formation of direct emulsions (oil in water) such as TWEEN 80 (HLB = 15).

Porous polymer microparticles may be obtained by the method of double water/oil/water emulsion in order to use them as macromolecular carriers for controlled/sustained delivery of hydrosoluble drugs.

In general, the multiple emulsions are prepared by a two-step process, this method being highly reliable and reproducible. For the water/oil/water double emulsions, the oil phase in the water/oil primary emulsion is represented by an organic solvent, the film forming polymer and the emulsifier. Strengthening of the particles is achieved by the diffusion of solvent and precipitation of the polymer (Mora-Huertas et al., 2010). There have been numerous studies in order to discover and develop efficient methods for encapsulation the hydrosoluble active principles. The low efficiency of encapsulation of the water-soluble substances may be improved by using the multiple emulsion techniques coupled with solvent evaporation. This technique is used for the preparation of microparticles or microcapsules with controlled delivery of hydrophilic active substances, proteins, or polypeptides (Zambaux et al., 1998).

Kwon et al. (2007) have obtained through the method of water/oil/water double emulsion porous microparticles based on poly(lactide-co-glycolide) (PLGA), using as porogenic agent and stabilizer for peptide drug the sulfobutyl ether β-cyclodextrins sodium salt (SBE-CD) and sucrose ethyl acetate (SAIB), respectively. Mass median aerodynamic diameter of the porous microparticles obtained is assessed at about 3 μm, ideal size for the local inhalation therapy. The inhalation therapy can treat patients with inflammatory diseases of the upper and lower respiratory tract. The advantages of therapy by inhalation are rapid absorption of drugs, their deposit in optimal dose directly into the affected region, and also avoidance of liver metabolism. Drug delivery systems used in inhalation therapy must encapsulate enough quantity of drug and have the ability to release drug over an extended period of time. There are also important the physical properties of particles such as their optimal size which must be within the range 1–5 μm.

The introduction of SAIB to obtain porous microparticles allowed the release of proteins (BSA) within 7 days, while the porous microparticles without SAIB showed a release profile of only 4 h. This finding suggests the possibility of applying these systems in the therapies of the respiratory tract.

Another study by Yang et al. (2009b) indicates the preparation of macroporous microparticles based on PLGA by the method of water/oil/water double emulsion, using the ammonium bicarbonate as a porogenic agent. The porosity of microparticles was controlled by simply adjusting the amount of ammonium bicarbonate, whereas their size has been achieved by reducing the homogenization speed. Thus, there has been obtained porous microparticles with the desired aerodynamic properties (medium aerodynamic diameter ≈ 4 μm) which are appropriate for the inhalation therapy. The polymer–drug systems were obtained by encapsulation of two biologically active principles (lysozyme and doxorubicin hydrochloride) into macroporous PLGA microparticles. The kinetic studies have shown that lysozyme release occurs in 6 h, while the release of doxorubicin takes place in 4 days.

A new type of dry powder aerosol consisting of PLGA porous particles with small mass density and large size was obtained by single and double emulsion methods combined with solvent evaporation method (Edwards et al., 1997). Testosterone was chosen as model drug. It is interesting that the large porous particles with a diameter of 20 μm provide an increased systemic bioavailability of testosterone and high aerosolization efficiency leading to the conclusion that this type of microparticles can be attractive for systemic inhalation therapies.

Badilli et al. (2011) were obtained a new topical delivery system based on PLGA microspheres by oil in water emulsion coupled with solvent evaporation technique, in order to provide the controlled release of clobetasol propionate which is able to offer a rapid healing of psoriatic lesions. From the in vitro release of clobetasol propionate studies the authors were observed that the emulgel formulation containing highly porous PLGA microspheres has a high efficiency in psoriasis treatment.

Also, Zhang et al. (2013a) have used hollow porous PLGA microspheres for controlled protein release and promotion of cell compatibility. The nonporous, porous, and hollow porous PLGA microparticles were prepared by a modified double emulsion method. The hollow porous microparticles present highest release rate of BSA compared to nonporous PLGA microparticles, while the porous microparticles showed an intermediate release rate.

The porous composite microparticles (PLGA porous microparticles/ PVA hydrogels) significantly promote cell adhesion, growth, and proliferation indicating that these systems have potential applications as controlled drug delivery systems or as tissue engineering scaffolds. For asthma treatment, Oh et al. (2011) were prepared PLGA porous microparticles by water/oil/water double emulsion method using ammonium bicarbonate as porogenic agent. In function of porogen concentration (1.5 and 3%), the diameters of particles were 8.2 and 9.2 μm, respectively. Also, the porous structure and release profile can be controlled by the porogen concentration.

By a combination of emulsion polymerization, solvent evaporation, and freeze-drying methods, Zhu et al. (2017a) were prepared PLGA large porous microparticles loaded with oridonin for in situ treatment of primary non-small cell lung cancer. Hollow porous PLGA microparticles can be fabricated as injectable scaffold systems in regenerative medicine by double emulsion method (Qutachi et al., 2014). These microparticles fuse together at 37°C into solid scaffold due to the ethanol-sodium hydroxide treatment leading to the formation of a new material that supports cell attachment and proliferation. The mesoporous PLGA microparticles were prepared by oil/water emulsion-solvent evaporation method and the in vitro release profiles of eprinomectin from these injectable microparticles were performed (Shang et al., 2015). The eprinomectin–microparticles system was administered to Japanese white rabbits by subcutaneous injection finding that the drug had a fast-acting and persistent effect. Accordingly, the mesoporous injectable PLGA microspheres can be used as the sustained release system of veterinary drugs. The growth factor loaded PLGA microparticles have received a great importance because they can induce angiogenesis when are directly injected to the site of damaged tissue. The heparin immobilized porous PLGA microparticles were prepared in three steps as follows:

1. PLGA microparticles with an open porous structure were obtained by oil/water single emulsion and solvent evaporation methods in presence of Pluronic F-127 as porogenic agent;
2. functionalization of porous PLGA microparticles with primary amine groups;
3. immobilization of heparin via covalent conjugation.

The basic fibroblast growth factor was loaded into heparin-conjugated and unconjugated microparticles using a solution dipping method. The in vitro and in vivo studies showed a sustained release profile of basic fibroblast growth factor and the ability to induce angiogenesis in an animal model (Chung et al., 2006).

6.2.3   PRECIPITATION AND DISPERSION POLYMERIZATION

Precipitation polymerization is a modified version of solution polymerization, which means that the polymerization takes place in a medium in which the formed polymer precipitates.

In the precipitation polymerization, the monomer and the initiator are soluble in the aqueous or nonaqueous dispersion medium (the continuous phase) and the process is initiated in the liquid phase. The mechanism of the particle formation includes the following steps:

1. The nucleation stage; in this stage, the transformation of the monomers in oligomeric radicals occurs followed by their cross-linking or aggregation leading to the formation of nuclei (Li and Stover, 2000).
2. The growth stage: The formation of spherical particles takes place either by the adsorption of the oligomeric radicals present in the reaction medium that interact with the residual unreacted vinyl groups situated on the surface of the nuclei or by capturing other polymeric nuclei (Downey et al., 1999). When using a porogen, it is initially adsorbed by the nuclei, and then extracted from the polymer matrix in order to form the pores. During the

polymerization, the particles can precipitate from the continuous phase in two ways:

a. Enthalpic precipitation when the interactions between the polymer and the solvent are unfavorable (e.g., bulk polymerization of AN and vinyl chloride) (Li et al., 2013). In this case, the oligomeric radicals continue to grow until they reach the solubility limit, separate from the solution, and are deposited on the polymer or on the surface of the particles present in the system. These nuclei and particles continue to grow either by capturing other nuclei or oligomeric radicals or by incorporating another monomer molecule.

b. Entropic precipitation occurs when the oligomeric radicals react with vinyl groups or with radicals situated on the surface of the particles (Chaitidou et al., 2008).

The precipitation polymerization has the great advantage that through it one can obtain monodispersed microparticles without the addition of stabilizer or surfactant (Xia et al., 2017; Pardeshi and Singh, 2016; Yoshimatsu et al., 2007). In good solvents, the polymerization leads to micro and macroscopic gels, while in bad solvents particles of micrometric size are obtained (Yan et al., 2009; Wang et al., 2007; Perrier-Cornet et al., 2008; Miura et al., 2016).

Based on improvements to the precipitation polymerization technique, new variants of this method have been developed in recent years:

1. Distillation–precipitation polymerization (Bai et al., 2004; Liu et al., 2007; Bai et al., 2006a, 2006b);

2. Photo-initiated precipitation polymerization (Joso et al., 2007; Lime and Irgum, 2007; Lime and Irgum, 2009);

3. Controlled/"living" radical precipitation polymerization is a combination between precipitation polymerization and living radical polymerization represented by atom-transfer radical polymerization (ATRP), reversible addition-fragmentation chain transfer polymerization (RAFT), and iniferter technique (Zhang, 2013b; Song et al., 2015).

The polymerization process is called dispersion polymerization if the particle coalescence is prevented by using the stabilizers or surfactants

(Arshady, 1992). A classic example of this type of polymerization is the polymerization of the vinyl acetate in water when water-soluble polymers are used as stabilizers (Elbert, 2011). The precipitation/dispersion polymerization has been used to obtain smart microparticles, sensitive to various stimuli which, due to their properties, can be used in various fields, particularly in pharmaceutical or biomedical fields. Through the precipitation polymerization technique of Kawaguchi, anionic microgels were obtained starting from the methacrylic acid and bisacrylamide. These acrylic microgels were subsequently loaded with doxorubicin hydrochloride (Kiser et al., 2000).

Also, glucose-sensitive nanoparticles were obtained by grafting the aminophenyl boronic acid on amphoteric poly(N-isopropyl acrylamide) microgels (Hoare and Pelton, 2008). The microgels exhibit glucose-dependent swelling responses according to the various parameters such as temperature, pH, and ionic strength.

The studies performed on these types of microgels showed that:

• Microgels may collapse or swell when glucose concentration increases. The glucose concentration can be adjusted in function of the pH and grafting efficiency of the acid groups, while the ionic groups located on the surface of the microgels can be modified from cationic to anionic to a certain critical value of the glucose concentration.
• The microgels have a high loading capacity of insulin, especially due to the electrostatic interactions between the microgel and the hormone.

The molecular imprinting technique is a method used for the preparation of the porous structures with specific molecular recognition properties for a particular compound. In general, molecular imprinting involves the following steps:

1. preorganization of the templates and monomers to achieve the complementary interactions between components;
2. polymerization of the imprint molecule (template)—functional monomer complex;
3. removing the template from the polymer by extraction (Ekberg and Mosbach, 1989).

The molecular imprinting can be classified as follows:

1. covalent molecular imprinting approach, where the template interacts with the suitable monomers by covalent bonds (Wulff, 1995);
2. noncovalent molecular imprinting approach, where the monomer–template complexes are formed by means of interactions such as hydrogen bonds, ionic, and π–π interactions, Van der Waals forces (Mosbach, 1994);
3. the semicovalent molecular imprinting approach, where, initially, the template interacts with the functional monomer through a covalent bond but the rebinding is achieved by noncovalent interactions (Whitcombe et al., 1995);
4. molecular imprinting mediated through metallic cations.

In the molecular imprinting, the macroporous structures can be obtained by phase separation that occurs during precipitation polymerization. The factors affecting the porous structure and the size of the microparticles are the balance between the solubility parameters of the formed polymer and the polarity of the porogenic agent (Wang et al., 2003), the nature of the cross-linker and the template (Yoshimatsu et al., 2007; Tamayo et al., 2005), the stirring speed, and the polymerization temperature (Yoshimatsu et al., 2009).

Although this method has many advantages, it also presents some disadvantages such as:

• The use of a large volume of porogenic agent. This disadvantage can be overcome if the initiating step of the polymerization process takes place by means of UV radiation and the porogen is a mixture of mineral oil and toluene (Jin et al., 2008);
• A long polymerization time that can be reduced by the use of distillation precipitation polymerization (Yang et al., 2009a; Chen et al., 2011a, 2011b).

Thus, microparticles based on methacrylic acid and trimethylolpropane trimetacrylate (TRIM) were obtained through the precipitation polymerization techniques, using the nalidixic acid as a template. Nalidixic acid from the quinolone group is a synthetic antibiotic with bactericidal action. The nalidixic acid is indicated in:

- urinary infections with one or more microbes (monomicrobial or polymicrobial, high or low, complicated or uncomplicated) caused by susceptible germs, especially Gram-negative bacteria;
- prolonged prophylaxis of recurrent urinary infections;
- prevention of urinary tract infections in transurethral surgery or transrectal biopsy of the prostate.

The most performing nanoparticles had the following characteristics: molar ratio of nalidixic acid:MAA:TRIM = 1:12:12, particle diameter of 94 nm, low polydispersity index, high specific surface area, mean pore diameter of 12 nm, and a selectivity factor of 10.4. Particle morphology has also been found to have played an important role in drug release kinetics and in the performance of the imprinted nanoparticles (Abouzarzadeh et al., 2012).

Ye et al. (2000) have used the precipitation polymerization technique of methacrylic acid and trimethylolpropane trimethacrylate, in the presence of suitable quantities of theophylline, caffeine, and 17β-estradiol as molecule template for the preparation of molecular imprinting porous microparticles.

The extraction of bioactive compounds derived from plants (alkaloids, flavonoids, and phenols) is a priority in the pharmaceutical industry due to the fact that these compounds present a series of properties with a major impact on the human health.

Imprinted microparticles are an alternative that has attracted attention in recent years. Thus, a number of alkaloids, terpenoids, and polyphenols were extracted from the plants using imprinted microparticles (Yin et al., 2011; Chen et al., 2011a, 2011b).

Vinca alkaloids are alkaloids derived from the plant *Catharanthus roseus*. Currently, *Vinca* alkaloids can be made synthetically and are used in chemotherapy to treat various cancers or as immunosuppressive drugs.

Among these, vinblastine is part of the treatment plan for the Hodgkin's disease, lung cancer, bladder, brain and testicles cancers, or various melanomas. In order to increase the efficiency of treatment of these diseases, nanoparticles based on MAA and TRIM were obtained through the imprinting technique, in which the templates was the vinblastine. Following the studies, it was found that:

- The drug can be resorbed by noncovalent interactions such as hydrogen bonds, ionic, and hydrophobic interactions.

- The release of vinblastine is performed in a sustained manner, according to the first order kinetic model (Zhu et al., 2017b).

Recently, the microporous molecularly imprinted microspheres based on 4-vinylpyridine and monoethylene glycol dimethacrylate with high selectivity to kaempferol are prepared by precipitation polymerization (Xia et al., 2017). Based on the adsorption studies, it can be concluded that the microparticles exhibit a higher selectivity and adsorption capacity to kaempferol. Kaempferol is a natural flavonoid found in many plants such as *Aloe vera, Coccinia grandis, Cuscuta chinensis, Rosmarinus officinalis, Glycine max*, etc. It is very well known that kaempferol exhibits many pharmacological properties such as antioxidant, anti-inflammatory, anticancer (breast and ovarian cancer, Leukemia, bladder, prostate and colorectal cancer, gastric, pancreatic, and lung cancer), antidiabetic, antimicrobian, cardioprotective, neuroprotective, antiosteoporotic, anxiolytic, analgesic, and antiallergic (Calderon-Montano et al., 2011).

6.2.4 SEED SWELLING POLYMERIZATION

Seed swelling polymerization (SSP) is a method used to obtain monodisperse porous polymeric microparticles. The SSP technique was subsequently modified in order to reduce the preparation time of the monodisperse porous microparticles. Thus, SSP can be classified as follows:

1. One-step seed swelling polymerization method (Ogino et al., 1995)

    Generally, in the reaction system we encounter the seed particles, the monomer, the disperse phase, the initiator, the cross-linker, and the stabilizer. The seed is usually obtained by suspension polymerization or by emulsion/dispersion polymerization. The seed particles swell directly into the mixture of the monomer and the porogen, and the pores are formed when the phase separation between polymer particles and seed particles occurs. The pore size can be controlled by the ratio between seed swelling and organic phase (monomer and porogenic agent) (Chai et al., 2000).

2. Two-step seed swelling polymerization method
   By this method, invented by Ugelstad et al. (1979) the monodisperse porous particles are obtained in two stages:

   a. swelling of the seed particles with an activating solvent which
      is a water-insoluble compound with low molecular weight
      such as 1-chlorododecane, hexadecane, dioctyl adipate, and
      octadecanol (Sugimoto, 2001).

   Two methods were used to incorporate the activating solvent into
   particles: (1) using a liquid compound that is incorporated into
   the particles or (2) using an oligomer compound that is obtained
   by polymerization inside the particle (Ugelstad et al., 1993). The
   obtaining of the oligomers by polymerization inside the particles
   can be realized as follows:

   a. The seed polymer particles are swollen in the presence
      of a monomer and of a chain transfer agent (carbon tetra-
      bromide, carbon tetrachloride, mercaptans, and toluene).
   b. The seed polymer particles are swollen in the presence of
      monomers and of an oil-soluble initiator (Ellingsen et al.,
      1990).

   b. The activated particles are subsequently swelled into a system
      containing a porogenic agent, an initiator, and the monomer. After
      a period of time adequate for swelling, the temperature of the reac-
      tion mixture rises, allowing the polymerization to take place.
   Through the two-step SSP a series of macroporous microparticles
   with different applications have been prepared:

   i. Polyacrylic macroporous particles used for antibody
      binding in order to perform the flow cytometric immuno-
      assays for carcinoembryonic antigen (CEA) and alpha-
      fetoprotein (AFP) (Ugelstad et al., 1993).
   ii. Porous microparticles based on styrene and DVB char-
       acterized by high specific surface area and subsequently
       modified with diazo resin in order to be used as stationary
       phase in chromatography (Yu et al., 2017).
   iii. Macroporous microparticles based on polystyrene, meth-
        acrylic acid, and DVB (Tuncel et al., 2002).
   iv. Microparticles imprinted with cyromazine (Zhang et
       al., 2013c) and phenobarbital (Hua et al., 2011), is

characterized by increased affinities and specificities for these two types of compounds. Cyromazine is an insecticide, whereas phenobarbital is a drug that is administered alone or in combination with other antiepileptics for the treatment of epilepsy, as hypnotic or sedative, and as hepatic choleretic and enzyme inducer.

3. Multistage seed swelling polymerization method
   To produce the uniform porous particles, a multistage seeded polymerization method was proposed (Ching Wang et al., 1994). In their method, the dibutyl phthalate was used as a low molecular weight compound and the seeded particles based on polystyrene were obtained by emulsifier free-emulsion polymerization.

4. Dynamic seed swelling polymerization method
   Okubo and Nakagawa (1992) have proposed a new type of monomer swelling method that has only a step and does not require a swelling agent. The porous particles were prepared as follows:

   a. the seed particles based on polystyrene (d = 1.9 μm) were obtained by dispersion polymerization;
   b. monodisperse polystyrene seed particles are dispersed in ethanol/water medium together with a mixture of monomer, hydrophobic initiator, and stabilizer. Then, the water was introduced continuously and slowly in the system to decrease the solubilities of the monomer and initiator and allow to be adsorbed into the seed particle.

6.2.5 SPRAY DRYING TECHNIQUE

"Spray drying" is a well-known technique for preparation of solid particles by atomizing/spraying a drop suspension, followed by a drying process (Masters, 1991). The spray drying technique allows the preparation of the microparticles with relatively small size (typically tens of nanometers to several micrometers) and very large surface areas (Boissiere et al., 2011; Friesen et al., 2008).

The "spray drying" process has 4 steps:

1. spraying the base solution in the form of drops (atomization);
2. contact between air and spray droplets;

3. drying the sprayed particles;
4. separation of the dried microparticles.

The polymer-drug solution and the compressed air are passed through two-fluid nozzles or ultrasonic nebulizers, in function of the required droplet size. Droplets and dry air passing through the drying chamber, heated at a temperature higher than the vaporization temperature of the solvent, thereby causing the formation of microparticles. The time required for drying the droplets depends on the droplet residence time in the gas phase as well as of the geometry of the chamber, the airflow rate, temperature, and pressure. The final step is accomplished by means of a separator. The air and microparticles formed in the drying chamber pass through the separator which can be a cyclone, a filter bag, or an electric field precipitator.

With this method, porous particles based on hyaluronic acid were successfully prepared by Iskandar et al. (2009) using polystyrene particles as template molecules. A suspension consisting of the aqueous solution of hyaluronic acid and polystyrene particles was spray-dried by means of a two-fluid nozzle system. The evaporation of water during spray drying process was carried out using air heated at a temperature of 120°C. The solid particles obtained were collected and washed with an organic solvent to remove the template particles. The porosity and the pore size of the particles were controlled by changing various parameters such as the mass ratio between hyaluronic acid and template, the concentration of the aqueous solution of hyaluronic acid, and the size of the polystyrene particles. The results of this study have shown that the porous particles with controlled aerodynamic properties have been obtained and these particles can be used in different regions of the respiratory system (tracheobronchial or pulmonary ways).

By spray drying, Straub et al. (2005) were prepared porous microparticles based on PLGA and AI-700 (perflubutane polymer microspheres) using ammonium bicarbonate as a porogenic agent. AI-700 is an imaging agent administered intravenously for the detection of coronary artery disease. The SEM micrographs have demonstrated that the resulting microparticles have a spherical shape and a porous internal structure. The average microparticle diameter of 2.3 μm makes them fit into the ideal size for intravenous administration, thus allowing the use of these systems in imaging evaluations.

Studies conducted by Lei et al. (2016) have explored the use of the microfluidic jet spray dryer technique (MFJSD) for obtaining the pH-sensitive microparticles based on hydroxypropyl methylcellulose phthalate (HPMCP) as drug delivery systems.

By changing the conditions of microparticles preparation such as the nature of the solvent and the drying temperature, it is possible to obtain particles with shapes and sizes that can be controllable. For example, the use of a highly volatile solvent and a high drying temperature leads to the formation of large, irregularly shaped microparticles with a denser matrix. Hydrocortisone and lysine were selected as model drugs. The results of this study have shown that using the MFJSD technique has led to obtain HPMCP-drug systems with the desired size, different morphologies, and structures.

6.2.6 OTHER TECHNIQUES

In addition to the abovementioned technique, there are other methods that have been reported in recent years such as sinter method (Cai et al., 2013), membrane/microchannel emulsification method (Vladisavljevic et al., 2007), electrospraying (Bohr et. al, 2012), etc.

The melt dispersion method was used by Baimark (2009) to produce the porous microspheres of methoxy poly(ethylene glycol)-b-poly(ε caprolactone-co-D,L-lactide). The advantage of this method is the absence of any organic solvent and surfactants. In this case, the porous structure is obtained either by the sublimation of the water in the microparticles or by using of PEG as porogen.

Another technique that can be used to produce the porous particles is microfluidic method (Wang et al., 2017). The microfluidic method that is a version of microchannel emulsification method (Gokmen and Du Prez, 2012) offers some important advantages (Lewis et al., 2005; Jeong et al., 2005):

- mass and heat transfer is facile and more efficient compared to the bulk system;
- allows the control of the stoichiometry of chemical reactions;
- produces the porous particles with high surface area;
- low reagent consumption.

The fabrication of the porous particles by microfluidic technique can be realized using the following methods: polymerization with/without porogens, sacrificial template, flow reaction, self-assembly, and flow lithography (Wang et al. 2017).

6.3 CONCLUSIONS AND FUTURE PERSPECTIVES

The microparticulate materials that possess external/internal pores or a combination of both have received a great attention in last decades due to their potential applications in healthcare as carriers for biomacromolecules and other substances, bone tissue engineering and regeneration, pulmonary drug delivery, cell therapy, and dental applications.

Porous particles can be obtained by different manufacturing processes like suspension, dispersion, precipitation, multistage, membrane/microchannel emulsification, and microfluidic polymerization.

The porous particles possess the excellent properties like greater surface area, lower mass density, superior cell attachment and cell proliferation, higher drug absorption, and appropriate drug release kinetics for desired applications.

Also, the use of particulate systems in biomedical applications confers a series of facilities: permits the election and formulation of different combinations polymer-active principle, offers a gradual release of the bioactive principle in such manner as to provide the desired results, and allows a diversification in the administration routes such as oral, transdermal, ophthalmic, nasal, anal, and vaginal administrations.

The discovery of new possibilities to control the pore characteristics like pore size, pore structure, and pore size distribution can provide various opportunities to realize the materials with desired properties which can be useful in biomedical applications. Also, the fabrication of ordered porous particles with multicompartmentalized and hierarchical porous structure is a challenge for the future research and an open way for the production of the porous materials that are difficult to obtain by classical methods.

ACKNOWLEDGMENTS

The authors thank for the financial support from the project PNII-PT-PCCA-2013-4-1570, financed by the Romanian National Authority for Scientific Research, CNCS-UEFISCDI.

KEYWORDS

- porous particles
- suspension polymerization
- emulsion polymerization
- precipitation polymerization
- seed swelling polymerization
- spray drying
- biomedical applications

REFERENCES

Abouzarzadeh, A., et al. Synthesis and Evaluation of Uniformly Sized Nalidixic Acid-imprinted Nanophres Based on Precipitation Polymerization Method for Analytical and Biomedical Applications. *J. Mol. Recognit.* **2012,** *25* (7), 404–413.

Ahmed, S. M. Effects of Agitation and the Nature of Protective Colloid on Particle Size during Suspension Polymerization. *J. Disper. Sci. Technol.* **1984,** *5* (3–4), 421–432.

Anton, N.; Jakhmola, A.; Vandamme, T. F. Trojan Microparticles for Drug Delivery. *Pharmaceutics* **2012,** *4* (1), 1–25.

Arshady, R. Suspension, Emulsion and Dispersion Polymerization: A Methodological Survey. *Colloid Polym. Sci.* **1992,** *270* (8), 717–732.

Asefa, T., et al. Controlling Adsorption and Release of Drug and Small Molecules by Organic-functionalization of Mesoporous Materials. *Adsorption* **2009,** *15* (3), 287–299.

Bai, F.; Yang, X.; Huang, W. Synthesis of Narrow or Monodisperse Poly(Divinylbenzene) Microspheres by Distillation-precipitation Polymerization. *Macromolecules* **2004,** *37* (26), 9746–9752.

Bai, F., et al. Preparation of Narrow-dispersion or Monodisperse Polymer Microspheres with Active Hydroxyl Group by Distillation-precipitation Polymerization. *Polym. Int.* **2006a,** *55* (3), 319–325.

Bai, F.; Yang, X.; Huang, W. Preparation of Narrow or Monodisperse Poly(Ethyleneglycol Dimethacrylate) Microspheres by Distillation-precipitation Polymerization. *Eur. Polym. J.* **2006b,** *42* (9), 2088–2097.

Badilli, U.; Sen, T.; Tarimci, N. Microparticulate Based Topical Delivery System of Clobetasol Propionate. *AAPS Pharm. Sci. Tech.* **2011,** *12* (3), 949–957.

Baimark, Y. Porous Microspheres of Methoxy Poly(Ethylene Glycol)-b-poly(-caprolactone-co-D,L-lactide) Prepared by a Melt Dispersion Method. *Polymer* **2009,** *50* (20), 4761–4767.

Bohr, A., et al. Particle Formation and Characteristics of Celecoxib-loaded Poly(Lactic-co-glycolic Acid) Microparticles Prepared in Different Solvents Using Electrospraying. *Polymer* **2012,** *53* (15), 3220–3229.

Boissiere, C., et al. Aerosol Route to Functional Nanostructured Inorganic and Hybrid Porous Materials. *Adv. Mater.* **2011,** *23* (5), 599–623.

Borwankar, R. P.; Chung, S. I.; Wasan, D. T. Drop Size in Turbulent Liquid-liquid Dispersions Containing Polymeric Suspension Stabilizers. I. The Breakage Mechanism. *J. Appl. Polym. Sci.* **1986,** *32* (7), 5749–5762.

Boyere, C., et al. Synthesis of Microsphere-loaded Porous Polymers by Combining Emulsion and Dispersion Polymerizations in Supercritical Carbon Dioxide. *Chem. Commun.* **2012,** *48* (67), 8356–8358.

Brena, B. M.; Batista-Viera, F. Immobilization of Enzymes. A Literature Survey. In *Methods in Biotechnology: Immobilization of Enzymes and Cells*; Humana Press Inc.: New York, NY, 2006; Vol. 22, pp 15–30.

Brodelius, P.; Mosbach, K. Immobilization Techniques for Cells/organelles. Overview. In *Methods in Enzymology*; Mosbach, K., Ed.; Academic Press: London, 1987; Vol. 135, pp 173–175.

Brooks, B. Suspension Polymerization Processes. *Chem. Eng. Technol.* **2010,** *33* (11), 1737–1744.

Buchholz, K.; Klein, J. Characterization of Immobilized Biocatalysts. In *Methods in Enzymology*; Mosbach, K., Ed.; Academic Press: London, 1987; Vol. 135, pp 3–30.

Buszello, K.; Muller, B. W. Emulsions as Drug Delivery Systems. In *Pharmaceutical Emulsions and Suspensions*; Nielloud, F., Marti-Mestres, G. R., Eds.; Marcel Dekker: New York, 2000; pp 191–228.

Cai, Y., et al. Porous Microsphere and its Applications. *Int. J. Nanomed.* **2013,** *8,* 1111–1120.

Calderon-Montano, J. M., et al. A Review on the Dietary Flavonoid Kaempferol. *Mini Rev. Med. Chem.* **2011,** *11* (4), 298–344.

Chai, Z., et al. Pore Size in One-step Swelling and Polymerization of Monodisperse Polymer Beads. *J. Polym. Sci. A Polym. Chem.* **2000,** *38* (18), 3270–3277.

Chaitidou, S., et al. Precipitation Polymerization for the Synthesis of Nanostructured Particles. *Mat. Sci. Eng. B* **2008,** *152* (1–3), 55–59.

Chen, F. F.; Wang, G. Y.; Shi, Y. P. Molecularly Imprinted Polymer Microspheres for Solid-phase Extraction of Protocatechuic Acid in *Rhizoma homalomenae. J. Sep. Sci.* **2011a,** *34* (19), 2602–2610.

Chen, L.; Xu, S.; Li, J. Recent Advances in Molecular Imprinting Technology: Current Status, Challenges and Highlighted Applications. *Chem. Soc. Rev.* **2011b,** *40* (5), 2922–2942.

Chen, Y. L.; Schweizer, K. S.; Fuchs, M. Phase Separation in Suspensions of Colloids, Polymers and Nanoparticles: Role of Solvent Quality, Physical Mesh, and Nonlocal Entropic Repulsion. *J. Chem. Phys.* **2003,** *118* (8), 3880–3890.

Ching Wang, Q.; Svec, F.; Frechet, J. M. J. Fine Control of the Porous Structure and Chromatographic Properties of Monodisperse Macroporous Poly(Styrene-co-divinylbenzene) Beads Prepared Using Polymer Porogens. *J. Polym. Sci. A Polym. Chem.* **1994,** *32* (13), 2577–2588.

Cho, Y. S.; Shin, C. H.; Han, S. Dispersion Polymerization of Polystyrene Particles Using Alcohol as Reaction Medium. *Nanoscale Res. Lett.* **2016,** *11* (1), 46.

Chung, H. J., et al. Heparin Immobilized Porous PLGA Microspheres for Angiogenic Growth Factor Delivery. *Pharm. Res.* **2006,** *23* (8), 1835–1841.

Cigu, T. A., et al. Adsorption and Release Studies of New Cephalosporin from Chitosan-g-poly(Glycidyl Methacrylate) Microparticles. *Eur. Polym. J.* **2016,** *82,* 132–152.

Conceicao, B. M., et al. A Study of the Initiator Concentration's Effect on Styrene-divinyl-benzene Polymerization with Iron Particles. *Polimeros* **2011,** *21* (5), 409–415.

Cui, Y. N., et al. Advances in Porous Polymer Particles. *Polym. Mat. Sci. Eng.* **2008,** *24* (8), 1–8.

Ding, L., et al. Synthesis of Macroporous Polymer Carrier and Immobilization of Papain. *Iran. Polym. J.* **2003,** *12* (6), 491–495.

Downey, J. S., et al. Growth Mechanism of Poly(Divinylbenzene) Microspheres in Precipitation Polymerization. *Macromolecules* **1999,** *32* (9), 2838–2844.

Dragan, S.; Nichifor, M.; Petrariu, I. Copolimeri Stiren-divinilbenzen Permanent Porosi Obtinuti in Prezenta Mediilor Inerte de Umflare Sau a Compusilor Macromoleculari. *Mat. Plast.* **1976,** *13* (3), 155–159.

Dusek, K. Phase Separation during the Formation of Three-dimensional Polymers. *J. Polym. Sci. Polym. Symp.* **1967,** *16* (3), 1289–1299.

Edwards, D. A., et al. Large Porous Particles for Pulmonary Drug Delivery. *Science* **1997,** *276* (5320), 1868–1872.

Ekberg, B.; Mosbach, K. Molecular Imprinting: A Technique for Producing Specific Separation Materials. *Trends Biotechnol.* **1989,** *7* (4), 92–96.

Elbert, D. L. Liquid-liquid Two Phase Systems for the Production of Porous Hydrogels and Hydrogel Microspheres for Biomedical Applications: A Tutorial Review. *Acta Biomater.* **2011,** *7* (1), 31–56.

Ellingsen, T., et al. Monosized Stationary Phases for Chromatography. *J. Chromatogr.* **1990,** *535,* 147–161.

Erbay, E., et al. Polystyrene Suspension Polymerization: The Effect of Polymerization Parameters on Particle Size and Distribution. *Polym. Plast. Technol. Eng.* **1992,** *31* (7–8), 589–605.

Esfandyari-Manesh, M., et al. The Control of Morphological and Size Properties of Carbamazepine-imprinted Microspheres and Nanospheres Under Different Synthesis Conditions. *J. Mater. Res.* **2013,** *28* (19), 2677–2686.

Fang, D.; Pan, Q.; Rempel, G. L. Preparation and Characterization of 2-hydroxyethyl Methacrylate-based Porous Copolymeric Particles. *J. Appl. Polym. Sci.* **2007,** *105* (5), 3138–3145.

Friesen, D. T., et al. Hydroxypropyl Methylcellulose Acetate Succinate-based Spray-dried Dispersion: An Overview. *Mol. Pharm.* **2008,** *5* (6), 1003–1019.

Galia, M.; Svec, F.; Frechet, J. M. J. Monodisperse Polymer Beads as Packing Material for High-performance Liquid Chromatography: Effect of Divinylbenzene Content on the Porous and Chromatographic Properties of Poly(Styrene-co-divinylbenzene) Beads Prepared in the Presence of Linear Polystyrene as a Porogen. *J. Polym. Sci. A Polym. Chem. Ed.* **1994,** *32* (11), 2169–2175.

Gao, Y., et al. Magnetic-responsive Microparticles with Customized Porosity for Drug Delivery. *RSC Adv.* **2016,** *6* (91), 88157–88167.

Garti, N.; Benichou, A. Double Emulsions for Controlled Release Applications: Progress and Trends. In *Encyclopedic Handbook of Emulsion Technology*; Sjoblom, J., Ed.; Marcel Dekker: New York, 2001; pp 377–407.

Gokmen, T. M.; Du Prez, F. E. Porous Polymer Particles—A Comprehensive Guide to Synthesis, Characterization, Functionalization and Applications. *Prog. Polym. Sci.* **2012,** *37* (3), 365–405.

Griffin, W. C. Classification of Surface Active Agents by HLB. *J. Cosmetic Chem.* **1949**, *1* (5), 311–326.

Griffin, W. C. Calculation of HLB Values of Non-ionic Surfactants. *J. Cosmetic Chem.* **1954**, *5* (4), 249–256.

Hancock, R. I. *Surfactants*; Tadros T. Th., Ed.; Academic Press: London, 1984; pp 287–321.

Hernan Perez de la Ossa, D., et al. Poly-ε-caprolactone Microspheres as a Drug Delivery System for Cannabinoid Administration: Development, Characterization and in Vitro Evaluation of Their Antitumoral Efficacy. *J. Control. Release* **2012**, *161* (3), 927–932.

Hoare, T.; Pelton, R. Charge-switching Amphoteric Glucose-responsive Microgels with Physiological Swelling Activity. *Biomacromolecules* **2008**, *9* (2), 733–740.

Horak, D., et al. Porous polyHEMA Beads Prepared by Suspension Polymerization in Aqueous Medium. *J. Appl. Polym. Sci.* **1993**, *49* (11), 2041–2050.

Hua, K., et al. Surface Hydrophilic Modification with a Sugar Moiety for a Uniform-sized Polymer Molecularly Imprinted for Phenobarbital in Serum. *Acta Biomater.* **2011**, *7* (8), 3086–3093.

Huichao, W., et al. The Application of Biomedical Polymer Material Hydroxyl Propyl Methyl Cellulose (HPMC) in Pharmaceutical Preparations. *J. Chem. Pharmaceut. Res.* **2014**, *6* (5), 155–160.

Ishizaki, K.; Komarneni, S.; Nanko, M. *Porous Materials. Process Technology and Applications*; Springer-Science + Business Media B. V: Netherlands, 1998.

Iskandar, F., et al. Production of Morphology-controllable Porous Hyaluronic Acid Particles Using a Spray-drying Method. *Acta Biomater.* **2009**, *5* (4), 1027–1034.

Jacobelli, H.; Bartholin, M.; Guyot, A. Styrene Divinyl Benzene Copolymers. I. Texture of Macroporous Copolymers with Ethyl-2-hexanoic Acid in Diluent. *J. Appl. Polym. Sci.* **1979**, *23* (3), 927–939.

Jangde, R., et al. *Monolithic Floating Tablets of Nimesulide*; The Pharmaceutical Magazine Institute of Pharmacy, Pt. Ravishankar Shukla University: Raipur, 2008; Vol. 1, pp 1–3.

Jeong, W. J., et al. Continuous Fabrication of Biocatalyst Immobilized Microparticles Using Photopolymerization and Immiscible Liquids in Microfluidic Systems. *Langmuir* **2005**, *21* (9), 3738–3741.

Jin, Y., et al. Narrowly Dispersed Molecularly Imprinted Microspheres Prepared by a Modified Preparation Polymerization Method. *Anal. Chim. Acta.* **2008**, *612*, 105–113.

Johnson, G. R. Effects of Agitation during VCM Suspension Polymerization. *J. Vinyl Technol.* **1980**, *2* (3), 138–140.

Joshi, R. V., et al. Dual pH and Temperature-responsive Microparticles for Protein Delivery to Ischemic Tissues. *Acta Biomater.* **2013**, *9* (5), 6526–6534.

Joso, R., et al. Ambient Temperature Synthesis of Well-defined Microspheres via Precipitation Polymerization Initiated by UV-irradiation. *J. Polym. Sci. Part A Polym. Chem.* **2007**, *45* (15), 3482–3487.

Kaneko, K. Determination of Pore Size and Pore Size Distribution.1. Adsorbents and Catalysts. *J. Membr. Sci.* **1994**, *96* (1–2), 59–89.

Kawaguchi, S.; Ito, K. Dispersion Polymerization. *Adv. Polym. Sci.* **2005**, *175*, 299–328.

Kiser, P. F.; Wilson, G.; Needham, D. Lipid-coated Microgels for the Triggered Release of Doxorubicin. *J. Control. Release* **2000**, *68* (1), 9–22.

Konishi, Y.; Okubo, M.; Minami, H. Phase Separation in the Formation of Hollow Particles by Suspension Polymerization for Divinylbenzene/toluene Droplets Dissolving Polystyrene. *Colloid Polym. Sci.* **2003**, *281* (2), 123–129.

Kwon, M. J., et al. Long Acting Porous Microparticles for Pulmonary Protein Delivery. *Int. J. Pharm.* **2007**, *333* (1–2), 5–9.

Langner, F.; Moritz, H. U.; Reichert, K. H. Reactor Scale-up for Polymerization in Suspension. *Chem. Eng. Sci.* **1980**, *35* (1–2), 519–525.

Leal-Calderon, F.; Schmitt, V.; Bibette, J. *Emulsion Science. Basic Principles*; Springer Science Business Media: Berlin, Germany, 2007.

Lei, H., et al. Aerosol-assisted Fast Formulating Uniform Pharmaceutical Polymer Microparticles with Variable Properties Towards pH-sensitive Controlled Drug Release. *Polymers* **2016**, *8* (5), 195–210.

Lewis, P. C., et al. Continuous Synthesis of Copolymer Particles in Microfluidic Reactors. *Macromolecules* **2005**, *38* (10), 4536–4538.

Li, G. L.; Mohwald, H.; Shchukin, D. G. Precipitation Polymerization for Fabrication of Complex Core-shell Hybrid Particles and Hollow Structures. *Chem. Soc. Rev.* **2013**, *42* (8), 3628–3646.

Li, J.; Zhang, Y. Porous Polymer Films with Size-tunable Surface Pores. *Chem. Mater.* **2007**, *19* (10), 2581–2584.

Li, L., et al. Synthesis and Characterization of Suspension Polymerized Styrene-divinylbenzene Porous Microspheres Using as Slow-release Active Carrier. *Chinese J. Chem. Eng.* **2006**, *14* (4), 471–477.

Li, W. H.; Stover, H. D. H. Monodisperse Cross-linked Core-shell Polymer Microspheres by Precipitation Polymerization. *Macromolecules* **2000**, *33* (12), 4354–4360.

Lime, F.; Irgum, K. Monodisperse Polymeric Particles by Photoinitiated Precipitation Polymerization. *Macromolecules* **2007**, *40* (6), 1962–1968.

Lime, F.; Irgum, K. Preparation of Divinylbenzene and Divinylbenzene-co-glycidyl Metacrilate Particle by Photoinitiated Precipitation Polymerization in Different Solvent Mixtures. *Macromolecules* **2009**, *42* (13), 4436–4442.

Liu, G.; Yang, X.; Wang, Y. Preparation of Monodisperse Hydrophilic Polymer Microspheres with N, N'-methylenediacrylamide as Crosslinker by Distillation Precipitation Polymerization. *Polym. Int.* **2007**, *56* (7), 905–913.

Lofgreen, J. E.; Ozin, G. A. Controlling Morphology and Porosity to Improve Performance of Molecularly Imprinted Sol-gel Silica. *Chem. Soc. Rev.* **2014**, *43* (3), 911–933.

Lungan, M. A., et al. Complex Microparticulate Systems Based on Glycidyl Methacrylate and Xanthan. *Carbohydr. Polym.* **2014**, *104*, 213–222.

Lungan, M. A., et al. Surface Characterization and Drug Release from Porous Microparticles Based on Methacrylic Monomers and Xanthan. *Carbohydr. Polym.* **2015**, *125*, 323–333.

Malmsten, M. *Surfactants and Polymers in Drug Delivery*; Marcel Dekker: New York, NY, 2002.

Mane, S.; Ponrathnam, S.; Chavan, N. Hyperhydrophilic Three Dimensional Crosslinked Beads as an Effective Drug Carrier in Acidic Medium: Adsorption Isotherm and Kinetics Appraisal. *New J. Chem.* **2015**, *39* (5), 3835–3844.

Masters, K. *Spray Drying Handbook*; Wiley Interscience: Cambridge, UK, 1991.

208																																																																																																																																																																																																						Smart Materials

Meenach, S. A., et al. Synthesis, Optimization and Characterization of Camptothecin-loaded Acetylated Dextran Porous Microparticles for Pulmonary Delivery. *Mol. Pharmaceut.* **2012,** *9* (2), 290–298.

Miura, C.; Matsunaga, H.; Haginaka, J. Molecularly Imprinted Polymer for Caffeic Acid by Precipitation Polymerization and its Application to Extraction of Caffeic Acid and Chlorogenic Agent from *Eucommia ulmodies* leaves. *J. Pharm. Biomed. Anal.* **2016,** *127,* 32–38.

Mora-Huertas, C. E.; Fessi, H.; Elaissari, A. Polymer-based Nanocapsules for Drug Delivery. *Int. J. Pharm.* **2010,** *385* (1–2), 113–142.

Mosbach, K. Molecular Imprinting. *Trends Biochem. Sci.* **1994,** *19* (1), 9–14.

Moustafa, A. B.; Faizalla, A. Preparation and Characterization of Porous Poly(Methacrylic Acid) Gel by Dispersion Polymerization. *J. Appl. Polym. Sci.* **1999,** *73* (9), 1793–1798.

Ogino, K., et al. Synthesis of Monodisperse Macroreticular Styrene-divinylbenzene Gel Particles by a Single-step Swelling and Polymerization Method. *J. Chromatogr. A.* **1995,** *699* (1–2), 59–66.

Oh, Y. J., et al. Preparation of Budesonide Loaded Porous PLGA Microparticles and Their Therapeutic Efficacy in a Murine Asthma Model. *J. Control. Release* **2011,** *150* (1), 56–62.

Okay, O. Macroporous Copolymer Networks. *Prog. Polym. Sci.* **2000,** *25* (6), 711–779.

Okay, O., et al. Phase Separation in the Synthesis of Styrene-divinylbenzene Copolymers with Di-2-ethylhexyl Phthalate as Diluent. *J. Appl. Polym. Sci.* **1985,** *30* (5), 2065–2074.

Okubo, M., et al. Preparation of Micron-size Monodispersed Polymer Particles by Seeded Polymerization Utilizing the Dynamic Monomer Swelling Method. *Colloid Polym. Sci.* **1991,** *269* (3), 222–226.

Okubo, M.; Nakagawa, T. Preparation of Micron-size Monodisperse Polymer Particles Having Highly Crosslinked Structures and Vinyl Groups by Seeded Polymerization of Divinylbenzene Using the Dynamic Swelling Method. *Colloid Polym. Sci.* **1992,** *270* (9), 853–858.

Pardeshi, S.; Singh, S. K. Precipitation Polymerization: A Versatile Tool for Preparing Molecularly Imprinted Polymer Beads for Chromatography Applications. *RSC Adv.* **2016,** *6* (28), 23525–23536.

Perez, R. A., et al. Therapeutic Bioactive Microcarriers: Co-delivery of Growth Factor and Stem Cells for Bone Tissue Engineering. *Acta Biomater.* **2014,** *10* (1) 520–530.

Perrier-Cornet, R., et al. Functional Crosslinked Polymer Particles Synthesized by Precipitation Polymerization for Liquid Chromatography. *J. Chromatogr. A* **2008,** *1179* (1), 2–8.

Philippova, O., et al. Magnetic Polymer Beads: Recent Trends and Developments in Synthetic Design and Applications. *Eur. Polym. J.* **2011,** *47* (4), 542–559.

Pinto, J. M.; Giudici, R. Optimization of Cocktail of Initiators for Suspension Polymerization of Vinyl Chloride in Batch Reactors. *Chem. Eng. Sci.* **2001,** *56* (3), 1021–1028.

Pratima, N. A.; Tiwari, S.; Kamble, S. Mucoadhesive: As Oral Controlled Gastroretentive Drug Delivery System. *Int. J. Res. Pharm. Sci.* **2012,** *2* (3), 32–59.

Qiu, J.; Charleux, B.; Matyjaszewski, K. Controlled/living Radical Polymerization in Aqueous Media: Homogeneous and Heterogeneous Systems. *Prog. Polym. Sci.* **2001,** *26* (10), 2083–2134.

Qutachi, O., et al. Injectable and Porous PLGA Microspheres that Form Highly Porous Scaffolds at Body Temperature. *Acta Biomater.* **2014,** *10* (12), 5090–5098.

Rabelo, D.; Coutinho, F. M. B. Porous Structure Formation and Swelling Properties of Styrene Divinylbenzene Copolymers. *Eur. Polym. J.* **1994,** *30* (6), 675–682.

Racovita, S., et al. Adsorption and Release Studies of Cefuroxime Sodium from Acrylic Ion Exchange Resin Microparticles Coated with Gellan. *React. Funct. Polym.* **2016,** *105*, 103–113.

Salonen, J., et al. Mesoporous Silicon Microparticles for Oral Drug Delivery: Loading and Release of Five Model Drugs. *J. Control. Release* **2005,** *108* (2–3), 362–374.

Saracoglu, B., et al. Synthesis of Monodisperse Glycerol Dimethacrylate-based Microgel Particles by Precipitation Polymerization. *Ind. Eng. Chem. Res.* **2009,** *48* (10), 4844–4851.

Schwarz-Linek, J., et al. Phase Separation and Rotor Self-assembly in Active Particle Suspensions. *Proc. Natl. Acad. Sci. USA* **2012,** *109* (11), 4052–4057.

Sene, F. F.; Martinelli, J. R.; Okuno, E. Synthesis and Characterization of Phosphate Glass Microspheres for Radiotherapy Applications. *J. Non Cryst. Solids* **2008,** *354* (42–44), 4887–4893.

Senel, S., et al. Nucleotide Adsorption-desorption Behaviour of Boronic Acid Functionalized Uniform-porous Particles. *J. Chromatogr. B* **2002,** *769* (2), 283–295.

Shang, Q., et al. Fabrication, Characterization and Controlled Release of Eprinomectin from Injectable Mesoporous PLGA Microspheres. *RSC Adv.* **2015,** *5* (92), 75025–75032.

Song, R. Y., et al. Synthesis of Glutathione Imprinted Polymer Particles via Controlled Living Radical Precipitation Polymerization. *Chinese J. Polym. Sci.* **2015,** *33* (3), 404–415.

Srikanth, M. V., et al. Ion Exchange Resins as Controlled Drug Delivery Carriers. *J. Sci. Res.* **2010,** *2* (3), 597–611.

Straub, J. A., et al. Porous PLGA Microparticles: AI-700, an Intravenously Administered Ultrasound Contrast Agent for Use in Echocardiography. *J. Control. Release* **2005,** *108* (1), 21–32.

Sugimoto, T. *Monodispersed Particles*; Elsevier: Amsterdam, Netherlands, 2001.

Sugiura, S., et al. Synthesis of Polymeric Microspheres with Narrow Size Distribution Employing Microchannel Emulsification. *Macromol. Rapid Commun.* **2001,** *22* (10), 773–778.

Sun, Q.; Deng, Y. In Situ Synthesis of Temperature-sensitive Hollow Microspheres via Interfacial Polymerization. *J. Am. Chem. Soc.* **2005,** *127* (23), 8274–8275.

Svec, F.; Frechet, J. M. New Designs of Macroporous Polymers and Supports: From Separation to Biocatalysis. *Science* **1996,** *273* (5272), 205–211.

Tamayo, F. G.; Casillas, J. L.; Martin-Esteban, A. Evaluation of New Selective Molecularly Imprinted Polymers Prepared by Precipitation Polymerisation for the Extraction of Phenylurea Herbicides. *J. Chromatogr. A* **2005,** *1069* (2), 173–181.

Tuncel, A., et al. Carboxyl Carrying-large Uniform Latex Particles. *Colloids Surf. A Physicochem. Eng. Asp.* **2002,** *197* (1–3), 79–94.

Ugelstad, J., et al. Absorption of Low Molecular Weight Compounds in Aqueous Dispersions of Polymer-oligomer Particles. 2) A Two Step Swelling Process of Polymer Particles Giving an Enormous Increase in Absorption Capacity. *Makromol. Chem. Phys.* **1979,** *180* (3), 737–744.

Ugelstad, J., et al. Preparation and Biochemical and Biomedical Applications of New Monosized Polymer Particles. *Polym. Int.* **1993,** *30* (2), 157–168.

Vasiliu, S., et al. Adsorption of Cefotaxime Sodium Salt on Polymer Coated Ion Exchange Resin Microparticles: Kinetics, Equilibrium and Thermodynamic Studies. *Carbohydr. Polym.* **2011**, *85* (2), 376–387.

Vasiliu, S.; Bunia, I.; Neagu, V. Core-shell Microparticles Based on Acrylic Ion Exchange Resin/polysaccharides as Drug Carriers. *Ion Exch. Lett.* **2009**, *2* (3), 27–30.

Vivaldo-Lima, E., et al. An Updated Review on Suspension Polymerization. *Ind. Eng. Chem. Res.* **1997**, *36* (4), 939–965.

Vlad, C. D., et al. Sinteza Schimbatorilor De Anioni Prin Policondensarea Epiclorhidrinei Cu Polialchilenpoliamine. *Mat. Plast.* **1978**, *15* (1), 75–79.

Vladisavljevic, G. T., et al. Shirasu Porous Glass Membrane Emulsifications: Characterisation of Membrane Structure by High-resolution X-ray Microtomography and Microscopic Observation of Droplet Formation in Real Time. *J. Membr. Sci.* **2007**, *302* (1–2), 243–253.

Wally, Z. J., et al. Porous Titanium for Dental Implant Applications. *Metals* **2015**, *5* (4), 1902–1920.

Wang, B., et al. Macroporous Materials: Microfluidic Fabrication, Functionalization and Applications. *Chem. Soc. Rev.* **2017**, *46* (3), 855–914.

Wang, C., et al. Intravitreal Controlled Release of Dexamethasone from Engineered Microparticles of Porous Silicon Dioxide. *Exp. Eye Res.* **2014**, *129*, 74–82.

Wang, J., et al. Monodisperse, Molecularly Imprinted Polymer Microspheres Prepared by Precipitation Polymerization for Affinity Separation Applications. *Angew. Chem. Int. Ed.* **2003**, *42* (43), 5336–5338.

Wang, J., et al. Synthesis and Characterization of Micrometer-sized Molecularly Imprinted Spherical Polymer Particulates Prepared via Precipitation Polymerization. *Pure Appl. Chem.* **2007**, *79* (9), 1505–1519.

Wang, R., et al. Modification of Poly(Glycidyl Methacrylate-divinylbenzene) Porous Microspheres with Polyethylene Glycol and Their Adsorption Property of Protein. *Colloids Surf. B Biointerface* **2006**, *51* (1), 93–99.

Watanabe, T., et al. Microfluidic Approach to the Formation of Internally Porous Polymer Particles by Solvent Extraction. *Langmuir* **2014**, *30* (9), 2470–2479.

Whitcombe, M. J., et al. A New Method for the Introduction of Recognition Site Functionality into Polymers Prepared by Molecular Imprinting: Synthesis and Characterization of Polymeric Receptors for Cholesterol. *J. Am. Chem. Soc.* **1995**, *117* (27), 7105–7111.

Wu, D., et al. Design and Preparation of Porous Polymers. *Chem. Rev.* **2012**, *112* (7), 3959–4015.

Wu, L.; Bai, S.; Sun, Y. Development of Rigid Bidisperse Porous Microspheres for High-speed Protein Chromatography. *Biotechnol. Prog.* **2003**, *19* (4), 1300–1306.

Wu, J., et al. Uniform-sized Particles in Biomedical Field Prepared by Membrane Emulsification Technique. *Chem. Eng. Sci.* **2015**, *125*, 85–97.

Wulff, G. Molecular Imprinting in Cross-linked Materials with the Aid of Molecular Templates—A Way Towards Artificial Antibodies. *Angew. Chem. Int. Ed. Engl.* **1995**, *34* (17), 1812–1832.

Xia, Q., et al. Preparation and Characterization of Monodisperse Molecularly Imprinted Polymer Microspheres by Precipitation Polymerization for Kaempferol. *Design. Monom. Polym.* **2017**, *20* (1), 201–209.

Yan, Q., et al. Precipitation Polymerization in Acetic Acid: Study of the Solvent Effect on the Morphology of Poly(Divinylbenzene). *J. Phys. Chem. B.* **2009,** *113* (10), 3008–3014.

Yang, K., et al. One-pot Synthesis of Hydrophilic Molecularly Imprinted Nanoparticles. *Macromolecules* **2009a,** *42* (22), 8739–8746.

Yang, P.; Gai S.; Lin, J. Functionalized Mesoporous Silica Materials for Controlled Drug Delivery. *Chem. Soc. Rev.* **2012,** *41* (9), 3679–3698.

Yang, Y., et al. Development of Highly Porous Large PLGA Microparticles for Pulmonary Drug Delivery. *Biomaterials* **2009b,** *30* (10), 1947–1953.

Ye, L.; Weiss, R.; Mosbach, K. Synthesis and Characterization of Molecularly Imprinted Microspheres. *Macromolecules* **2000,** *33* (22), 8239–8245.

Yin, X., et al. Development of Andrographolide Molecularly Imprinted Polymer for Solid-phase Extraction. *Spectrochim. Acta Part A* **2011,** *79* (1), 191–196.

Yoo, J. W.; Lee, J. S.; Lee, C. H. Characterization of Nitric Oxide-releasing Microparticles for the Mucosal Delivery. *J. Biomed. Mater. Res.* **2010,** *92* (4), 1233–1243.

Yoshimatsu, K., et al. Uniform Molecularly Imprinted Microspheres and Nanoparticles Prepared by Precipitation Polymerization: The Control of Particle Size Suitable for Different Analytical Applications. *Analyst. Chim. Acta.* **2007,** *584* (1), 112–121.

Yoshimatsu, K., et al. Peptide-imprinted Polymer Microspheres Prepared by Precipitation Polymerization Using a Single Bi-functional Monomer. *Analyst* **2009,** *134* (4), 719–724.

Yu, B., et al. Preparation of Porous Poly(Styrene-divinylbenzene) Microspheres and Their Modification with Diazoresin for Mix-mode HPLC Separations. *Materials* **2017,** *10* (4), 440–453.

Yuan, Q., et al. Preparation of Particle-stabilized Emulsions Using Membrane Emulsification. *Soft Matter* **2010,** *6* (7), 1580–1588.

Zambaux, M. F., et al. Influence of Experimental Parameters on the Characteristics of Poly(Lactic Acid) Nanoparticles Prepared by a Double Emulsion Method. *J. Control. Release* **1998,** *50* (1–3), 31–40.

Zhang, G. H., et al. Fabrication of Hollow Porous PLGA Microspheres for Controlled Protein Release and Promotion of Cell Compatibility. *Chinese Chem. Lett.* **2013a,** *24* (8), 710–714.

Zhang, H. Controlled/"living" Radical Precipitation Polymerization: A Versatile Polymerization Technique for Advanced Functional Polymers. *Eur. Polym. J.* **2013b,** *49* (3), 579–600.

Zhang, Y. J., et al. Synthesis and Evaluation of Cyromazine Molecularly Imprinted Polymeric Microspheres by Two-step Seed Swelling Polymerization. *Asian J. Chem.* **2013c** *25* (15), 8329–8332.

Zhu, L., et al. Inhalable Oridonin-loaded Poly(Lactic-co-glycolic)Acid Large Porous Microparticles for In Situ Treatment of Primary Non-small Cell Lung Cancer. *Acta Pharm. Sin. B.* **2017a,** *7* (1), 80–90.

Zhu, Y., et al. Molecularly Imprinted Nanoparticles and Their Releasing Properties, Biodistribution as Drug Carriers. *Asian J. Pharm. Sci.* **2017b,** *12* (2), 172–178.

CHAPTER 7

AMINO-SILICONES AS ACTIVE COMPOUNDS IN THE DETECTION AND CAPTURE OF CO_2 FROM THE ENVIRONMENT

ALEXANDRA BARGAN* and MARIA CAZACU

Inorganic Polymers Department, "Petru Poni" Institute of Macromolecular Chemistry, Gr. Ghica Voda Alley 41A, 700487 Iasi, Romania

Corresponding author. E-mail: anistor@icmpp.ro

CONTENTS

ABSTRACT

Because anthropogenic carbon dioxide (CO_2) emissions in the atmosphere have greenhouse effect contributing to global warming, worldwide efforts are being made to reduce them, developing techniques for separating and capturing CO_2 being a priority. One of the most effective technologies for CO_2 capture consists of chemical absorption in a liquid medium containing the amine (alkanolamine and ammonia) with the formation of carbamate or bicarbonate. Because the reaction is reversible, CO_2 can then be removed by heating with the amine regeneration and reuse it. Silicone materials have also been studied as means of capturing CO_2, among them amino-silicones recently proved to be highly efficient absorber of this. For such use, amines containing siloxanes has several advantages over the classic organic amines such as high thermal stability, low volatility, and low viscosity, which allows their use as such, without the need for dissolution/dilution with water or organic solvents. This makes the heat energy needed to release CO_2 and absorber regeneration to be reduced. The effectiveness of the amino-silicones in retaining CO_2 is extended in their use as sensors for this gas. This chapter critically reviewed and analyzed the results of the authors and those reported in the literature on both these directions.

7.1 INTRODUCTION

Global warming caused by the emissions of greenhouse gases, particularly carbon dioxide (CO_2), has been a widespread preoccupation in the last years. CO_2 contributes more than 60% to global warming due to its great emission amount. Many efforts are being made for reducing these emissions amounts by developing new or improving the existent techniques for separating and capturing of CO_2 (Yu et al., 2012; Sanz-Perez et al., 2016; Perry, 2016). A very simple definition of the global warming is the increase in the average global temperature at the fastest rate in recorded history for the last 50 years. With one exception all of the 16 hottest years in National Aeronautics and Space Administration (NASA) records had happened since 2000. Some scientists discussed about a "pause" or a "slow down" in this global warming, in the increase of the global temperatures, but several recent studies disapproved this affirmation (Karl et al., 2015). Karl et al. in the journal *Science*, in 2015 reported an update global surface temperature analysis that does not support the notion of a "slowdown" in the increase of global surface temperature.

CO_2, other air pollutants, and greenhouse gasses are collected in the atmosphere and absorb the sunlight and solar radiation. The radiation normally should escape into space but because of these pollutants, the heat is trapped causing the increase in the planet temperatures. It is getting hotter and hotter. This phenomenon is known as greenhouse effect. One of the major problems is that these pollutants can last for many years, even centuries in the planet's atmosphere. This increase in the global atmospheric CO_2 concentration is caused by the continuous increase of the global population, thus conducing to a continued increase in fossil fuel use (Stone et al., 2009). If the emissions do not decrease and measures for reducing them are not taken, the average US temperatures could grow up to 10°F (Fahrenheit) in the next century. In the United States, where the electricity is mainly made by burning the fossil fuels, that is being the largest source of heat-trapping pollution, about two billion tons of CO_2 are produced every year. The second-largest source of carbon pollution is the transportation sector which produces about 1.7 billion tons of CO_2 emissions a year.

According to the data obtained from the Mauna Loa station (Institute of Oceanography, San Diego University), where the average concentration of CO_2 is monitored and registered, the value of this parameter has increased at an annual rate of about 2 ppm, from 401.33 ppm in April 2014 to 403.01 ppm in December 2015 and 404.44 ppm in December 2016. Despite all the policies to moderate them, the emissions of gases from the greenhouse effect increased to levels unequaled until now. The main cause of global warming is this increase in CO_2 emissions that why many efforts have been dedicated to the capture and storage of CO_2 from atmosphere. To mitigate the environmental impact from enhanced emission of CO_2 is imperative to develop improved removal technologies (IPCC, 2005). Through the other capturing processes, the adsorption became the chosen one because of its simplicity, less energy demand, and ease to use.

Carbon capture and storage or carbon capture and sequestration (CSS) is the process of capturing CO_2 from many sources (fossil fuel power plants), transporting it to a storage place, and depositing it in a way that will not enter in the atmosphere. The main reason is to prevent the release of large quantities of CO_2 into the atmosphere. The long-term storage of CO_2 is a new concept. Some commercial examples were: the Weyburn-Midale Carbon Dioxide Project developed in 2000, Sask Power's Boundary Dam, and Mississippi Power's Kemper Project. CCS applied to a modern power plant can reduce the CO_2 emissions in the atmospheres with almost 80–90%. The CO_2 can be captured using adsorption, membrane gas separation, or adsorption technologies. Capturing and compressing the CO_2 increase the energy needs with 25–40%, this and the other system costs increasing the cost per watt energy obtained by 21–91% for fossil fuel power plants. Using these technologies would be more expensive, mainly if those are far from a capturing site. The CO_2 is stored either in deep geological formations or like mineral carbonates. Storing the CO_2 in the ocean is not yet possible because of the ocean acidification. The most promising sealing sites are considered to be the geological formations. Even there are places with enough storage capacity of CO_2, also remains the problem of long-term predictions about these underground and submarine storage security which are very difficult and not secure. The main risk is that CO_2 could leak in the atmosphere. Monitoring the CO_2 emission is very important aim because of its association with potential climate change.

For reducing the emissions of CO_2, a promising manner to control the emission of CO_2 in coal-fired power plants is CO_2 capture and storage. After the CO_2 is captured, it can be used for obtaining different products, as can be seen in Table 7.1. The reuse of the CO_2 is another important aspect considered for decreasing the atmospheric CO_2 level.

TABLE 7.1   Use of the CO_2 Captured.

	Use	Products
	Food/products	Carbonated beverages
	Biological conversion	Algae/greenhouse gases
	Extractant	Flavors/fragrances
	Mineralization	Carbonates
	Chemicals	Methanol, urea, CO, methane, and salicylic acid (liquid fuels, fertilizers, and secondary chemicals)
Captured CO_2	Refrigerant	Refrigeration and dry ice
	Miscellaneous	Aerosol, dry ice pellets injected into metal castings, with medical O_2 as a respiratory stimulant
	Inerting agent	Protect carbon powder shield gas in welding blanket products
	Fire suppression	Fire extinguishers
	Plastics	Polymers and polycarbonate
	Enhanced fuel recovery	Oil and gas

Reprinted with permission from MacDowell, N., et al. An Overview of CO_2 Capture Technologies. *Energy Environ. Sci.* **2010**, *3*, 1645–1669. © 2010 Royal Society of Chemistry.

7.2 TECHNIQUES AND MATERIALS FOR SEPARATING AND CAPTURING CO_2, REGENERATION, AND REUSE

CO_2 can be captured out of air or flue gas using the appropriate techniques, making in this way possible the reuse of the CO_2 captured and decreasing in that manner the air pollution.

Until now many methods/technologies have been proposed in order to remove and recover CO_2 from the air, industrial wastes, mine gases, and human respiration and these technologies are presented in Table 7.2. They can be divided as we just mentioned in absorption and adsorption (physical or chemical), cryogenics, membrane, and microbial/algal systems. For the chemical absorption, the reagents most widely used are based principally on aqueous solutions of alkanolamines (monoethanolamine (MEA), diethanolamine, and mixed alkanolamines) but these reagents, highly efficient for CO_2 adsorption have some disadvantages: drawbacks-bugs during the application, such as corrosion, oxidative degradation, and pollution caused by their high volatilities; large and expensive equipment; high power consumption, etc. For physical adsorption, the solvents used are Selexol, Purisol, Rectisol, etc.

Absorption with amines is the most used technology, the other ones being still in the developmental research stage. The chemical absorption in a liquid medium containing the amine with the formation of carbamate or bicarbonate is one of the most effective technologies for CO_2 capture. Because the reaction is reversible, CO_2 can then be removed by heating with the amine regeneration and reuse it.

The CO_2 capture can also be realized physically or chemically by solid sorbents. Due to the fact that the first process, physically sorption of CO_2, can be reduced too much by the presence of water vapors, the chemisorption seems to be a better choice.

The solid adsorbents as compared to the aqueous solutions of amine are easy to handle, can save energy, and are not causing corrosion difficulties. In the last time, many solid adsorbents have been studied as possible CO_2 capture candidates: mesoporous silica, zeolites, activated carbon, metal-organic frameworks (MOFs), and amine-silica hybrid/composite materials. Although some of these materials, MCM-41 and SBA-16, can improve the drawbacks of chemisorbents due to their good ordered structures and mesoporous characteristics, their CO_2 adsorption capacities are poor.

TABLE 7.2 CO_2 Separation and Capture Technologies.

CO₂ separation and capture		
Absorption	Chemical solvents	MEA, TEA, MDEA, Benfield, Caustic, Sulfinol, and others
	Physical solvents	Selexol, Purisol, Rectisol, fluorinated solvents, ionic liquids, and others
Adsorption	Physical	Mesoporous silica, zeolites, alumina-activated carbon, MOFs
	Chemical	CaO, Ca(OH)₂, NaOH, MgO, Li₂ZrO₃ and Li₄SiO₄
Cryogenics	Ryan–Holmes process, liquefaction distillation	
Membrane	Gas separation	Polyphenylene oxide and polydimethylsiloxanes
	Gas absorption	Polypropylene
Microbial/ algal system	Algae, cyanobacteria, β-proteobacteria, clostridia, and archaea	

MEA, monoethanolamine; TEA, triethanolamine; MDEA, methyl diethylamine denfield-potassium carbonate; sulfinol, tetrahydrothiophene 11'-dioxide, an alkanolamine and water; Rectisol, methanol; Purisol, N-methyl-2-pyrrolidone; Selexol, dimethyl ethers of polyethylene glycol (DMPEG).

Most technologies for CO_2 capture are based on liquid solvents, generally consisting of amines (Esam and Amana, 2013). Amines have been used since 1930 to separate CO_2 from natural gas and hydrogen, the most commonly used amine for this purpose being MEA. The amine reaction with CO_2 is well known, and this is the basis of many functional technologies to retain it. Primary and secondary amines by reaction with CO_2 form carbamates through a zwitterion mechanism; under anhydrous conditions, two amine groups are required to capture one mole of CO_2 (Perry et al., 2010; Wang and Li, 2015). Carbamates themselves represent an important class of organic compounds, used in a variety of applications including polyurethanes, pesticides, fungicides, medicinal drugs, and synthetic intermediates (Bates et al., 2002).

When subjected to heating under vacuum or in presence of moisture, the carbamates can decompose to the initial amine (Dumitriu et al., 2012). The amine is regenerated by water vapor stripping at 100–120°C, which condenses leaving CO_2 which is compressed to 100–150 bar and can be sequestered geologically (Wang and Li, 2015).

Among the amine-based solvents, MEA is one of the most favored solvents for CO_2 capture due to its high CO_2 capture capacity and fast reaction kinetics (Park et al., 2012).

Thus, first commercial CO_2 sequestration facility started in Norway in 1996, followed by the processes developed by Fluor Daniel Inc., Dow Chemical Co., and several other companies, and they are based on chemical absorption using MEA-containing solvents. The disadvantages of using this compound consist in: high cost (around \$40 per ton); high amount of heat to regenerate the solvents; partial loss of solvents during regeneration; stable salt formation with certain impurities, thus reducing the absorption capacity; degradation over time with the production of corrosive compounds that require certain precautions in handling and storage (Esam and Amana, 2013); MEA high volatility, its corrosive fume being a concern for the process design and operation; the concentration of MEA has to be limited to 15–30 wt% and this makes the CO_2 capture and the solvent regeneration processes complicated and costly (Park et al., 2012).

Therefore, a number of innovative organic and inorganic materials including amine functionalized solid mesoporous sorbents and liquid solvents, that is, ionic liquids and organic solvents such as amino-silicones, are being developed to capture CO_2. Ionic liquids have the advantage of negligible vapor pressure even at elevated temperatures but as drawbacks their complex synthesis and purification steps and high cost (Park et al., 2012).

7.3   SILICONE MATERIALS FOR CO_2 CAPTURING

Silicone materials have also been studied as means of capturing CO_2, and among these, amino-silicones (mono-, di-, or tetra-primary or secondary amines) recently proved to be highly efficient absorber of this. For such use, amine-containing siloxanes have several advantages over the classic organic amines, such as, high thermal stability of the siloxane bond, low volatility, and low viscosity, which allows their use as such, without the need for dissolution/dilution with water or organic solvents (Noll, 1968). This makes the heat energy needed to release CO_2 and absorber regeneration to be reduced. The effectiveness of the amino-silicones in retaining CO_2 is extended in their use as sensors for this gas.

In recent years, many efforts were made for developing new adsorbents and to optimize their CO_2 capture and storage. Since 1990s mesoporous silica materials were extensively investigated in many areas including environment, energy, biomedicine, and catalysis. The most used silica material is MCM-41, due to its combination of superior properties, as high surface area, thermal stability, and pore volume which can vary depending on the surfactant used in the synthesis reaction. The pore sizes can have dimensions between 2 and 50 nm. The mesoporous silica has a high adsorption capacity for CO_2, CH_4, N_2, H_2, and O_2 and this capacity can be improved for CO_2 adsorption by different methods: adjusting the synthesis parameters, functionalizing the mesoporous silica, or impregnation of the material.

Amine–silica hybrid/composite materials are made of amines and silica. These materials can exhibit high and fast CO_2 adsorption, low energy consumption, and good selectivity. A very important fact is that these materials are tolerant to moisture. The amine–silica hybrid/composite materials can be obtained by grafting or impregnation. Impregnation is a physical process, while the grafting means to synthesize an amine–silica hybrid material through the formation of a covalent bond between organic-amines and silica. Amine-functionalized silica materials were investigated for their use as CO_2 sorbents by Li et al. (2007, 2014a, 2014b), Gunathilake et al.(2016), Sanz-Perez et al. (2017), Loganathan and Ghoshal (2017), Leal et al.(1995), Kishor and Ghoshal (2015), Gholami et al. (2016), and others from different point of view. For example, Sanz-Perez et al. studied the reuse and recycling of amine–silica materials as CO_2 adsorbents after their lifespan. The conclusion of their work was that the materials obtained using the impregnation of the calcined samples maintained their CO_2 adsorption properties more than six cycles without changes in their efficiency for CO_2 adsorption, and the materials obtained by grafting the calcined samples yield to smaller amine efficiency during CO_2 capture.

CO_2 reversibly adsorbed on a silica gel containing 3-aminopropyl groups bonded to surface atoms of silicon. These act as the active sites for the chemisorption of CO_2 at room temperature which is liberated by temperature programmed desorption at about 100°C. The material is capable of adsorbing about 10 STP cm³ (standard temperature and pressure) of dry CO_2 per gram and can be regenerated upon heating. It might be used as a scrubber for CO_2 from industrial gaseous streams. Adsorption of humid CO_2 produces a small amount of formaldehyde which suggests activation of the CO_2 (Leal et al., 1995).

7.3.1 SOLID AMINE SORBENTS

CO_2 capturing on solid sorbents is a very up-and-coming technology in the last decade. Until now around 11.436 papers have been published on solid CO_2 sorbents, between 2011 and 2014, around 2000 approaching different aspects on this field. The CO_2 sorbents were organized according to their working temperatures: low-temperature (<200°C), intermediate-temperature (200–400°C), and high-temperature (>400°C) and taking into account: sorption capacity, kinetics, recycling stability, cost, etc. (Wang et al., 2014).

Various mesoporous solid adsorbents impregnated with polyamines and grafted with aminosilanes were used in CO_2 capture technologies for postcombustion power plants. Studies concerning their CO_2 adsorption capacity, adsorption rate, cost, thermal stability, etc., were performed (Yu et al., 2012).

CO_2 can be captured by adsorption onto the surface of a specific sorbent like, activated carbons (Pino et al., 2016), zeolites amine immobilized mesoporous silica types of MCM-41 (Costa et al., 2014; Costa et al., 2015; Hao et al., 2015; Melendez-Ortiz et al., 2013; Xu et al., 2002; Thi Le, 2014), SBA-15 (Huang et al., 2013; Yoo et al., 2015), MOFs, microporous organic polymers (MOPs), and amine-grafted sorbents (Chang et al., 2009; Nigar et al., 2016; Niu et al., 2016; Soto-Cantu et al., 2012; Qi et al., 2011).

The ideal sorbent for CO_2 should have a high CO_2 adsorption capacity, high selectivity for CO_2, should be economically regenerable and without any cyclic performance loss. The postcombustion capture of CO_2 is made at ambient temperature that is why chemical adsorbents with basic amine groups are the most suitable. These adsorbents can be developed using different techniques like surface modification of low-cost carbons and mesoporous silica using a polymeric amine, such as tetraethylene-pentamine (TEPA) or polyethylenimine (PEI), and synthesis of a high nitrogen content carbon adsorbent.

The solid amine sorbents for CO_2 capture obtained by amination of different types of supports as mesoporous silica (MCM-41, SBA-15, and

KIT-6) and two polymeric supports, PMMA and PS, were studied and developed as chemical sorbents and their CO_2 sorption performance and stability were reviewed (Unveren et al., 2016).

Some very useful materials for CO_2 adsorption are amine-functionalized adsorbents. Many types of molecules and macromolecules containing different types of amines (primary, secondary, tertiary, or combinations of them) were investigated for CO_2 adsorption. These adsorbents can be arranged into three classes based on the function of their physical and chemical characteristics.

- *First class* are adsorbents containing amine molecules (with short alkyl chains, such as diethylenetriamine, TEPA, and branched polymers (poly(ethylenimine)), physically bounded (impregnated) on the solid support.
- *Second class* are adsorbents with molecules having amines chemically grafted onto the support material or bounded in another manner (covalent bonds): amine-containing alkoxysilanes (3-aminopropyltrialkoxysilane, 3-(2-aminoethylamino)propyltrimethoxysilane, and 3-[2-(2-aminoethylamino)ethylamino]propyltrimethoxysilane) are bound to presynthesized oxide supports or co-condensed with silica forming molecules like tetraethyl orthosilicate.
- *Third class* is obtained by in situ polymerization reactions of amine monomers in the pores of support materials (Yoo et al., 2015).

7.3.2 AEROGELS IN CO₂ SORPTION

Aerogels represent any material derived from organic, inorganic, or hybrid molecular starting compounds which were prepared by sol–gel process and dried with a special technology in order to keep the three dimensional and highly porous networks (Maleki, 2016). This term was introduced for the first time by S. Kistler in 1930 when he obtained an air-filled solid material with almost the same dimensions as the original gel. He used a supercritical drying technology. Many studies were focused on silica type aerogels (SiO_2). The most important reasons for its popularity are its special look (like "frozen smoke") and its well-known applications in space experiments (as insulate electronics and collectors for space-dust and comet particles for analysis at the return on Earth).

Even the silica aerogel was obtained for the first time in 1930, this domain saw an increase in development only after 1985 due to the attempts of Stanislaus Teichner from the University Claud Bernard to produce silica aerogel for the storage of rockets fuels. This type of aerogel possesses a number of special properties and it is the lightest and lower-density solid material known. It has a sponge-like structure with high porosity (80–99.8%) and high specific surface area (500–1200 m^2/g). Silica aerogel is 1000 times less dense than glass and it can support more than 10,000 times of its weight. As a function of its density, the silica aerogel can have a very large surface area. Also, it is much known for its low thermal conductivity, low density, low refractive index, ultralow dielectric constant, low sound speed, and very high transparency (Dorcheh and Abbasi, 2008; Cui et al., 2011; Czaun et al., 2013). With these special properties, silica aerogels exhibit an amazing range of characteristics which are not observed in other types of materials, making it is the perfect candidates for the use as sensors for gas detection, pollution filters, nuclear particle detectors, and many other interesting application.

The major part of the aerogels is made by sol–gel reaction, but with different starting precursors and different operating conditions. The synthesis of silica aerogels has three steps:

- Gel preparation by sol–gel process from a silica source solution and by addition of catalyst. The gels are usually classified according to the dispersion medium: water, alcohol, and air (hydrogel/aquagel, alcogel, and aerogel);
- Aging of the gel in its mother solution;
- Drying of the gel when all liquids should be eliminated. For preventing the collapse of the gel structure, the drying is made in special conditions: supercritical dying: at high temperature (HTSCD) and low temperature (LTSCD). The potential applications of aerogels depend on their microstructure and surface groups that being determined by the parameters of the sol–gel reaction. This is the reason why it is important to select the appropriate type of synthesis and processing for the aerogels for target applications.

Figure 7.1 indicates the sol–gel reaction for obtaining a silica aerogel. This reaction is a versatile process, which allows tailoring the nanostructure by adjusting the reaction's parameters: concentration of the precursors, solvent type, ratios of water to silica precursors, temperature, and pH. Using it, a molecular compound or other phase could be incorporated for giving special characteristics to the gel network (at molecular and at nanoscale level). This can be achieved physically (by using dopants or additives in the porous network) or chemically (with proper organofunctional alkoxide derivatives).

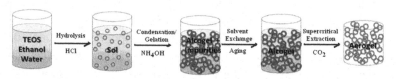

FIGURE 7.1   Synthesis of silica aerogels. Reprinted with permission from Sanli, D.; Erkey, C. Effect of Polymer Molecular Weight and Deposition Temperature on the Properties of Silica Aerogel/Hydroxy-Terminated Poly(Dimethylsiloxane) Nanocomposites Prepared by Reactive Supercritical Deposition. *J. Sup. Fluids* **2015**, *105*, 99–107. © 2015 Elsevier.

Recent studies have the principal purpose, the preparation of new aerogels as new CO_2 capture materials, varying the chemophysical properties of different aerogels. Amine-modified SiO_2 aerogels were explored as CO_2 capture materials. The SiO_2 aerogels were modified with 3-aminopropyltriethoxysilane (APTES) (Cui et al., 2011), PEI, and TEPA (Qi et al., 2011; Linneen et al., 2013) for increasing their CO_2 sorption performance (5.1 mmol/g) and for having a good cycling stability (over 10 cycles).

Minju et al. (2015), in their study about amine impregnated porous silica gel sorbents, have used 3-aminopropyltrimethoxysilane (APTMS) and PEI to modify the silica aerogels derived from water glass. The conclusions of this study were that the effect of amino loading is essential for having good results: meaning good selectivity and an improved sorption capacity (Minju et al., 2015). The CO_2 sorption capacity is also determined by the temperature, partial pressure of CO_2, sorbent particle size, density of sorbent, equipment, etc. The presence of water in the gas flow, during the CO_2 sorption experiments at low concentrations improved the sorbent stability because the water blocks the formation of urea during the regeneration process (Veneman et al., 2014). A large amount of water affects the process due to the increase in the demand for heat energy.

Scientists at Aspen Aerogels (Begag et al., 2013) developed amino-functionalized aerogels starting from amino-alkyltrialkoxysilane co-precursors and methyltrimethoxysilane (MTMS) precursor and obtained materials with a CO_2 sorption capacity of almost 1.8 mmol/g and a cycling stability over 2000 cycles.

7.3.3   LIQUID SILICONE AMINE SORBENTS

The silicones properties as thermal stability, low vapor pressures, low-viscosity fluids, and low heat capacities make them the candidate molecules for CO_2 capture.

For the first time, in 1972, Kammermeyer and Sollami patented the silicone oils as absorbents for CO_2 from a gaseous fluid (Kammermeyer and Sollami, 1972), at ambient or lower temperatures (Avila et al., 2016). A variety of silicone-based materials have been tested for this purpose. It has been found that the gas solubility of the polymer can be influenced by changing the side chains. Thus, the substitution with diphosphate increases the solubility of CO_2, while OH-terminal polydimethylsiloxanes are superior to those with trimethylsiloxyl groups with respect to the solubility of CO_2. The silica mesoporous and the amine functionalized aerogel were also tested (Feron, 2016). After absorption, the silicone oils full with CO_2 are circulated through a series bends in contact with a heat generating equipment. The silicone oils are transported after that to a desorption chamber where are agitated and heated. Using vacuum applied to the surfaces of the absorbents the gas is eliminated and transported to a storage chamber and the silicone oils are then recirculated by a pump and brought back to the absorption chamber at ambient or lower temperature (Perry, 2016).

The high reactivity of the amine group with CO_2, make as the silicones functionalized with this group to be between the main candidates for CO_2 capture. The reaction of CO_2 with an amino-silicone is presented in Figure 7.2 and illustrates the formation of a carbamate salt, this reversible reaction near fast reaction rates and low cost being a key chemical criterion.

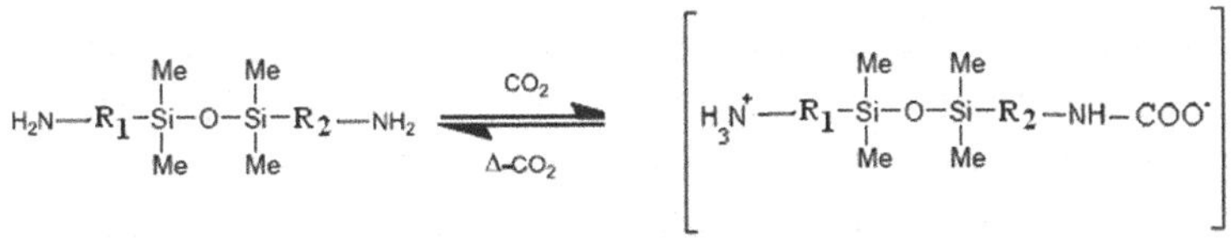

FIGURE 7.2   The reaction scheme for the formation of carbamate derivative of the siloxane compound.

Because of the good thermal stability of the liquid silicones, the process of capturing the CO_2 can take place at high temperatures releasing the gas after that with elevated pressures. This fact determines a decrease in the energy consumption. Table 7.3 presents some of the possible silicone derivatives as candidates for CO_2 capture, studied by Perry (2016). He focused especially on primary or secondary amino functionality, and small silicone backbone. For the practical experiments, dry gas (CO_2) was inserted in a mixed volume of neat amino-silicone at 40°C, for 2 h and 1 bar of CO_2.

TABLE 7.3 CO_2 Absorption of Amino-silicones.

Compound	Structure of the compound
	CH₃ CH₃ structure: R_1—Si—O—Si—R_2 with CH₃ groups
1	$R_1=R_2=C_3H_6-NH_2$
2	$R_1=R_2=C_4H_8-NH_2$
3	$R_1=R_2=C_5H_{10}-NH_2$
4	$R_1=R_2=C_3H_6-NH-C_5H_{10}-NH_2$
5	$R_1=R_2=C_6H_{12}-NH_2$
6	$R_1=R_2=CH_2-NH-C_2H_4-NH_2$
7	$R_1=R_2=C_3H_6-NH-C_2H_4-NH_2$
8	$R_1=R_2=C_3H_6-NH-C_3H_7$

Reprinted with permission from Perry, R. J.; O'Brien, M. J. Amino Disiloxanes for CO_2 Capture. *Energy Fuels* **2011**, *25*, 1906–1918. © 2011 American Chemical Society.

Figure 7.3 represents the percent weight gained by every amino-silicone in comparison with the percent of theoretical CO_2 uptake that the weight gain corresponded to. Some of the amino-silicones exhibit values near the theoretical uptake (1, 2, and 3), whereas others do not achieve their potential (7 and 8). The simplest amino-silicone is 1,3-bis(3-aminopropyl) tetramethyldisiloxane (APDS) (Fig. 7.4).

This is the reagent used in different chemical reaction for post-functionalization or inserting as segment in copolymer structure with different organic partners. At the same time, when left in the air it is able to absorb CO_2 to form a solid precipitate that is deposited (Fig.7.5a). The process is favorized by bases such as imidazolecarboxaldehyde, present in the mixture for the purpose of synthesizing the corresponding Schiff base but being impeded by the separation of carbamate (Bargan et al., 2014).

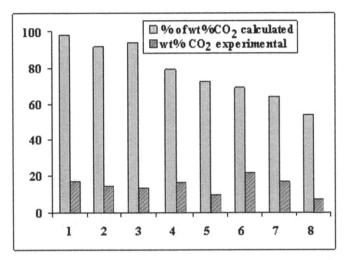

FIGURE 7.3   Theoretical estimated and experimental values for CO_2 capturing capacity for the silicone derivatives presented in Table 7.3. Reprinted with permission from Perry, R. J.; O'Brien, M. J. Amino Disiloxanes for CO_2 Capture. *Energy Fuels* **2011**, *25*, 1906–1918. © 2011 American Chemical Society.

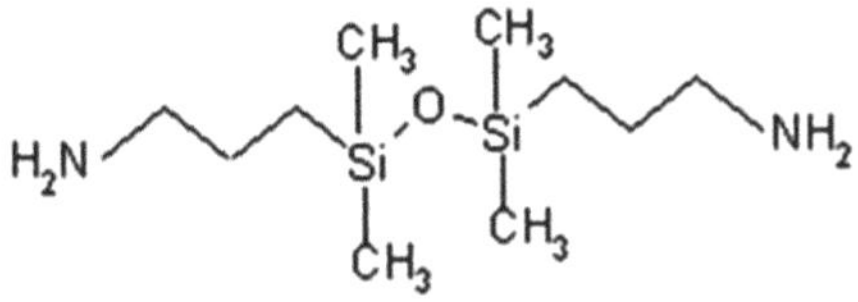

FIGURE 7.4   Structure of 1,3-bis(3-aminopropyl) tetramethyldisiloxane.

The reaction of the $-NH_2$ group with CO_2 is a complex process and there may be several unstable species in transition. The reaction products may be a mixture of ammonium bicarbonates, carbonate, and carbamate which may coexist with the reactants. However, under anhydrous conditions at room temperature, the main reaction product is ammonium carbamate. Under the influence of humidity, it turns into ammonium carbonate (Meng, 2014). As a result, IR spectrum is complex and its pattern in the region specific for C=O, C—O, and N—H absorption bands may change depending on the conditions in which it was recorded but with certainty, it is different from that of the starting amine as in Figure 7.5.

In the case of silicone derivatives, taking into account that the absorption of CO_2 occurs in bulk into a relative hydrophobic compound, due to the tetramethyldisiloxane presence in the structure is presumed that

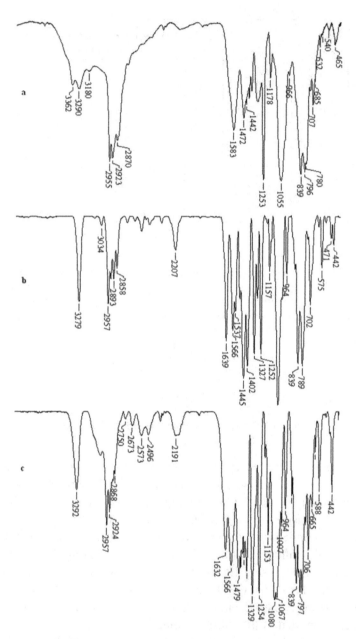

FIGURE 7.5   Comparative IR spectra for (a) 1,3-bis(3-aminopropyl) tetramethyldisiloxane; (b) carbonated compounds formed in basic medium in presence of imidazolecarboxaldehyde; and (c) carbonated compounds as a result of long exposure in air.

carbamate is the main CO_2 chemisorption product. The band at 1566 cm^{-1} in the spectra of the two carbonated compounds (Fig. 7.5b,c) is assigned to asymmetric stretching in formed carbamate, while the band at 1639 cm^{-1} is considered the proton band (Park et al., 2008; Dibenedetto et al., 2008). The reaction leading to this product is similar to that shown in Figure 7.3, where $R_1=R_2=CH_2-CH_2-CH_2$ (Bargan et al., 2014).

In fact, the product's spectrum is shown in Figure 7.5b. It was isolated in crystalline state and X-ray single crystal diffraction analysis confirmed the carbamate form (Fig. 7.6).

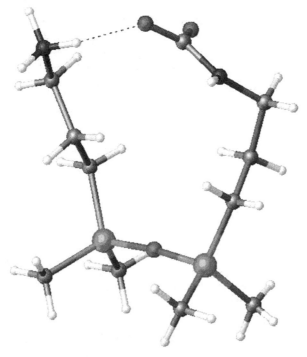

FIGURE 7.6 (See color insert.) Structure of carbamate derivative of 1,3-bis(3-aminopropyl)tetramethyldisiloxane, as determined crystallographically. Reprinted with permission from Bargan, A., et al. A New ZwitterionicSiloxane Compound: Structural Characterization, the Solution Behavior and Surface Properties Evaluation. *J. Mol. Liquids* **2014**, *196*, 319–325. © 2014 Elsevier.

Low viscosity (4 Cp at 25°C) makes APDS usable as such without solvent addition while its low volatility (boiling point 265°C) simplifies CO_2, stripping process is no longer necessary to separate the solvents evaporated. This, together with the low heat required in the stripping process due to its lower specific heat than MEA (i.e., 2.3 vs. 3.7 kJ/(kg K), makes the APDS-based process requires 50% less energy than 30% MEA-based solution (Budzianowski, 2016).

A formulation used for CO_2 capture consists in a 60/40 wt/wt mixture of α,ω-bis(3-aminopropyl)polydimethylsiloxane (AP) with tri-ethylene glycol (TEG) as a cosolvent. Its behavior and performance are compared with those of mono-ethanolamine (MEA). It was found that the former is a much cheaper ($46.04/t of CO_2) version than the latter ($60.25/t of CO_2). This efficiency is attributed to several factors, such as the higher working capacity of the amino-silicone solvent compared the MEA, which reduces the solvent flow rate required, reducing equipment sizes.

The affinity of silicone amines was also highlighted when the APTMS was reacted with triethanolamine, in molar ratio 1:1, in bulk, in air when the silatrane formed resulted in carbamate form (Fig. 7.7).

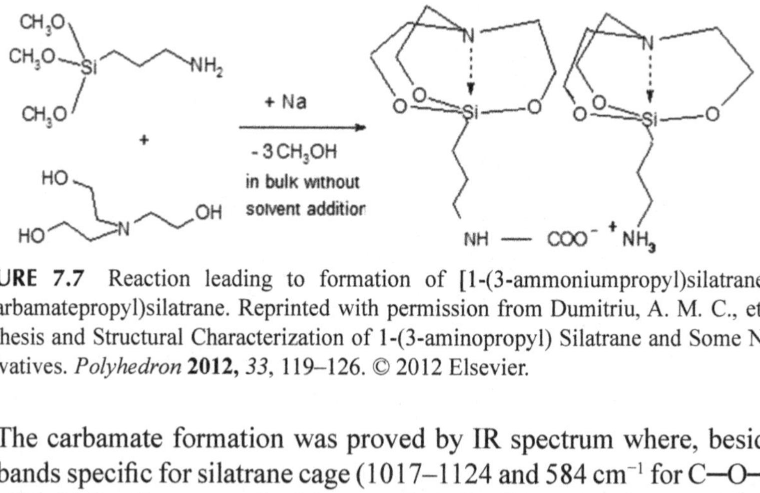

FIGURE 7.7 Reaction leading to formation of [1-(3-ammoniumpropyl)silatrane]1-(3-carbamatepropyl)silatrane. Reprinted with permission from Dumitriu, A. M. C., et al. Synthesis and Structural Characterization of 1-(3-aminopropyl) Silatrane and Some New Derivatives. *Polyhedron* **2012**, *33*, 119–126. © 2012 Elsevier.

The carbamate formation was proved by IR spectrum where, besides the bands specific for silatrane cage (1017–1124 and 584 cm^{-1} for C—O—Si and Si←N bonds, respectively), a well-defined band assigned to C═O symmetric stretching of carbamate appears at 1534 cm^{-1}. Because the compound could be isolated in a suitable crystalline state, its structure was determined by X-ray single crystal diffraction (Fig. 7.8).

Due to the fact that from the reaction of the amino-silicone with the CO_2, results the carbamate in form of solid, gum, or a viscous liquid, the transport of the CO_2 through the unreacted amino-silicone decrease. For improving the transport and reducing the viscosity of the carbamate products, the triethylene glycol (TEG) is often added to the reaction. This compound has similar characteristics with the amino-silicones (low vapor pressure, low toxicity, thermal stability, and low specific heat). With the TEG present in reaction, the CO_2 absorption improved or surpassed the limit (Perry, 2016).

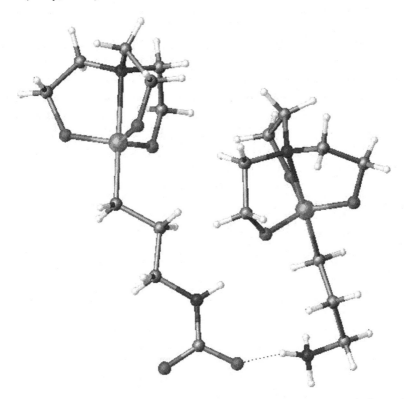

FIGURE 7.8   (See color insert.) X-ray structure of [1-(3-ammoniumpropyl)silatrane]1-(3-carbamatepropyl)silatrane. Reprinted with permission from Dumitriu, A. M. C., et al. Synthesis and Structural Characterization of 1-(3-aminopropyl) Silatrane and Some New Derivatives. *Polyhedron* **2012,** *33*, 119–126. © 2012 Elsevier.

7.4   AMINO-SILICONES—HIGH EFFICIENT ABSORBER FOR CO_2: ADVANTAGES AND DISADVANTAGES

The practical technologies for the capture and sequestration of CO_2 use as sorbents: MOFs, and various absorption solvents and solids, such as aqueous ethylenediamine solutions, amino-silicone mixtures, amino acid salt solutions, amidines, polyethylene glycol-stabilized silica-supported tetraethylenepentamine adsorbents, high surface area carbon molecular sieves, and metal-substituted layered double hydroxides.

From this large diversity of silicon-based materials, a part of them have been studied as possible candidates for CO_2 capture media. It is known that the solubility for CO_2 of the silicone oils, this fact making them suitable as CO_2 absorber.

Relative to the MEA CO_2 capture process, one of the most used, the use of the amino-silicones have some advantages: lower liquid-absorbent volatility (materialize in an decrease in the energy for evaporating liquid absorbent in the desorption process), an easy separation of the CO_2 from the liquid absorbent, a high thermal stability than MEA, a good working capacity, a small number of compression stages for transportation and storage of the CO_2. The advantages of silicone CO_2 sorbents as compared with their drawbacks are shown in Table 7.4.

TABLE 7.4   Silicone Materials as CO_2 Sorbents.

Silicone materials as CO_2 sorbents	
Advantages	**Disadvantages**
• Thermal stability	• The generally higher silicon price
• Low vapor pressures	• The amino-silicones absorbents react with sulfur compounds in the flue gas causing an increase of the costs associated with the absorbent loss
• Low viscosity fluids	
• Low heat capacities	
• High CO_2 loading	• Costs in removing the heat-stable salts formed
• Fast reaction rates	
• Low costs	
• Low liquid-absorbent volatility	
• Good work capacity	
• High desorption pressures	
• Small number of compression stages needed for transportation and storage of the CO_2 recovered	

7.5   APPLICATIONS AS GAS SENSORS

A CO_2 sensor represents an instrument for measuring the CO_2 gas. The most used principle for these types of sensors is infrared gas sensors and chemical sensors. Monitoring the quality of indoor air quality by measuring the CO_2 (e.g., in air-conditioning systems) is very important in our days.

An example of CO_2 sensor designed by Fleischer et al., in 2009, and published in 2014 has a APTMS/PTMS layer as sensor layer. The gas sensor projected by these scientists is based on field effect transistor construction and has gas-sensitivity to CO_2 layer, a polymer-based material. An impediment of this type of sensor is its sensitivity to moisture but that is compensated by adding a humidity sensor (Fleischer et al., 2014).

Different siloxane polymers and copolymers were tested as sensitive layers for CO_2 by Steigmeier et al.(2009). They proved to have good performances, such as fast response time, a high long-term stability and a high sensitivity between 400 ppm CO_2 and 4000 ppm CO_2 (Steigmeier et al., 2009). Chemical CO_2 gas sensors having the sensitive layers based on heteropolysiloxane or polymer (fluoropolymers) have the most important advantages: very low energy consumption and they can be reduced in size sufficiently to fit into microelectronic-based systems. As the CO_2 is absorbed by the polymer layer, its capacitance is modified. This change can be determined and measured in order to know the CO_2 concentration. There are also important drawbacks: low overall lifetime and short- and long-term drift effects in comparison with the nondispersive infrared (*NDIR*) measurement principle (Singh, 2014). Most CO_2 sensors are fully calibrated prior dispatching from the factory but over time, zero point of the sensor needs to be calibrated for maintaining its long-term stability. A complex of lanthanum with Schiff base derived from APDS and 2,4-dihydroxybenzaldehyde was tested by Telipan et al.(2008) an active layer in a CO_2 gas detector. In this case not the formation of carbamate or carbonate species occurs, but probably the involvement of CO_2 in the coordination of the metal leads to changes in film conductivity.

The sensor was made by thin films technologies using an alumina plate $4 \times 4 \times 0.6$ mm as substrate, two plate electrodes. The complex was dissolved in chloroform and coated as 200 nm thickness sensitive layer on the substrate by dip coating method. The sensor was mounted on the transistor. The gases detection was performed in NO_x and CO_2 atmospheres. The sensor proved to be more sensitive for NO_x detection than for CO_2 (Telipan et al., 2008).

Other types of CO_2 sensors are the optical ones based on room temperature of ionic liquids ((RTILs)-1-butyl-3-methylimidazolium salts) encapsulated in silicone matrix (Borisov et al., 2007). The materials are also used as fluorescent and absorption-based pH indicator. By varying the nature of ionic liquid, it is possible to tune the dynamic range. Semi-quantitative determination of CO_2 is achieved by dissolving an absorption-based pH indicator [thymol blue (TB) or bromthymol blue (BTB)] in the RTIL. A quantitative fluorimetric sensor makes use of 8-hydroxypyrene-1,3,6-trisulfonate (HPTS). The response of the fluorimetric sensors to CO_2 can be linearized making the calibration of the sensor very simple.

A reference inert fluorescent dye (4-dicyanomethylene-2-methyl-6-(4-(dimethylamino)styryl)-4H-pyran) can be added for radiometric measurements. It is also shown that stable inorganic salts such as sodium phosphate can be used instead of quaternary ammonium hydroxides. Sensitivity of the material can be tuned by varying the pK_a of an indicator (Borisov et al., 2007). The sensors can be applied in environmental monitoring.

7.6 CONCLUDING REMARKS AND OUTLOOKS

The development of efficient, low-cost materials for capturing CO_2 from the flow of gases emitted by fossil fuel power plants is a time domain. CSS has the potential to lower the greenhouse gas emissions in the next few decades, cutting emissions and providing the opportunity to transition to a low carbon economy. The use of CSS would have different effects on the amount of pollutants emitted in the atmosphere, while CO_2 and sulfur dioxide emissions would decrease substantially as these gases would be removed from the flow of gases resulted from burning fossil fuels, emissions of ammonia would increase as the traditional techniques for capturing CO_2 (Niu et al., 2016). Developing new materials (based on silicone materials) that will be used as substrates for capturing CO_2 gas is a very important issue for the future of our world. The new materials for carbon capture would be designed to allow a larger number of cycles of sorption–desorption of CO_2 as compared with liquid amines that were previously tested for CO_2 capture.

From a scientific standpoint, the materials created in further studies will allow controlled capture of CO_2 gas and the possibility to install economically viable devices for carbon capture at all power plants and industrial sites that use fossil fuels. This in turn will allow humankind to avoid an increase of the CO_2 concentration in the atmosphere above 450 ppm, thus avoiding the most negative effects of climate change due to accumulation of greenhouse gases in the atmosphere, creating the conditions for continued economic growth around the world.

ACKNOWLEDGMENTS

This work was supported by a grant from the Ministry of National Education, CNCS-UEFISCDI, project number PN-III-P4-ID-PCE-2016-0642.

KEYWORDS

- CO_2 capture and storage
- silicones
- amino-silicones
- silica
- functionalized-silica

REFERENCES

Avila, S. G., et al. Incorporation of Monoethanolamine (MEA), Diethanolamine (DEA) and Methyldiethanolamine (MDEA) in Mesoporous Silica: An Alternative to CO_2 Capture. *J. Environ. Chem. Eng.* **2016,** *4* (4), 4514–4524.
Bargan, A., et al. A New Zwitterionic Siloxane Compound: Structural Characterization, the Solution Behavior and Surface Properties Evaluation. *J. Mol. Liquids* **2014,** *196*, 319–325.
Bates, E. D., et al. CO_2 Capture by a Task-specific Ionic Liquid. *J. Am. Chem. Soc.* **2002,** *124* (6), 926–927.
Begag, R., et al. Superhydrophobic Amine Functionalized Aerogels as Sorbents for CO_2 Capture. *Greenhouse Gas Sci. Technol.* **2013,** *3*, 30–39.
Borisov, S. M., et al. Optical Carbon Dioxide Sensors Based on Silicone-encapsulated Room-temperature Liquids. *Chem. Mater.* **2007,** *19* (25), 6187–6194.
Budzianowski, W. M. *Energy Efficient Solvents for CO_2 Capture by Gas-liquid Absorption: Compounds, Blends and Advanced Solvent Systems*; Springer: Cham, Switzerland, 2016.
Chang, F. Y., et al. Adsorption of CO_2 onto Amine-grafted Mesoporous Silicas. *Sep. Purif. Technol.* **2009,** *70*, 87–95.
Costa, C. C., et al. Synthesis Optimization of MCM-41 for CO_2 Adsorption Using Simplex-centroid Design. *Mater. Res.* **2015,** *18* (4), 714–722.

Costa, J. A. S., et al. A New Functionalized MCM-41 Mesoporous Material for Use in Environmental Applications. *J. Braz. Chem. Soc.* **2014,** *25* (2), 197–207.

Cui, S., et al. Mesoporous Amine-modified SiO_2 Aerogel: A Potential CO_2 Sorbent. *Energy Environ. Sci.* **2011,** *4,* 2070–2074.

Czaun, M., et al. Organoamines-grafted on Nano-sized Silica for Carbon Dioxide Capture. *J. CO_2 Util.* **2013,** *1,* 1–7.

Dibenedetto, A., et al. Hybrid Materials for CO_2 Uptake from Simulated Flue Gases: Xerogels Containing Diamines. *Chem. Sus. Chem.* **2008,** *1,* 742– 745.

Dorcheh, A. S.; Abbasi, M. H. Silica Aerogels; Synthesis, Properties and Characterization. *J. Mater. Proc. Technol.* **2008,** *199* (1–3), 10–26.

Dumitriu, A. M. C., et al. Synthesis and Structural Characterization of 1-(3-aminopropyl) Silatrane and Some New Derivatives. *Polyhedron* **2012,** *33,* 119–126.

Esam, O. A. CO_2 Capture on Porous Adsorbents Containing Surface Amino Groups. Electronic Theses and Dissertations, Paper 2304, 2013. http://dc. etsu.edu/etd/2304.

Feron, P. *Absorption-based Post-combustion Capture of Carbon Dioxide*; Woodhead Publishing: Duxford, UK, 2016.

Fleischer, M.; Pohle, R.; Steigmeier, S. Carbon Dioxide Sensor and Associated Method for Generating a Gas Measurement Value. U.S. Patent 8683845 B2, 2014.

Gholami, M.; Talaie, M. R.; Aghamiri, S. F. CO_2 Adsorption on Amine Functionalized MCM-41: Effect of Bi-modal Porous Structure. *J. Taiwan Inst. Chem. Eng.* **2016,** *59,* 205–209.

Gunathilake, C., et al. Amine-modified Silica Nanotubes and Nanospheres: Synthesis and CO_2 Sorption Properties. *Environ. Sci. Nano.* **2016,** *3,* 806–817.

Hao, N., et al. One Step Synthesis of Amine-functionalized Hollow Mesoporous Silica Nanoparticles as Efficient Antibacterial and Anticancer Materials. *ACS Appl. Mater. Interfaces* **2015,** *7,* 1040–1045.

Huang, C. H.; Klinthong, W.; Tan, C. S. SBA-15 Grafted with 3-aminopropyl Triethoxysilane in Supercritical Propane for CO_2 Capture. *J. Supercrit. Fluids* **2013,** *77,* 117–126.

IPCC. *IPCC Special Report on Carbon Dioxide Capture and Storage, Prepared by Working Group III of the Intergovernmental Panel on Climate Change*; Metz, B., Davidson, O., de Coninck, H. C., Loos, M., Meyer, L. A., Eds.; Cambridge University Press: Cambridge, UK and New York, NY, 2005; pp 442.

Kammermeyer, K.; Sollami, B. J. Method of Recovering Carbon Dioxide from a Fluid. U.S. Patent 3665678, May 30, 1972.

Karl, T. R., et al. Possible Artifacts of Data Biases in the Recent Global Surface Warming Hiatus. *Science* **2015,** *348* (6242), 1469–1472.

Kishor, R.; Ghoshal, A. K. APTES Grafted Ordered Mesoporous Silica KIT-6 for CO_2 Adsorption. *Chem. Eng. J.* **2015,** *262,* 882–890.

Leal, O., et al. Reversible Adsorption of Carbon Dioxide on Amine Surface-bonded Silicagel. *Inorg. Chim. Acta.* **1995,** *240* (1–2), 183–189.

Li, S.; Li, Y.; Wang, J. Solubility of Modified Poly(Propylene Oxide) and Silicones in Supercritical Carbon Dioxide. *Fluid Ph. Equilibria* **2007,** *253,* 54–60.

Li, K. M., et al. Influence of Silica Types on Synthesis and Performance of Amine-silica Hybrid Materials Used for CO_2 Capture. *J. Phys. Chem. C.* **2014a,** *118,* 2454–2462.

Li, K., et al. The Influence of Polyethyleneimine Type and Molecular Weight on the CO_2 Capture Performance of PEI-nano Silica Adsorbents. *Appl. Energy* **2014b,** *136,* 750–755.

Lineen, N.; Pfeffer, R.; Lin, Y. S. CO_2 Capture Using Particulate Silica Aerogel Immobilized with Tetraethylenepentamine. *Micropor. Mesopor. Mater.* **2013**, *176*, 123–131.

Loganathan, S.; Ghoshal, A. K. Amine Tethered Pore-expanded MCM-41: A Promising Adsorbent for CO_2 Capture. *Chem. Eng. J.* **2017**, *308*, 827–839.

MacDowell, N., et al. An Overview of CO_2 Capture Technologies. *Energy Environ. Sci.* **2010**, *3*, 1645–1669.

Maleki, H. Recent Advances in Aerogels for Environmental Remediation Applications: A Review. *Chem. Eng. J.* **2016**, *300*, 98–118.

Melendez-Ortiz, H. I., et al. Hydrothermal Synthesis of Mesoporous Silica MCM-41 Using Commercial Sodium Silicate. *J. Mex. Chem. Soc.* **2013**, *57* (2), 73–79.

Meng, L. Development of an Analytical Method for Distinguishing Ammonium Bicarbonate from the Products of an Aqueous Ammonia CO_2 Scrubber and the Characterization of Ammonium Bicarbonate. Masters Theses & Specialist Projects, Western Kentucky University, Paper 243, 2004. tp://digitalcommons.wku.edu/theses/243.

Minju, N., et al. Amine Impregnated Porous Silica Gel Sorbents Synthesized from Waterglass Precursors for CO_2 Capturing. *Chem. Eng. J.* **2015**, *269*, 335–342.

Nigar, H., et al. Amine-functionalized Mesoporous Silica: A Material Capable of CO_2 Adsorption and Fast Regeneration by Microwave Heating. *AIChE J.* **2016**, *62*, 547–555.

Niu, M., et al. Amine-impregnated Mesoporous Silica Nanotube as an Emerging Nanocomposite for CO_2 Capture. *ACS Appl. Mater. Interfaces* **2016**, *8*, 17312–17320.

Noll, W. Chemistry and Technology of Silicones. In *Properties of Technical Products*; Noll, W., Ed.; Academic Press: New York, NY, 1968; pp 439–440, Chapter 9.

Park, H., et al. Analysis of the CO_2 and NH_3 Reaction in an Aqueous Solution by 2D IR COS: Formation of Bicarbonate and Carbamate. *J. Phys. Chem. A.* **2008**, *112*, 6558–6562.

Park, A. H. A., et al. Methods and Systems for Capturing Carbon Dioxide and Producing a Fuel Using a Solvent Including a Nanoparticle Organic Hybrid Material and a Secondary Fluid. US Patent 20150014182A1, 2012.

Perry, R. J., et al. Aminosilicone Solvents for CO_2 Capture. *Chem. Sus. Chem.* **2010**, *3*, 919–930.

Perry, R. J.; O'Brien, M. J. Amino Disiloxanes for CO_2 Capture. *Energy Fuels* **2011**, *25*, 1906–1918.

Perry, R. J. Aminosilicone Systems for Post-combustion CO_2 Capture. In *Absorption-based Post-combustion Capture of Carbon Dioxide*; Elsevier: Amsterdam, Netherlands, 2016; pp 121–144.

Pino, L., et al. Sorbents with High Efficiency for CO_2 Capture Based on Amine-supported Carbon for Biogas Upgrading. *J. Environ. Sci.* **2016**, *48*, 138–150.

Qi, G., et al. High Efficiency Nanocomposite Sorbents for CO_2 Capture Based on Amine-functionalized Mesoporous Capsules. *Energy Environ. Sci.* **2011**, *4*, 444–452.

Sanli, D.; Erkey, C. Effect of Polymer Molecular Weight and Deposition Temperature on the Properties of Silica Aerogel/Hydroxy-Terminated Poly(Dimethylsiloxane)Nanocomposites Prepared by Reactive Supercritical Deposition. *J. Sup. Fluids* **2015**, *105*, 99–107.

Sanz-Perez, E. S., et al. Reuse and Recycling of Amine-functionalized Silica Materials for CO_2 Adsorption. *Chem. Eng. J.* **2017**, *308*, 1021–1033.

Sanz-Perez, E. S., et al. Direct Capture of CO_2 from Ambient Air. *Chem. Rev.* **2016**, *116*, 11840–11876.

Singh, M. *Introduction to Biomedical Instrumentation*; Prentice Hall: Upper Saddle River, NJ, 2014; p 248.

Soto-Cantu, E., et al. Synthesis and Rapid Characterization of Amine-functionalized Silica. *Langmuir* **2012**, *28*, 5562–5569.Steigmeier, S.; Fleischer, M.; Tawil, A. Optimization of the Work Function Response of the CO_2 Sensing. Polysiloxane Layers by Modification of the Polymerization. *Sensors 2009 IEEE*, **2009**, 1742–1746. INSPEC Accession Number 11119247, DOI 10.1109/ICSENS.2009.5398480.

Stone, E. J.; Lowe, J. A.; Shine, K. P. The Impact of Carbon Capture and Storage on Climate. *Energy Environ. Sci.* **2009**, *2*, 81–91.

Telipan, G., et al. Lanthanum Complex for Gas Sensing. *J. Optoelectron. Adv. Mater.* **2008**, *10* (12), 3409–3412.

Thi Le, M. U.; Lee, S. Y.; Park, S. J. Preparation and Characterization of PEI-loaded MCM-41 for CO_2 Capture. *Int. J. Hydrogen Energy* **2014**, *39*, 12340–12346.

Unveren, E. E., et al. Solid Amine Sorbents for CO_2 Capture by Chemical Adsorption: A Review. *Petroleum* **2016**, *3*, 1–14.

Veneman, R., et al. Adsorption of CO_2 and H_2O on Supported Amine Sorbents. *Energy Procedia* **2014**, *63*, 2336–2345.

Wang, J., et al. Recent Advances in Solid Sorbents for CO_2 Capture and New Developments Trends. *Energy Environ. Sci.* **2014**, *7*, 3478–3518.

Wang, X.; Li, B.; Phase-change Solvents for CO_2 Capture. In *Novel Materials for Carbon Dioxide Mitigation Technology*; Elsevier: Amsterdam, Netherlands, 2015.

Xu, X., et al. Novel Polyethylenimine-modified Mesoporous Molecular Sieve of MCM-41 Type as High-capacity Adsorbent for CO_2 Capture. *Energy Fuels* **2002**, *16*, 1463–1469.

Yoo, C. J.; Lee, L. C.; Jones, C. W. Probing Intramolecular Versus Intermolecular CO_2 Adsorption on Amine-grafted SBA-15. *Langmuir* **2015**, *31*, 13350–13360.

Yu, C. H.; Huang, C. H.; Tan, C. S. A Review of CO_2 Capture by Absorption and Adsorption. *Aerosol Air Qual. Res.* **2012**, *12*, 745–769.

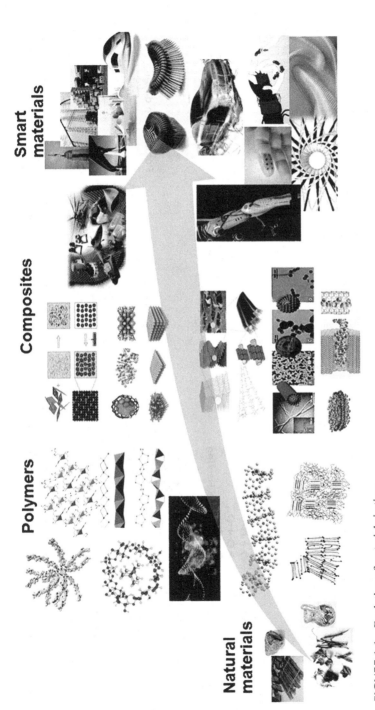

FIGURE 1.1 Evolution of materials in time.

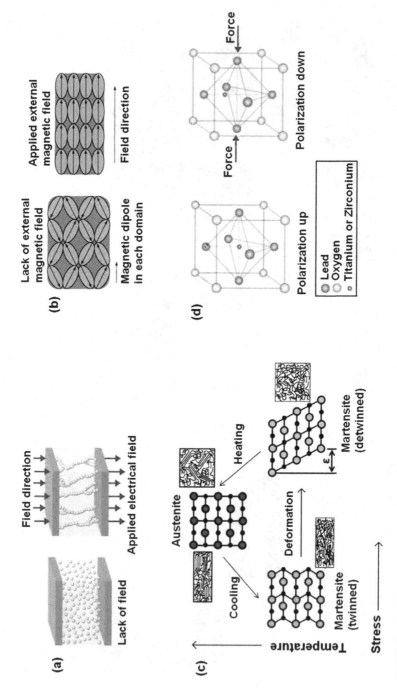

FIGURE 1.3 Schematic representations of shape fixing and recovery mechanisms for some types of smart materials: (a) magneto- and electro-rheological, (b) magnetostriction (electrostriction), (c) shape–memory alloys, and (d) piezo materials.

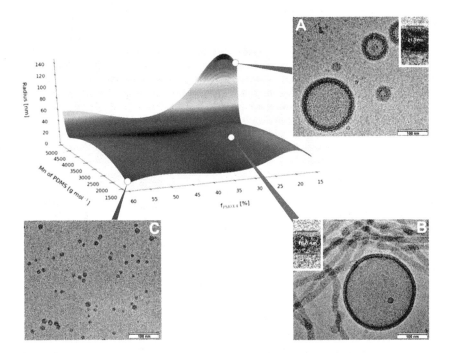

FIGURE 2.3 3D phase diagram of various amphiphilic diblock copolymers based on PDMS and PMOXA. The three white points with the corresponding cryo-TEM micrographs represent the supramolecular architectures formed by self-assembly of PDMS$_{65}$-b-PMOXA$_{14}$ (A), PDMS$_{39}$-b-PMOXA$_8$ (B), and PDMS$_{16}$-b-PMOXA$_7$ (C). As inset, the enlarged view of the membrane thickness for PDMS$_{65}$-b-PMOXA$_{14}$ (A; 21.3 nm), and PDMS$_{39}$-b-PMOXA$_8$, respectively (B; 16 nm), is inserted. Reprinted with permission from Wu, D., et al. Effect of Molecular Parameters on the Architecture and Membrane Properties of 3D Assemblies of Amphiphilic Copolymers. *Macromolecules* **2014**, *47* (15), 5060−5069. © 2014 American Chemical Society.

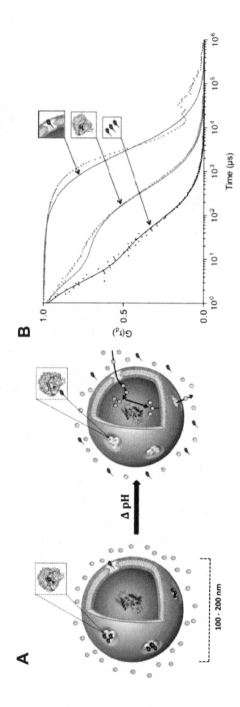

FIGURE 2.4 Schematic representation of a nanoreactor with a pH-triggered enzymatic activity controlled by chemically modified porin "gates" reconstituted in a polymersome membrane based on PMOXA$_6$–*b*–PDMS$_{44}$–*b*–PMOXA$_6$ triblock copolymer. (A) The pH change induces the cleavage of molecular cap (green dots) opening the "gate" for both the entrance of substrates (red dots) and release of the products resulted from enzymatic reaction (yellow dots). (B) Autocorrelation curves of pH-sensitive fluorescent cap, cyanine5-hydrazide (Cy5-hydrazide), in phosphate buffer (black), Cy5-modified OmpF in 3% octyl-glucopyranoside micelles (red), and polymersomes with reconstituted Cy5-modified OmpF (blue). The significant increase of diffusion time for micelles and polymersomes containing the capped porin compared to the freely diffusing dye indicates the chemical modification of channel protein and its insertion into the polymer membrane. Reprinted with permission from Einfalt, T., et al. Stimuli-triggered Activity of Nanoreactors by Biomimetic Engineering Polymer Membranes. *Nano Lett.* **2015**, *15* (11), 7596–7603. © 2015, American Chemical Society.

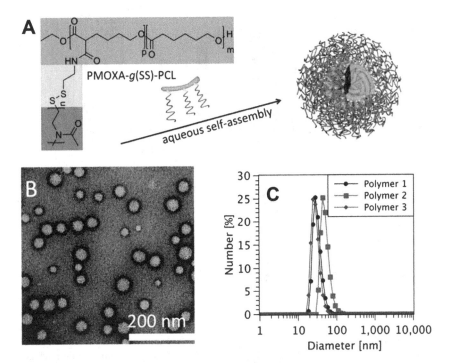

FIGURE 2.7 (A) Chemical structure of reduction-responsive amphiphilic graft copolymer, PMOXA-*g*(SS)-PCL, and the schematic representation of NPs self-assembled in aqueous medium; (B) TEM micrographs of NPs based on PMOXA-*g*(SS)-PCL copolymers; and (C) hydrodynamic diameter of NPs and their number distribution in phosphate buffer determined by DLS. Reprinted with permission from Najer, A., et al. An Amphiphilic Graft Copolymer-based Nanoparticle Platform for Reduction-responsive Anticancer and Antimalarial Drug Delivery. *Nanoscale* **2016,** *8* (31), 14858–14869. © 2016 Royal Society of Chemistry.

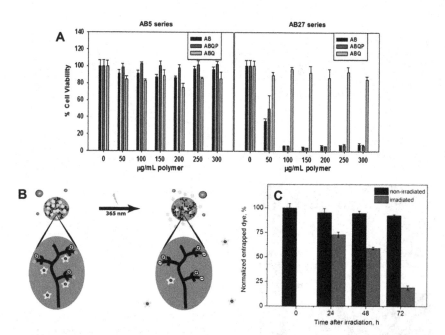

FIGURE 2.8 (A) HeLa cell viability vs. polymer concentration after 24 h for both Quaternized and non-quaternized NPs; (B) schematic representation of the photo-triggered release mechanism for the quaternized self-assembled NPs loaded with anionic payload; and (C) the content of entrapped dye (%), evaluated by fluorimetry, at various time points of dye release from the irradiated ABQ27 NPs compared to non-irradiated NPs. Reprinted with permission from Dinu, I. A., et al. Engineered Non-toxic Cationic Nanocarriers with Photo-triggered Slow-release Properties. *Polym. Chem.* **2016,** *7* (20), 3451–3464. © 2016 Royal Society of Chemistry.

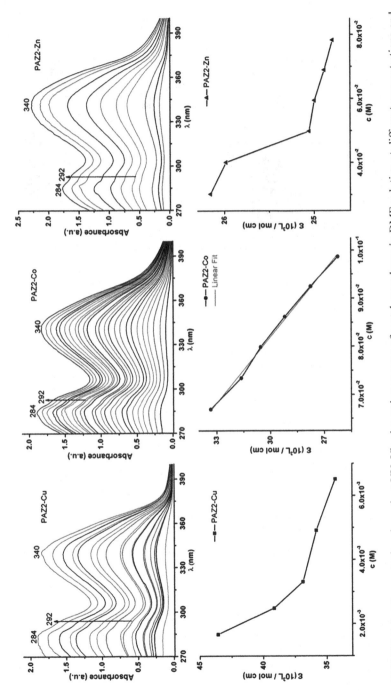

FIGURE 5.4 Concentration-dependent UV–Vis absorption spectra of metal complexes in DMF solution at different concentrations and plots of extinction coefficient of absorbance maximum (ε) at 340 nm vs. molar concentration (c (M)).

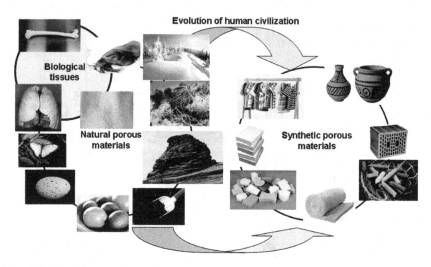

FIGURE 6.1   The evolution of porous materials.

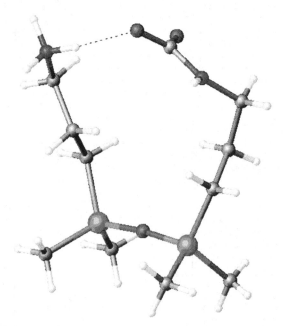

FIGURE 7.6   Structure of carbamate derivative of 1,3-bis(3-aminopropyl)tetramethyld-isiloxane, as determined crystallographically. Reprinted with permission from Bargan, A., et al. A New ZwitterionicSiloxane Compound: Structural Characterization, the Solution Behavior and Surface Properties Evaluation. *J. Mol. Liquids* **2014**, *196*, 319–325. © 2014 Elsevier.

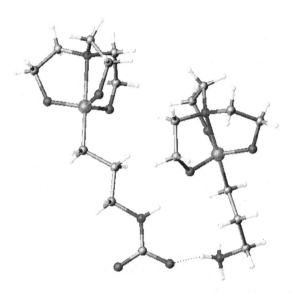

FIGURE 7.8   X-ray structure of [1-(3-ammoniumpropyl)silatrane]1-(3-carbamatepropyl) silatrane. Reprinted with permission from Dumitriu, A. M. C., et al. Synthesis and Structural Characterization of 1-(3-aminopropyl) Silatrane and Some New Derivatives. *Polyhedron* **2012**, *33*, 119–126. © 2012 Elsevier.

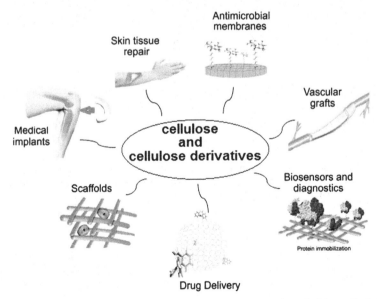

FIGURE 8.1   Applications of cellulose and cellulose derivatives in biomedical fields. Adapted from Rojas et al. (2015).

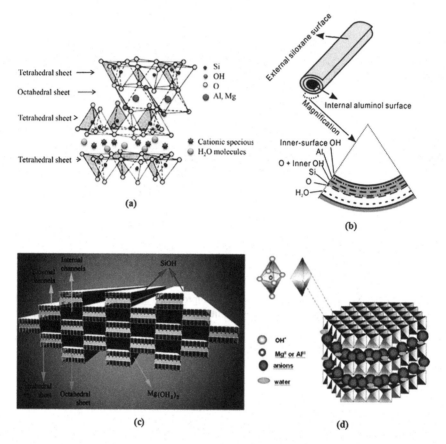

FIGURE 9.2   Schematic representation of (a) 2:1 layered clay, (b) halloysite (1:1 layered clay), (c) sepiolite (chain silicate), and (d) layered double hydroxide structures.

a: Reprinted with permission from Ghadiri, M.; Chrzanowski, W.; Rohanizadeh, R. Biomedical Applications of Cationic Clay Minerals. *RSC Adv.* **2015**, *5* (37), 29467–29481. © 2015 Royal Chemical Society.

b: Reprinted with permission from Yuan, P.; Tan, D.; Annabi-Bergaya, F. Properties and Applications of Halloysite Nanotubes: Recent Research Advances and Future Prospects. *Appl. Clay Sci.* **2015**, *112–113*, 75–93. © 2015 Elsevier.

c: Reprinted with permission from Soheilmoghaddam, M., et al. Characterization of Bio Regenerated Cellulose/Sepiolite Nanocomposite Films Prepared via Ionic Liquid. *Polym. Test.* **2014**, *33*, 121–130. © 2014 Elsevier.

d: Reprinted with permission from Mohapatra, L.; Parida, K. A Review on the Recent Progress, Challenges and Perspective of Layered Double Hydroxides as Promising Photocatalysts. *J. Mater. Chem. A.* **2016**, *4* (28), 10744–10766. © 2016 Royal Chemical Society.

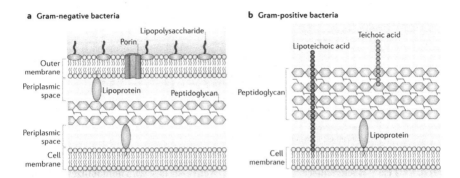

FIGURE 11.2 Cell wall structures of Gram-negative bacteria (a) and Gram-positive bacteria (b). Reprinted with permission from Brown, L., et al. Through the Wall: Extracellular Vesicles in Gram-positive Bacteria, *Mycobacteria* and Fungi. *Nat. Rev. Microbiol.* **2015**, *13* (10), 620–630. © 2015 Nature Publishing Group.

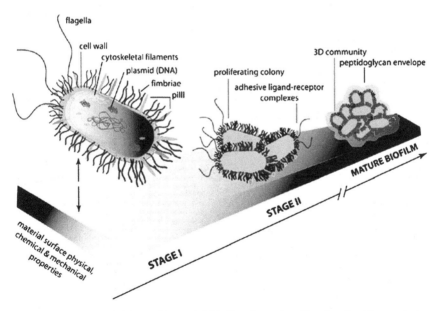

FIGURE 11.3 Mechanism of bacterial biofilm formation. Reprinted with permission from Lichter, A. J., et al. Design of Antibacterial Surfaces and Interfaces: Polyelectrolyte Multilayers as a Multifunctional Platform. *Macromolecules* **2009**, *42* (22), 8573–8586. © 2009 American Chemical Society.

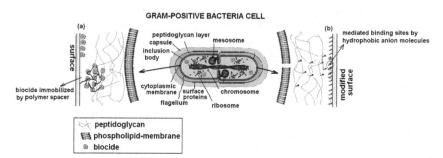

FIGURE 11.5 Bactericidal action mode for (a) Active-surface with immobilized biocides via polymeric spacer and (b) active-modified surface inducing the removal of hydrophobic anion from microbial cell wall.

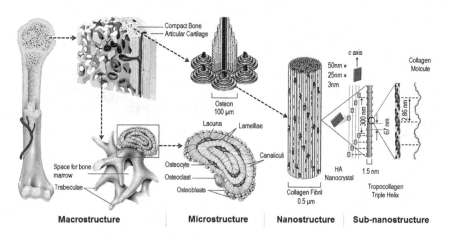

FIGURE 12.1 The seven levels of hierarchy in bone structure. Reprinted from permission from Wang, X., et al. Topological Design and Additive Manufacturing of Porous Metals for Bone Scaffolds and Orthopaedic Implants: A Review. *Biomaterials* **2016**, *83*, 127–141.© 2016 Elsevier.

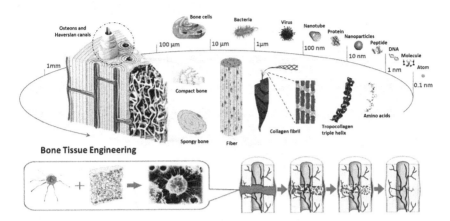

FIGURE 12.2 Bone tissue engineering concepts of biomimetic microstructure and microcomposition. Reprinted with permission from Hao, Z., et al. The Scaffold Microenvironment for Stem Cell Based Bone Tissue Engineering. *Biomater. Sci.* **2017**, *5*, 1382–1392. © 2017 Royal Society of Chemistry.

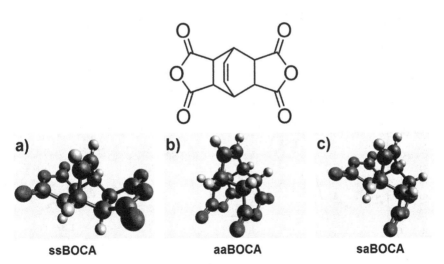

FIGURE 13.1 Spatial conformation of bicyclo-2,2,2-oct-7-ene-2,3,5,6-tetracarboxylic dianhydride—BOCA: (a) *sin*-conformation; (b) *anti*-conformation; (c) *sin-anti* conformation.

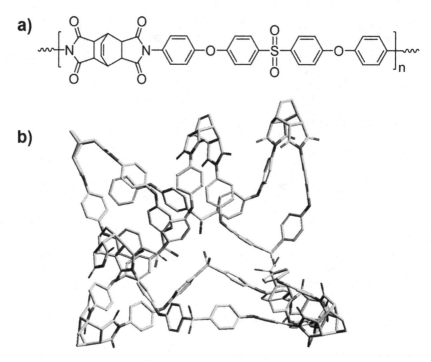

FIGURE 13.2 Chemical composition of BOCA—pBAPS structural unit (a) and 3D conformation of 7SU chain (b).

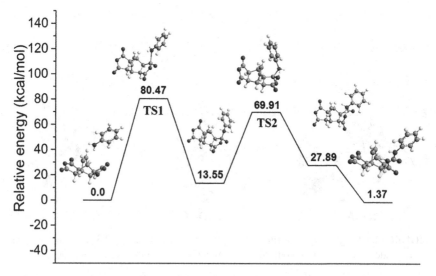

FIGURE 13.5 Energetic profile of addition of one aniline molecule to BOCA.

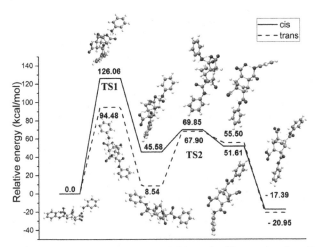

FIGURE 13.7   Energetic profile for second addition of aniline molecule to BOCA.

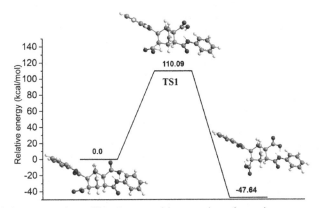

FIGURE 13.8   Energetic profile of transposition reaction of one site.

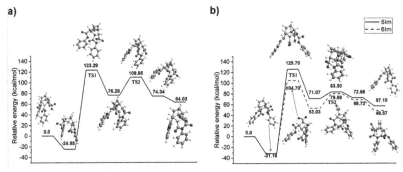

FIGURE 13.11   Energetic profile of the first step of dehydration reaction: (a) dehydration of *cis* form; (b) dehydration of *trans* form.

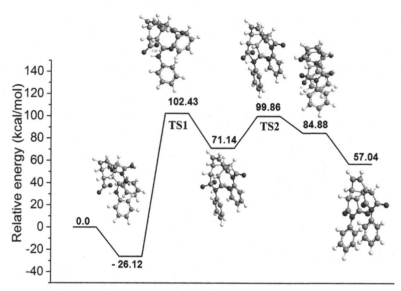

FIGURE 13.12 Energetic profile of second step of dehydration reaction for *cis* conformation.

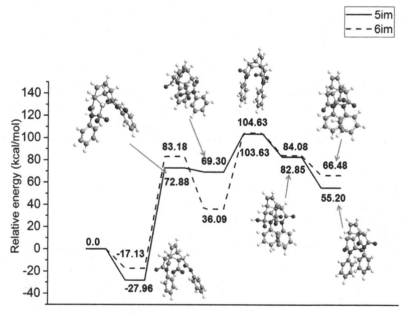

FIGURE 13.13 Energetic profile of second step of dehydration reaction for *trans* conformation.

CHAPTER 8

SMART BEHAVIOR OF CELLULOSE MATERIALS IN PHARMACEUTICAL INDUSTRIES

MIHAELA-DORINA ONOFREI*

Department of Physical Chemistry of Polymers, "Petru Poni" Institute of Macromolecular Chemistry, 41A Grigore Ghica Voda Alley, 700487 Iasi, Romania

E-mail: myha1976@yahoo.com

CONTENTS

ABSTRACT

Cellulose and cellulose derivatives have provided a very attractive area of research due to their possible applications in different economic sectors, especially pharmaceutical industries. Such materials are among the excipients frequently used in pharmaceutical compounded and industrialized products and, become more and more important in this area due to production of the new derivatives, and finding applications with different purposes. The fact that their intelligent behavior as reaction to environmental stimuli—temperature, pH, and ionic strength—has made materials based on cellulose attractive for various pharmaceutical applications (controlled-release drug delivery systems using matrices, binding agents, film formation during tablet coating, suspending agents for suspensions, thickening agents for ointments and creams, and hydrogels), discovery of their properties becomes essential. In this context, the chapter provides information on the basic properties of the cellulose and cellulose derivatives and how these polymers can be optimized to generate specific properties. The synthesis routes of novel materials and specific design techniques are examined, and the significant impact of the smart materials based on cellulose in the pharmaceutical industries is demonstrated.

8.1 INTRODUCTION

Over the last decade, importance of smart polymers is increasing day by day, because these polymers can undergo large reversible, chemical, or physical changes in response to small changes in the environmental conditions—pH, temperature, dual-stimuli, light, and phase transition (Mahajan and Aggarwal, 2011). As is known, cellulose has been the subject of vast investigations in macromolecular chemistry and the smart behavior has been observed among large number of cellulosic materials, which have provided a very attractive area of research in designing and synthesis smart behavior in biochemical sciences in many ways.

Cellulose is the most abundant resource in nature, present in herbal cells and tissues, which possess some promising properties—high swelling capacity, biocompatibility, biodegradability, and biological functions, able of broad chemical modification and a high mechanical stiffness and strength (Billah, 2015; Qiu and Hu, 2013; Rojas et al., 2015). Furthermore, due to the interesting properties and new functionalities, cellulose is utilized in varied applications—smart membranes, drug delivery systems, hydrogels, electro-active papers, and sensors. Such natural polysaccharides have been widely applied in different fields, like, wastewater treatment, food industry, wood and paper, fibers and clothes, cosmetic, biomedical, an excipient in pharmaceutical industries, and tissue engineering applications (Qiu and Hu, 2013; Duan et al., 2014; Kim et al., 2006). Moreover, this smart material based on cellulose show great interest, particularly their intelligent behavior in reaction to environmental stimuli (temperature, pH, and ionic strength), and they can be utilized for many conditions, mostly as biomaterials and drug carriers (Billah, 2015; Rojas et al., 2015; Qiu and Hu, 2013; Kim et al., 2006).

Withal, chemical modification of cellulose for various applications such as food, cosmetics, printing, textile, and pharmaceutical, has been an interesting subject of research in recent years, and new applications are continuously sought. Thus, chemical modification of cellulose is carried out to improve processability and to produce strong, reproducible, recyclable, and biocompatible cellulose derivatives with different physicochemical and mechanical properties which can be adapted for specific industrial applications, in particular for pharmaceutical applications (Fig. 8.1) (Rojas et al., 2015; Lavanya et al., 2011).

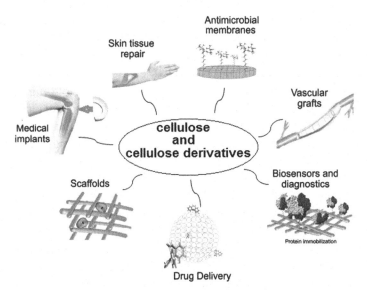

FIGURE 8.1   (See color insert.) Applications of cellulose and cellulose derivatives in biomedical fields. Adapted from Rojas et al. (2015).

The most common cellulose derivatives produced by chemical modification are cellulose esters and ethers, which are obtained by the reaction of some hydroxyl groups on the repeating unit of cellulose. Indeed, cellulose and cellulose derivatives behave as smart materials and offer a great diversity of properties that can be employed in various applications (Onofrei et al., 2015). Thanks to biodegradability and biocompatibility of cellulose, its derivatives are utilized in various applications—pharmaceuticals, immobilization of proteins and antibodies, coatings, optical films, laminates, textiles, and foodstuffs, as well as the formation of cellulose composites—synthetic polymers and biopolymers (Khan et al., 2016). Cellulose and cellulose derivatives have been widely used in medical applications such as wound dressing, tissue engineering, controllable drug delivery system, blood purification, etc., due to its biocompatibility, hydrophilicity, biodegradability, nontoxicity, and antimicrobial properties. Furthermore, it has also found an application in the treatment of renal failure, as well as in a variety of more recent and evolving clinical applications such as for scaffolds in tissue engineering, temporary skin substitute, hemostatic agent, postoperative adhesion barrier, and as a culture material for hepatocytes (Hoenich, 2006).

For successful application of cellulose-based materials in different fields, as hydrophobic coatings, oleophobic or antibacterial functionalities, and technology and biomaterials, it is necessary to adjust the surface properties for water repellence, hydrophilicity for long time, antibacterial assets, enhanced adhesion, etc. (Cortese et al., 2015). In addition, cellulose and its derivatives, have played an integral part in the evolution of drug delivery technology by assuring controlled release of therapeutic agents in constant doses (over long periods) and by gradual dosage of drugs (hydrophilic and hydrophobic). Moreover, these are employed as excipients for oral, topical/parenteral administration in pharmaceutical products.

From cellulose ethers can be prepared a very wide range of polymer products, which vary from each other with respect to type of substituent, substitution level, viscosity, molecular weight, and particle size. Thus, cellulose ethers are biocompatible and therefore, can be used for food, cosmetics and especially, pharmaceutical scope (as binders, coating agents, emulsifying, stabilizing, agents, and tablet disintegrants) (Mahanta and Mahanta, 2015). Also, water-insoluble cellulose esters are polymers with good film-forming characteristics which find a variety of applications, as material coatings and controlled-release systems, hydrophobic matrices, and semipermeable membranes—for applications in pharmacy, agriculture, and cosmetics. In recent years, there has been observed a significant interest for cellulose esters compounds in the field of biomedical materials, which combine natural and synthetic polymers. These polymers can be used in a range of applications, including drug delivery systems, wound closure, novel vascular grafts, or scaffolds for tissue engineering (in vitro or in vivo). (Mahanta and Mahanta, 2015; Messersmith and Giannelis, 1995; Gorrasi et al., 2003; Kunioka et al., 2007; Erceg et al., 2008; Bordes et al., 2009; Botana et al., 2010; D'Amico et al., 2012; Bruno Rocha e Silva et al., 2013).

Nowadays, cellulose and cellulose derivatives (ether and ester) are among the excipients frequently used in pharmaceutical compounded and industrialized products, become more and more important in this area due to production of the new derivatives, and finding applications with different purposes. This chapter reviews the main researches of smart materials based on cellulose, including the preparations, classification, properties, and applications of these materials in the pharmaceutical industries.

8.2 CLASSIFICATION OF CELLULOSE MATERIALS

Cellulose is a natural polymeric polysaccharide constitutes long chains of anhydro-D-glucopyranose units (AGU), with each cellulose molecule possessing three hydroxyl groups per AGU (excepting of the terminal ends). Cellulose derivatives belong to the general class of hydrophilic colloids and have in common that they are hydrophilic, semisynthetic linear macromolecules obtained through chemical modification of cellulose (Fig. 8.2).

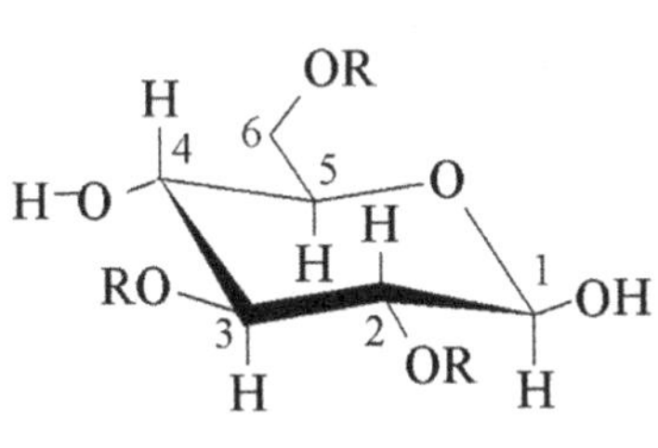

FIGURE 8.2   General chemical structure of cellulose derivatives.

In this context, a detailed classification of cellulose derivatives can be made on the basis of different criteria (chemical modification of cellulose—etherification or esterification, water solubility, and electrolytic dissociation). Such classifications are useful because they facilitate the discussion of cellulose derivatives, which were designed and fine-tuned to obtain certain desired properties. These hydroxyl groups of cellulose can be modified chemically to form esters or ethers which differ in physicochemical properties, offering a wide range of applications (Shanbhag et al., 2007; Kamel et al., 2008; Vlaia et al., 2016). Due to the strong intramolecular hydrogen bonding, pure cellulose is insoluble in cold or hot water. When is converted to cellulose ethers or esters cellulose, the derivatives become water-soluble, used in wide range of applications. Therefore, modified cellulose derivatives expand water retention capacity, pseudoplastic behavior, film-forming properties, and complexation (Mahanta and Mahanta, 2015).

- Based on the type of chemical modification of cellulose, cellulose derivatives can be separated in:
    - cellulose ethers derivatives—polymers formed by hydroxyl etherification with the appropriate alkyl halide of previously alkalinized cellulose;
    - cellulose esters derivatives—polymers formed by hydroxyl esterification with various organic acids in the presence of a strong acid as catalyst.
- Based on their water solubility, cellulose derivatives can be classified into two groups:
    - water-soluble polymers, including most of the cellulose ethers;
    - water-insoluble polymers, including the cellulose esters.
- According to their electrolytic dissociation or charge, cellulose derivatives can be classified as:
    - nonionic (uncharged) polymers that do not have an electric charge;
    - ionic (anionic and cationic) polymers with electric charge (Lowman and Peppas, 2000).

8.2.1   CELLULOSE DERIVATIVES ETHERS

Different cellulose ethers were achieved by etherification of the three hydroxyl groups on anhydroglucose units of cellulose, which produce water-soluble derivatives and finally cellulose ethers (Clasen and Kulicke, 2001; Marques-Marinho and Vianna-Soares, 2013). Cellulose ethers are water-soluble polymers derived from cellulose, which differ from each other with respect to type of substituent, molecular weight, chemical structure, degree of substitution, and molar substitution, properties that include solubility, viscosity, surface activity, thermoplastic film characteristics, and stability against biodegradation, heat, hydrolysis, and oxidation (Mahanta and Mahanta, 2015). Due to their multifunctional properties, methyl cellulose (MC), ethyl cellulose (EC), hydroxyethyl cellulose (HEC), carboxymethyl cellulose (CMC), sodium carboxymethyl cellulose (NaCMC), hydroxypropyl cellulose (HPC), and hydroxypropyl methylcellulose (HPMC) are the most common types of cellulose ethers (Clasen and Kulicke, 2001; Marques-Marinho and Vianna-Soares, 2013) (see Table 8.1).

TABLE 8.1   Cellulose Ethers Structure.

R	Cellulose ethers
$-CH_3$	Methyl cellulose
$-CH_2CH_3$	Ethyl cellulose
$-CH_2CH_2CH_2OH$	Hydroxypropyl cellulose
$-CH_3$ or $-CH_2CH(OH)CH_3$	Hydroxypropylmethyl cellulose
$-CH_2CH_2OH$	Hydroxyethyl cellulose
$-C_2H_5$ or $-CH_2CH_2OH$	Ethyl hydroxyethyl cellulose
$-CH_2COONa$	Carboxymethyl cellulose

Source: Adapted from Clasen and Kulicke (2001).

A very wide range of polymer products can be prepared using different cellulose ethers, which present a good solubility, high chemical resistance, and nontoxic nature. They are utilized in drilling technology and building materials, as additives for drilling fluids and processing of plaster systems, as stabilizers in food, pharmaceutical, and cosmetics formulations as the fundamental component, as well as important excipients for designing matrix tablets (Lowman and Peppas, 2000).

Furthermore, cellulose ethers start to swell on contact with water, and the hydrogel layer starts to increase around the dry core of the tablet. Thus, the hydrogel represents a diffusional barrier for water molecules, able to penetrate the polymer matrix and to release the drug molecules (Poersch-Parcke and Kirchner, 2003; Kamel et al., 2008; Mahanta and Mahanta, 2015; Yilmaz et al., 2015). In addition, cellulose ethers are gaining considerable importance because they are biocompatible and can be used for pharmaceutical purposes. Due to their hydrophilicity and good gel-forming characteristics, cellulose ethers (soluble and insoluble) have been utilized in hydrophilic polymeric matrices.

HPMC which comprises both hydrophobic and hydrophilic structural units is water-soluble cellulose ether, with excellent swelling ability, good compressibility, and a fast hydration characteristic. Thus, HPMC is a pharmaceutically significant natural polymer which is mainly used in the preparation of controlled-release tablets. HPC is a nonionic water-soluble, used in hydrophilic matrix systems as thickening agent, tablet binder, and modified release, and film coating polymer. Also, HEC manifests the highest swelling ability and hydration rate. EC is a nonionic cellulose ether, insoluble in water, but soluble in some polar organic solvents which can be used in hydrophobic matrix system. NaCMC is used as an emulsifying agent in pharmaceuticals and in cosmetics, presenting a wide range of functional properties like binding, thickening, and stabilizing agent. Moreover, mixtures of cellulose ethers with other cellulose ethers or different polymers are used in controlled release systems (Shokri et al., 2011; Mahanta and Mahanta, 2015; Yilmaz et al., 2015).

8.2.2   CELLULOSE DERIVATIVES ESTERS

Cellulose ester's properties are adapted to the requirements of pharmaceutical applications, which permit construction of drug delivery systems that address patient needs. These properties comprise a low toxicity, stability, dietary decomposition products, high water permeability, film strength, compatibility with a wide range of actives, and capacity to form microparticles and nanoparticles. Moreover, these polymers are frequently used with cellulose ethers for preparation of microporous delivery membranes (Mahanta and Mahanta, 2015; Edgar, 2007). In this context, cellulose esters behave as smart materials and offer a variety of properties that can be exploited in several applications.

Table 8.2 displays different cellulose esters which are formed by hydroxyl esterification with acetic, trimellitic, dicarboxylic phthalic, or

TABLE 8.2 Cellulose Esters Structure.

Ester groups		R	
Acetate	H, I		I
Acetate trimellitate	H, I, II		I
			II
Acetate phthalate	I, III		I
			III
Hydroxypropyl methyl phthalate	H, CH$_3$, CH$_2$CH(OH)CH$_3$, III, IV		III
			IV
Hydroxypropyl methyl phthalate acetate succinate	H, CH$_3$, CH$_2$CH(OH)CH$_3$, I, V		I
			V

Source: Adapted from Marques-Marinho and Vianna-Soares (2013).

succinic acids, or a combination of these compounds. In addition, all these reactions occur in the presence of a strong acid that supports the acid catalysis (Marques-Marinho and Vianna-Soares, 2013). Cellulose esters such as cellulose acetate (CA), cellulose acetate phthalate (CAP), cellulose acetate butyrate (CAB), cellulose acetate trimellitate (CAT), hydroxypropyl methylcellulose phthalate (HPMCP), hydroxypropyl methylcellulose acetate succinate (HPMCAS), and many others were used as matrix materials. They are used extensively as binders, fillers, and laminate layers in composites and laminates, as excellent material for photographic films, and as membrane-forming materials applicable for gas separation, water purification, food and beverage processing, medicinal and bioscientific fields. Also, depending on the film application, cellulose esters properties can be modified by varying degree of substitution, chain length, or hydroxyl content (Colombo et al., 1999; Kusumocahyo et al., 2005; Edgar, 2007; Shanbhag et al., 2007; Malhotra et al., 2015).

8.3 DESIGNING CELLULOSES AS SMART MATERIALS FOR PHARMACEUTICAL APPLICATIONS

Smart materials based on cellulose present significant roles in various biomedical applications such as clinical applications—as wound dressings, carriers for drug delivery, osmotic drug delivery systems, granules and tablets as binders, membranes for prevention of postoperative adhesions, meshes for hernia repair, materials for hemostasis, membranes for hemodialysis, preparations for treatment of ophthalmological disorders, and also as materials for plastic, reconstructive and aesthetic surgery. Moreover, cellulose materials were tested as cell carriers for tissue engineering, some of these results were included in the practice (Bacakova et al., 2015).

In this context, pharmaceutical applications of different cellulose comprise matrices in controlled-release drug delivery systems using matrices, binding agents, film formation during tablet coating, suspending agents for suspensions, thickening agents for ointments and creams, and hydrogels (Guo et al., 1998). Literature data reveal that the most natural polymers form a gel phase with decrease of the temperature. Moreover, aqueous solutions of some cellulose derivatives present reverse thermogelation at higher temperatures (Franz, 1989; Klouda et al., 2008).

Recently, a number of significant advances are made in the development of new technologies for drug delivery systems. Drug delivery systems refer to systems for transporting a pharmaceutical compound in the body as needed to obtain its desired therapeutic effect, which enhances patient compliance and offers various advantages—extend the time of pharmacological action and reduce the adverse effects (Gong et al., 2008; Pardeshi et al., 2015). In addition, oral controlled release systems continue to be the most common of all the drug delivery systems thanks to their several advantages. Thus, literature data present different gastrointestinal mucoadhesive dosage forms such as discs, microspheres, and tablets (Ahuja and Khar, 1997; Jadhav et al., 2013). Also, cellulose ethers are highly used as major excipients for designing matrix tablets. These compounds start to swell on contact with water and the hydrogel coating begin to increase around the dry core of the tablet. Thus, the hydrogel, which exhibits a diffusional barrier for water molecules, penetrate the polymer matrix and the drug molecules are released (Baumgartner et al., 1998; Siepmann et al., 1999; Colombo et al., 1999; Lowman and Peppas, 2000). Moreover, due to their excellent properties, cellulose esters present a great importance to controlled drug delivery in the pharmaceutical industry. In this context, most of them, in combination with other polymers are commonly used for enteric coating. For example, cellulose acetate was one of the first used materials for the production of semipermeable membranes in primary osmotic pumps (Siepmann et al., 2005; Shanbhag et al., 2007).

Indeed, as smart polymers, cellulose and its derivatives are flexible and can accomplish different conformations such as membrane, microparticle, capsule, hydrogel, and so on (Xie and Li, 2017). For instance, cellulose derivatives (such as MC, EC, HPC, HPMC, MEC, and CMC) can be used as thickening agents and stabilizers in food products and disintegrants in pharmaceuticals (Trygg, 2015).

Due to interesting features, modified and/or composite cellulose can be used to fabricate environmental stimuli-responsive smart materials, which will display intelligent comportment when exposed to proper environmental stimulus. Thus, in response to the introduction of a predetermined external stimulus, smart cellulose composites, environmentally stimuli-responsive, manifest a change of features in a controlled manner. Furthermore, these are fully biodegradable in a wide variety of environmental conditions and are interesting because of their excellent mechanical and thermal performance. Usually, these stimuli-responsive materials are used to afford controllable changes on the cellulose composite and to control the modification of different properties (like shape, mechanical rigidity and flexibility, opacity, and porosity). Therefore, environmentally stimuli-responsive smart cellulose composites have been used in many applications, mostly as biomaterials and drug carriers (Doelker, 1993; Kalia et al., 2009; Mashkour et al., 2011; Belgacem and Gandini, 2005; Reid et al., 2008; Edgar et al., 2001; Klemm et al., 2005; Kontturi et al., 2006; Spence et al., 2011).

Literature data reveal that some of cellulose derivatives such as CMC, HPC, HEC, CA, etc., exhibit stimuli-responsive behavior (Doelker, 1993; Klemm et al., 1998a; Klemm et al., 1998b; Kalia et al., 2009; Tan et al., 2010; Ekici, 2011; Zhang et al., 2011; Bai et al., 2012). The possible applications of various cellulose derivatives used in the pharmaceutical industry are summarized in Table 8.3 (Jones, 2004).

For examples, EC and HPC are extensively used in the pharmaceutical industry as coatings and shells of microspheres. In this regard, there are many reports on the use of these materials to fabricate stimuli-responsive materials as coatings and shells (Edgar et al., 2001; Kamel et al., 2008; Edgar, 2007; Rogers et al., 2011; Murtaza, 2012; Rogers et al., 2012). Besides, in the prospect of working as protecting shells, EC and CAP were used to coat drugs and then dispersed in stimuli-responsive matrices (Lecomte et al., 2005; Tripathi et al., 2012; Josephine et al., 2012).

Moreover, cellulose and cellulose derivatives-based biocomposites and nanocomposites present a very wide range of possible applications. Stimuli-responsive cellulose-based nanocomposites (responsive to pH, temperature, redox potential, light, magnetic field, and electrical field) have many applications in drug delivery systems, which include controlled delivery in specific intracellular locations or to targeted tissues and promotion in controlled drug release. Thus, in recent years have been intensively investigated the stimuli-responsive smart drug delivery systems (Cheng et al., 2011; Delcea et al., 2011; Wohl et al., 2012; Manchun et al., 2012; Fleige et al., 2012; Khan et al., 2016; Wang et al., 2016; Xie and Li, 2017).

Cellulose and cellulose derivatives-based wound dressings have been achieved for treating acute and chronic skin wounds such as burns, leg venous ulcers, and immune disorders (Lagus, et al., 2013; Dini et al., 2013; Araújo et al., 2013; Jaeger et al., 2015).

For preventing secondary infection of wounds, and to accelerate healing, cellulose materials have been mixed with antimicrobial agents (such as silver ions or antibiotics) (Duteille et al., 2012; Albaugh et al., 2013).

Another important field for the clinical application of cellulose-based materials is ophthalmology and also, in preventing postoperative adhesions after gynecological, abdominal, and lumbar surgery (Wander et al., 2011; Babi-zhayev et al., 2009; Robertson et al., 2010).

TABLE 8.3 Pharmaceutical Applications of Smart Cellulose Derivatives.

Cellulose derivatives	Applications
Methylcellulose (MC)	Controlled release
(Methocel A, Dow Chemicals)	Tablet coating and granulation
	Water-soluble thermoplastics
	Thickeners
Ethylcellulose (EC)	Microencapsulation
(Ethocel, Dow Chemicals)	Sustained release tablet coating
	Tablet coating
	Water-insoluble films
Hydroxypropyl cellulose (HPC)	Controlled release matrix
(Klucel, Hercules)	Film coating
	Tablet binder
Hydroxyethylcellulose (HEC)	Ophthalmic formulations
(Natrosol, Hercules)	Thickener
	Stabilizer
	Water binder
	Topical formulations
Hydroxypropyl methylcellulose (HPMC)	Ophthalmic preparations
(Methocel E,K, Dow Chemicals)	Tablet binder
	Stabilizing agent
	Film coatings
	Viscosity increasing agent
Hydroxyethyl methylcellulose (Culminal, Hercules)	Suspending and a thickening agent

Source: Adapted from Jones (2004).

Due to their good mechanical, physical, and biological properties, mainly biodegradability, biocompatibility, and the reduced cytotoxicity, nanocelluloses have become essential for the designing of new biomaterials. Furthermore, nanocelluloses present special functions such as tissue repair, regeneration and healing, sensors, antimicrobial nanomaterials, shape memory materials, and smart membranes (Fig. 8.3) (Kim et al., 2006; Shanmuganathan et al., 2010; Qiu and Hu, 2013; Lin and Dufresne, 2014; Jorfi and Foster, 2015; Xie and Li, 2017). In this context, due to structural properties of the backbone and the adaptability of its different derivatives, cellulose is extensively used in applications. Thus, it will explain in detail, the possible ways in which cellulose and its derivatives can be utilized in pharmaceutical applications.

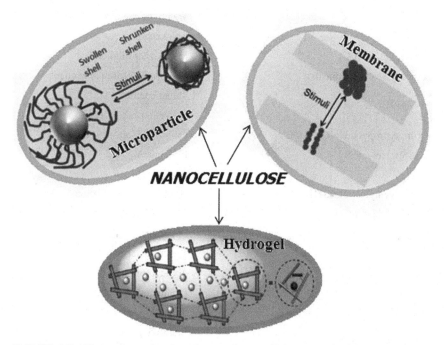

FIGURE 8.3   Illustrations of various forms of nanocellulose organization with potential application as drug delivery systems. Adapted from Xie and Li (2017).

Over the last decade, controlled release solid oral dosage forms have been widely used in pharmaceutical industry. These allow more comfortable application of drugs and at the same time assuring a sustained and reproducible method of release (Gallardo et al., 2008). Generally, the word "enteric" indicates small intestine, which prevents the release of medication before it reaches the small intestine. Thus, an enteric coating is a barrier which controls the location of oral medication in the body. The excellent properties of enteric coating material are the strength to gastric fluids, sensitivity to intestinal fluid, capacity of continual film formation, compatibility with most coating solution components, cheap, nontoxic, and ease of application (Singh et al., 2012). Due to their ease of administration, production, and identification, solid dosage forms such as capsules and tablets are the most used dosage forms. To eliminate the degradation in the gastric environment of acid labile actives and to protect the stomach from slight inflammation or other discomfort in the body, frequently, is applied an enteric coating to a solid dosage form (Agyilirah and Banker, 1991; Liu et al., 2011). Thus, tablets are solid dosage forms prepared by compressing a therapeutic agent which has been mixed or granulated before with pharmaceutical excipients, mainly polymers.

Generally, capsules are solid dosage forms which depend on the composition and are composed of gelatin. These can be classified in hard capsules composed of gelatin, sugar, water, and colorant and soft capsules—filled with the active ingredients and additionally, diluents or filler (Jones, 2004). Frequently, enteric-coated solid dosage forms constitute the basic groups of delayed release drug delivery systems designed for release of the drugs in the gastrointestinal tract. Moreover, enteric dosage forms can be considered as a type of oral site-specific pharmaceuticals that initiate drug release after switching from stomach. For example, the enteric oral dosage forms are adequate for drugs with irritancy potential for internal protective layer of stomach, this category of drugs belonging to the nonsteroidal anti-inflammatory drugs (NSAIDs). Habitually, the materials used in enteric-coated formulations are pH-dependent polymers which comprise carboxylic acid groups. These polymers present the good film-forming properties to produce smooth coats with good integrity. Also, with increasing of pH toward natural and light alkaline zone similar to the small intestine condition, these polymers become ionized and remain unionized in low pH conditions (environment of stomach) (Liu et al., 2011).

The most used materials for enteric coatings include different cellulose derivatives such as CAT, CAP, carboxymethylethyl cellulose (CMEC), HPMCP, and hydroxypropyl methylcellulose acetate succinate (HPMCAP) (Williams III and Liu, 2000; Singh et al., 2012). In addition, for obtaining resistant and uniform enteric films, it is very important to use different plasticizers (diethyl phthalate, glyceryl triacetate, triethyl citrate, and glyceryl monocaprylate). Thus, due to its very good film-forming properties and suitable polymer-to-polymer adhesion with enteric coating polymers, HPMC is used in enteric coating process as precoating or subcoating polymer, particularly with cellulose ester derivatives—CAT, CAP, CMEC, HPMCP, and HPMCAS (Williams III and Liu, 2000).

Hydroxyethyl methylcellulose, a partially O-methylated and partly O-(hydroxyethylated) cellulose, is used as an excipient in a wide range of pharmaceutical formulations mainly as a coating for solid dosage forms and granules and as a suspending agent for dispersing systems. Moreover, hydroxyethyl cellulose is used in a series of pharmaceutical applications such as a viscosity-modifying agent in ophthalmic and topical formulations, as a matrix for controlled release in solid dosage forms and as a film coating agent in solid dosage forms (Jones, 2004).

For improving physicochemical and mechanical characteristics of coating, literature has investigated aqueous nanodispersions of enteric coating HPMC. Thus, it was found that the ion-exchange process for the protonation of dissociated carboxylic acid group of the HPMCP nanoparticles, introduced for controlling the release rate of drug and hydrophobic nature of HPMCP coating layer, is necessary for pharmaceutical domain (Kim et al., 2003).

Recently, hydrophilic polymers are commonly used as rate-controlling polymers for extended release matrix-type dosage forms. In this context, HPMCs are cellulose ethers which are widely used as the basis for sustained release hydrophilic matrix tablets, their properties as gelling agents being important in the formulation. Furthermore, due to this property, they are responsible for the formation, by hydration, of a diffusion, and erosion-resistant gel layer which is capable to control drug release. For instance, HPMC matrix tablets may be affected by several formulation variables such as polymer concentration molecular weight drug levels and solubility, type of excipient and tablet shape, and size (Skoug et al., 1991).

Cellulose derivatives such as HEC and HPC, are used in hydrophilic matrix systems, while EC can be used in hydrophobic matrix system. In addition, hydrophobic EC was employed as matrix carriers for sustained release solid dosage. For instance, literature data show that EC is a hydrophobic polymer very used for the preparation of different pharmaceutical dosage forms—including film-coated tablet, microspheres, microcapsule, and matrix tablets (for soluble and poorly soluble drugs) (Eldrige et al., 1990; Akbuga, 1991; Rowe, 1992; Thies and Kleinebudde, 1999; Voinovich et al., 2000; Lavanya et al., 2011).

Literature data reveal that CMC sodium salt is used in oral pharmaceutical formulations as a disintegrant for capsules, tablets, and granules. It can be mention that, swelling power of this cellulose derivative is reduced at low pH and in concentrated solution of salts mainly in the existence of divalent and trivalent cations. Different celluloses such as NaCMC and HPMC were evaluated by in vitro release studies with respect to ibuprofen extended release from hard gelatin capsules. It was showed that different grades of these polymers could control ibuprofen release (nonsteroidal anti-inflammatory) to a substantial degree when used as diluents (Ford et al., 1987; Ojantakanen, 1992; Omidian and Park, 2012).

Furthermore, literature data have reported that the hydrophilic polymer matrix swells as water diffuses into the tablet. Cellulose derivatives, like HPMC, upon contact with aqueous media, start to hydrate, swell, coalesce, and form a viscous phase around the surface of the tablet. So, drug release is dependent on the relative contribution of diffusion and erosion release mechanisms. In addition, an important factor for drug releases from extended-release dosage forms is the matrix geometry (Ritger and Peppas, 1987; Witt et al., 2000; Siepmann et al., 2000).

Another study, mentions the antimicrobial and antiviral properties of CAP with applications in the pharmaceutical domain for enteric coating film of oral tablets or capsules. Due to the pH-dependent solubility of the aqueous medium, enteric coatings of this cellulose derivative are resistant to the gastric acid and easily soluble in the slightly alkaline environment of the intestine. Also, micronized CAP, used for tablet coating from water dispersions, mention its possibility to adsorb and to inhibit infections by human immunodeficiency type 1 virus, several herpes viruses in vitro, and other sexually transmitted disease pathogens. Moreover, literature shows that a gel formulation has potential as a topical microbicide for preventing sexually transmitted diseases including acquired immunodeficiency syndrome (Neurath et al., 2001; Neurath et al., 2003; Dobos et al., 2012; Dobos et al., 2013).

Polysaccharides are frequently used in the ocular mucoadhesive delivery system. Recently, most of ophthalmic preparations are usable in the form of aqueous solutions, because homogeneous solution dosage form affords a lot of advantages (e.g., the simplicity of large-scale fabrication). The development of new products-based smart cellulose derivatives for treatment of ophthalmic diseases shows a special interest for pharmaceutical industry (Kumar and Himmelstein, 1995; Jain et al., 2008; Mohan et al., 2009; Gupta et al., 2010a, 2010b; Abdelkader et al., 2014; Avinash and Ajay, 2015).

Various studies have shown that the inclusion of viscosity-increasing agents in the formulation such as HPMC, HEC, CMC, MC, and EC is used to increase the tear viscosity. Thus, decreases drainage and therefore, prolonging precorneal retention of the drops in the eye. For example, the ocular concentration of timolol was enhanced in the presence of the sodium carboxymethylcellulose vehicle, compared with nonviscous eye drops (Kyyrönen and Urtti, 1990). Furthermore, acetazolamide formulated in CMC was compared with the saline solution of the drug in patients with glaucoma. The time of action was longer than 8 and less than 24 h, and the results were appreciable when using high drug concentrations (Tous and El Nasser, 1992). Baeyens et al. (2009) prepared bioadhesive ophthalmic drug inserts of gentamycin using different cellulose derivatives—HPMC, EC, HPC, and CAP.

Recently, Nanjawade et al. (2009) investigated a combination of HPMC and polyacrylic acid as a vehicle for the formulation of eye drops of ketorolac tromethamine. It was found that the clarity, pH, and drug content of the developed formulation are satisfactory. In addition, the formulation was therapeutically efficacious, nonirritant, stable, and offered sustained drug release over a large period (8 h). Also, Bain et al. (2009) studied methylcellulose-based in situ fast gelling vehicle for ophthalmic drug delivery. Literature data show in situ gelling system of gatifloxacin prepared using HPMC and carbopol, evaluating its clarity, temperature, pH, tonicity, sterility, rheological behavior, in vitro release, transcorneal permeation, and ocular irritancy. The obtained results have demonstrated that the HPMC/ carbopol mixture can be used as an in situ gelling vehicle for improving the ocular bioavailability of gatifloxacin (Abdul Malik and Satyananda, 2014).

Attwood et al. (2001) presented a new in situ gum-based ophthalmic drug delivery system of Linezolid by using hydroxypropyl guar and xanthum in combination with HEC, sodium alginate, and carbopol as viscosity enhancing agents. All the formulations were evaluated for gelling ability, clarity, drug content evaluation, antibacterial activity, rheological, and in vitro diffusion study, respectively, for eye irritation testing. The results showed that sustained release of drug obtained, increase residence time of the drug over a long period of time (6 h).

Usually, in most of the in situ gelling systems are required high concentrations of polymers. For instance, to form strong gel upon administration in the eye is necessary to use a concentration of CAP (30% w/v) and Pluronics (25% w/v). Due to the increases of carbopol concentration in the vehicle, its acidic nature may lead to the stimulation of the eye tissue (Lin and Sung, 2000).

There are a lot of pharmaceutical grade derivatives of cellulose widely used in various administration routes. Over the last decades, different cellulose derivatives have demonstrated to be efficient on increasing the intranasal absorption of drugs, including soluble and insoluble cellulose derivatives such as HPMC, MC, HPC, CMC, HEC, EC, and microcrystalline cellulose. Thus, due to their mucoadhesive feature, the residence time of drugs in the nasal cavity may be prolonged, and high viscosity following hydration in the nasal cavity determines the drugs release of these derivatives. In this context, utilization of cellulose-based materials as absorption enhancer may lead to enlarge intranasal absorption and increased bioavailability (Kapoor et al., 2015; Zaki et al., 2007; Wang et al., 2008; Ugwoke et al., 2000). Also, a complete bioavailability (90.77%) could be realized for ketorolac tromethamine administered with microcrystalline cellulose (Quadir et al., 1999). In addition, by intranasal administration of apomorphine with CMC, can achieve a suitable bioavailability (102%) compared with subcutaneous injection in rabbits (Ugwoke et al., 1999).

Literature data reveal that a combination of the cellulose derivatives with other absorption enhancer produces a higher effectiveness than using only the polymer. It is observed that the intranasal bioavailability of ciprofloxacin at rabbits using HEC and MC alone as enhancer is small (18.2 and 19.46%), and increase when is in combination with Tween 80 (22.35 and 25.39%) (Paulsson, 2001).

For example, dopamine which is inefficacious by oral administration, due to first pass metabolism, was applied to the animals through rectal, dermal, buccal, and nasal routes. Thus, Ikeda et al. (1992) show that the effects of the addition of HPC and azone on the nasal absorption led to an ideal bioavailability of about 100% and, 25% for using only HPC. Furthermore, bacterial cellulose has been used in surgery of the lateral wall of the nose preventing nasal bleeding, surgical wound infections, clotting, and local pain (Kalia et al., 2011).

Over the last decade, significant progresses were made in improving the properties of smart cellulose-based materials used in the fields of drug delivery. Usually, cellulose and its derivatives are used as coating materials for drugs, additives of pharmaceutical products, supports for immobilized enzymes, blood coagulant, artificial kidney membranes, in wound treatment, as implant material, respectively, and scaffolds in tissue engineering.

Thus, utilization of cellulose and its derivatives as biomaterials for the design of tissue engineering scaffold has received a special attention, due to the excellent biocompatibility of cellulose and its good mechanical properties. In this context, cellulose and its derivatives are well supported by cells and tissues, inducing a moderately powerful foreign body reaction in the tissue (Miyamoto et al., 1989).

Besides vascular tissue engineering, cellulose and its derivatives have also been utilized for engineering a diversity of other tissues, like skin, cardiac muscle, skeletal muscle, heart valves, bone, and cartilage (Entcheva et al., 2004; Kingkaew et al., 2010; Andersson et al., 2010; Mohammadi, 2011; Shi et al., 2012; Dugan et al., 2013a, 2013b). Furthermore, cellulose-based materials have been used for building carriers for delivery and differentiation of mesenchymal stem cells and neural stem cells for neural tissue regeneration, for encapsulating and immunoisolating Langerhans islets, for constructing nanofibrous three-dimensional carriers for liver cells, and also for creating tubes for regenerating damaged peripheral nerves (Risbud et al., 2003; Gu et al., 2010; Bhattacharya et al., 2012; Kowalska-Ludwicka et al., 2013; Mothe et al., 2013).

Bacterial cellulose, which is identical to plant cellulose in chemical structure, is remarkable for its mechanical strength and biocompatibility, and it has often been applied in tissue engineering (Backdahl et al., 2006; Petersen and Gatenholm, 2011). For instance, microporous scaffolds made from bacterial cellulose and inoculated with human urine-derived stem cells maintained the formation of a multilayered urothelium. In this way, these constructions ensure the creation of tissue-engineered urinary conduits for urinary reconstruction (Bodin et al., 2010). Due to its special structure and biocompatibility, bacterial cellulose exhibit a higher mechanical strength and water retention, which afford it to serve as a natural scaffold material for the regeneration of a broad diversity of tissues (Czaja, 2006; MacNeil, 2007; Siró, 2010).

Literature reveals that nontoxic and biocompatible-silated hydroxypropylmethyl cellulose (Si-HPMC), present a large use in biomedical areas such as implanted in bone defects, scaffold for cell culture and cartilage model (Bourges et al., 2002).

Moreover, Si-HPMC hydrogel represents a possible basis for an innovative bone repair material, being appropriate to maintain survival, osteoblastic, proliferation, and differentiation when it is used as a new scaffold (Trojani et al., 2005).

Furthermore, cellulose-based materials have been widely used in medicine for application as membranes in the treatment of renal failure (Waron, 1986; Grassmann et al., 2005; Sato et al., 2005; Hoenich, 2006). Also, for potential urinary bladder reconstruction was applied cellulose acetate and other cellulose materials in the shape of porous membranes, for building a bioartificial renal tubule system, and in the form of electrospun porous microfibrous three-dimensional scaffolds (Sato et al., 2005; Han and Gouma, 2006).

Sannino et al. (2003) have proposed the use of cellulose derivatives-based hydrogels (CMC and HEC) as water absorbents in treating edemas. As well, cellulose derivatives can be utilized for the therapeutic practicability of superoxide dismutase enzyme, presented as hydrogels of CMC carrying the enzyme for its controlled release (Chiumiento et al., 2006).

Literature data reveal that in vitro and in vivo applications of smart cellulose-based materials have presented just negligible foreign body and inflammatory response reactions. Therefore, tissue biocompatibility of MC, EC, HEC, aminoethyl cellulose, and cellulosic polyion complexes was investigated by two in vivo tests (one for foreign body reaction and one for absorbance by living tissue). The foreign body reaction was relatively slow for all the samples examined, showing that cellulose can be changed to biocompatible materials (chemical and/or physical transformation), while for absorbance by living tissue has been found to be depended on the degree of crystallinity and the chemical structure of the sample (Miyamoto et al., 1989).

Generally, scaffolds represent a fundamental structural component in tissue engineering of a vascular substitute for small grafts by playing an important role in integrating the total tissue constructs. Thus, for potential application in vascular tissue engineering of small diameter grafts, Pooyan et al. (2012) designed a completely bio-based fibrous porous scaffold constitute of cellulose nanocrystals and reinforced cellulose acetate propionate matrix, where the resulted scaffold gave excellent mechanical performance.

Wound dressings form an important part of the medical and pharmaceutical wound care market. The most important characteristic for repair materials is their capacity to block exudate during the dressing process, but also their elimination from a wound surface after recuperation. Cellulose and its derivatives are highly used as a functional part of diverse wound dressing material (e.g., as hydrogels, fiber, and nonwoven). Literature data reveal that bacterial cellulose is an optimal material for skin tissue repair. This material showed that can provide a constant moist environment to a wound, and applied as a wound dressing shows good cytocompatibility and histocompatibility (Jeong et al., 2010; Petersen and Gatenholm, 2011). Furthermore, literature shows that the local applications of microbial cellulose membranes improve the healing process of burns and chronic wounds. Thus, Czaja et al. (2006) have utilized a microbial cellulose membrane as a wound-healing device for severely damaged skin and as a substitute of blood vessel with small diameter. Another study proved an absolute closure of the facial wound (44 days) with no considerable signs of extensive scarring (Czaja et al., 2007).

Also, Peršin et al. (2014) enhanced the hydrophilicity and microorganism inhibition of cellulose-based wound dressing materials, comparing two different procedures to increase the performance of attainment desired wound dressing characteristics. For example, it has been developed a membrane with bacterial cellulose and propolis extract, leading to antimicrobial and anti-inflammatory activities in chronic wounds that absorbing purulent exudates (Dugan et al., 2013a, 2013b). In addition, literature data showed that bacterial cellulose—dressed animals present more rapid wound healing within 14 days without any evidence of toxicity (Fu et al., 2012).

Generally, in pharmaceutical and dermatological practice, most of the drug products applied to the skin is semisolid preparations (like ointments, creams, and gels). From the category of antifungal agents were selected—clotrimazole, bifonazole, and fluconazole—as model drugs, solubilized in emulsion or microemulsion systems. In this context, the main studies on smart cellulose-materials as vehicles for dermal delivery of various drugs are summarized in Table 8.4 (Vlaia et al., 2016). All these were loaded in a cellulose derivative-based hydrogel for improving the viscosity of the emulsion or microemulsion, making it appropriate for cutaneous practice. Moreover, model drugs used in different studies exploring the effects of the structure and the components of the vehicle on drug release and penetration from cellulose derivative-based hydrogels were nonsteroidal anti-inflammatory agents (e.g., diclofenac sodium, oxicams (piroxicam, meloxicam), and etoricoxib).

TABLE 8.4   Applications of Cellulose Materials as Vehicles for Cutaneous Delivery of Drugs.

Cellulose derivative	Drug	Pharmaceutical dosage form	References
HPMC	Aminophylline	Hydrogel	Kouchak and Handali (2014)
HPMC	Bifonazole	Microemulsion-loaded	Sabale and Vora (2012)
HPMC	Clarithromycin	Emulgel	Baibhav et al. (2012)
HPMC	Chlorphenesin	Emulgel	Mohamed (2004)
HPMC	Clotrimazole	Emulgel	Shahin et al. (2011)
HEC, HPC, MC	Alaptide	Hydrogel	Sklenář et al. (2012)
CMCNa	Diclofenac sodium	Hydrogel	Gupta et al. (2010a, 2010b)
HPMC	Ketorolac tromethamine	Hydrogel	Hosny et al. (2013)
HPC	Lidocaine	Hydrogel	Sawant et al. (2010)
HPC	Meloxicam	Hydrogel	Jantharaprapap and Stagni (2007); Chang et al. (2007)
HPMC	Propranolol hydrochloride	Hydrogel	Vlaia et al. (2014)
HPMC	Piroxicam	Emulgel	Shokri et al. (2012)
MC, HPMC, CMC		Microemulsion-loaded hydrogel	Abd-Allah et al. (2010)
CMC HPMC	Fluconazole	Emulgel, microemulsion-loaded hydrogel Hydrogel	Salerno et al. (2010) Abdel-Mottaleb et al. (2007)
HPMC	Etoricoxib	Hydrogel	Prakash et al. (2010)

Source: Adapted from Vlaia et al. (2016).

These studies have proved that the release properties of topical formulations are particularly controlled by the thermodynamic activity of drug, diffusion, and particle size through the preparation.

8.4   CONCLUSIONS

Smart polymers exhibit an immense potential in different applications. From the broad group of macromolecular compounds, so-called smart polymers/stimuli-sensitive polymers are cellulose, a versatile polymer used in pharmaceutical application obtained from varied sources. In pharmaceutical technology, cellulose and its derivatives constitute an important class of excipients of continuously increasing relevance, so discovering of their properties becomes vital. These compounds, in different forms, have been utilized to prepare smart materials by chemical modifications (homogeneous or heterogeneous conditions) or by physical incorporation.

This chapter assembles the current knowledge on the structure and chemistry of celluloses, and in the development of innovative cellulose derivatives (esters and ethers) for pharmaceuticals. Cellulose and its derivatives are often used to modify the release of drugs in tablet and capsule formulations, also as tablet binding, thickening, and rheology control agents, for film formation, water retention, and improving adhesive resistance. This aspect is even more relevant when these products need specific pharmaceutical features (such as delayed release) and, therefore, proper techniques for obtaining drug therapy success. Furthermore, stimuli-responsive materials based on cellulose have large potential in tissue engineering-based applications due to their biocompatibility and biodegradability, where pH, temperature, and magnetic responses are frequently applied. Although excellent designs of smart materials based on cellulose have already been applied successfully in this field, more work still perforce to be done to make them more practical.

KEYWORDS

- cellulose and cellulose derivatives
- smart behavior
- biocompatibility
- biodegradability
- pharmaceutical applications

REFERENCES

Abdelkader, H.; Pierscionek, B.; Alany, R. G. Novel In Situ Gelling Ocular Films for the Opioid Growth Factor Receptor Antagonist Naltrexone hydrochloride: Fabrication,

Mechanical Properties, Mucoadhesion, Tolerability and Stability Studies. *Int. J. Pharm.* **2014,** *477* (1–2), 631–642.

Abd-Allah, F. I.; Dawaba, H. M.; Ahmed, A. M. S. Preparation, Characterization, and Stability Studies of Piroxicam Loaded Microemulsions in Topical Formulations. *Drug. Discov. Ther.* **2010,** *4* (4), 267–275.

Abdel-Mottaleb, M. M. A.; et al. Preparation and Evaluation of Fluconazole Gels. *Egypt. J. Biomed. Sci.* **2007,** *23* (1), 35–41.

Abdul Malik, P. H.; Satyananda, S. pH-induced In Situ Gelling System of an Anti-infective Drug for Sustained Ocular Delivery. *J. Appl. Pharm. Sci.* **2014,** *4* (1), 101–104.

Albaugh, K. W.; Biely, S. A.; Cavorsi, J. P. The Effect of a Cellulose Dressing and Topical Vancomycin on Methicillin-resistant Staphylococcus Aureus (MRSA) and Gram-positive Organisms in Chronic Wounds: A Case Series. *Ostomy Wound Manage.* **2013,** *59* (5), 34–43.

Agyilirah, G. A.; Banker, G. S. Polymers for Enteric Coating and Applications, In *Polymers for Controlled Drug Delivery*; Tarcha, P. J., Ed.; CRC Press: Boca Raton, 1991; pp 39–66.

Ahuja, R. P. K.; Khar, J. A. Mucoadhesive Drug Delivery System. *Drug Dev. Ind. Pharm.* **1997,** *2* (5), 489–515.

Akbuga, J. Furosemide-loaded Ethyl Cellulose Microspheres Prepared by Spherical Crystallization Technique: Morphology and Release Characterization. *Int. J. Pharm.* **1991,** *76* (3) 193–198.

Araújo, C. P., et al. Aproposal for the Use of New Silver-seaweed-cotton Fibers in the Treatment of Atopic Dermatitis. *Cutan. Ocul. Toxicol.* **2013,** *32* (4), 268–274.

Andersson, J., et al. Behavior of Human Chondrocytes in Engineered Porous Bacterial Cellulose Scaffolds. *J. Biomed. Mater. Res.* A. **2010,** *94* (4), 1124–1132.

Attwood D., et al. In Situ Gelling Xyloglucan Formulations for Sustained Release Ocular Delivery of Pilocarpine Hydrochloride. *Int. J. Pharm.* **2001,** *299* (1–2), 29–36.

Avinash, K. T.; Ajay, S. Formulation and Evaluation of Thermoreversible In Situ Ocular Gel of Clonidine Hydrochloride for Glaucoma. *Pharmacophore* **2015,** *6* (5), 220–232.

Babizhayev, M. A., et al. N-Acetylcarnosine Sustained Drug Delivery Eye Drops to Control the Signs of Ageless Vision: Glare Sensitivity, Cataract Amelioration and Quality of Vision Currently Available Treatment for the Challenging 50,000-patient Population. *Clin. Interv. Aging* **2009,** *4*, 31–50.

Bacakova L., et al. Cell Interaction with Cellulose-based Scaffolds for Tissue Engineering: A Review, In *Biochemistry Research Trends*; Mondal I. H., Ed.; Nova Science Publishers: New York, NY, 2015; Chapter 13, pp 341–374.

Backdahl, et al. Mechanical Properties of Bacterial Cellulose and Interactions with Smooth Muscle Cells. *Biomaterials* **2006,** *27* (9), 2141–2149.

Bai, Y., et al. Novel Thermo- and pH-responsive Hydroxypropyl Cellulose- and Poly (L-glutamic Acid)-based Microgels for Oral Insulin Controlled Release. *Carbohydr. Polym.* **2012,** *89* (4), 1207–1214.

Baibhav, J., et al. Development and Characterization of Clarithromycin Emulgel for Topical Delivery. *Int. J. Drug Dev. Res.* **2012,** *4* (3), 310–323.

Baeyens, V., et al. Evaluation of Soluble Bioadhesive Ophthalmic Drug Inserts (BODIR) for Prolonged Release of Gentamicin: Lachrymal Pharmacokinetics and Ocular Tolerance. *J. Ocul. Pharmacol. Ther.* **2009,** *14* (3), 263–272.

Bain, M. K.; Bhowmik, M.; Ghosh, S. N.; Chattopadhyay, D. In Situ Fast Gelling Formulation of Methyl Cellulose for in Vitro Ophthalmic Controlled Delivery of Ketorolac Tromethamine. *J. Appl. Polym. Sci.* **2009,** *113* (2), 1241–1246.

Baumgartner S., et al. Physical and Technological Parameters Influencing Floating Properties of Matrix Tablets Based on Cellulose Ethers. *S. T. P. Pharma Pratiques* **1998,** *8* (5), 182–187.

Belgacem, M. N.; Gandini, A. The Surface Modification of Cellulose Fibers for Use as Reinforcing Elements in Compostite Materials. *Compos. Interface* **2005,** *12* (1–2), 41–75.

Bhattacharya M., et al. Nanofibrillar Cellulose Hydrogel Promotes Three-dimensional Liver Cell Culture. *J. Control Release* **2012,** *164* (3), 291–298.

Billah Reduwan, S. M. Environmentally Responsive Smart Cellulose Composites. In *Cellulose and Cellulose Composites: Modification, Characterization and Applications*; Mondal, I. H., Ed.; Biochemistry Research Trends, Nova Science Publishers: New York, NY, 2015; pp 211–243.

Bodin, A., et al. Tissue-engineered Conduit Using Urine-derived Stem Cells Seeded Bacterial Cellulose Polymer in Urinary Reconstruction and Diversion. *Biomaterials* **2010,** *31* (34), 8889–8901.

Bordes, P., et al. Effect of Clay Organomodifiers on Degradation of Polyhydroxyalkanoates. *Polym. Degrad. Stab.* **2009,** *94* (5), 789–796.

Botana, A., et al. Effect of Modified Montmorillonite on Biodegradable Phb Nanocomposites. *Appl. Clay Sci.* **2010,** *47* (3–4), 263–270.

Bourges, X., et al. Synthesis and General Properties of Silated-hydroxypropyl Methylcellulose in Prospect of Biomedical Use. *Adv. Colloid. Interface. Sci.* **2002,** *99* (3), 215–228.

Bruno Rocha e Silva, M., et al. Dynamic and Structural Evaluation of Poly(3-hydroxybutyrate) Layered Nanocomposites. *Polym Test.* **2013,** *32* (1), 165–174.

Chang, J. S., et al. Formulation Optimization of Meloxicam Sodium Gel Using Response Surface Methodology. *Int. J. Pharm.* **2007,** *338* (1–2), 48–54.

Cheng, R., et al. Glutathione-responsive Nano-vehicles as a Promising Platform for Targeted Intracellular Drug and Gene Delivery. *J. Control. Release* **2011,** *152* (1), 2–12.

Chiumiento, A., et al. Anti-inflammatory Properties of Superoxide Dismutase Modified with Carboxymetil-cellulose Polymer and Hydrogel. *J. Mater. Sci. Mater. Med.* **2006,** *17* (5), 427–435.

Clasen C.; Kulicke, W. M. Determination of Viscoelastic and Rheooptical Material Functions of Water-soluble Cellulose Derivatives. *Prog. Polym. Sci.* **2001,** *26* (9), 1839–1919.

Colombo, P.; Bettini, R.; Peppas, N. A. Observation of Swelling Process and Diffusion Front Position during Swelling in Hydroxypropylmethyl Cellulose (HPMC) Matrices Containing a Soluble Drug. *J. Control. Release* **1999,** *61* (1–2), 83–91.

Cortese B., et al. Strategies and Applications for Incorporating Physical and Chemical Modifications for Functional Cellulose Based Materials. In *Cellulose and Cellulose Derivatives Biochemistry Research Trends*; Mondal, I. H., Ed.; Nova Science Publishers: New York, 2015; Chapter 2, pp 35–67.

Czaja, W., et al. Microbial Cellulose—The Natural Power to Heal Wounds. *Biomaterials* **2006,** *27* (2), 145–151.

Czaja, W. K., et al. The Future Prospects of Microbial Cellulose in Biomedical Applications. *Biomacromolecules* **2007,** *8* (1), 1–12.

D'Amico, D. A.; Manfredi, L. B.; Cyras, V. P. Crystallization Behavior of Poly(3-hydroxy-butyrate) Nanocomposites Based on Modified Clays: Effect of Organic Modifiers. *Thermochim. Acta* **2012**, *544* (20), 47–53.

Delcea, M.; Moehwald, H.; Skirtach, A. G. Stimuli-responsive LbL Capsules and Nanoshells for Drug Delivery. *Adv. Drug Deliv. Rev.* **2011**, *63* (9), 730–747.

Dini, V., et al. Improvement of Periulcer Skin Condition in Venous Leg Ulcer Patients: Prospective, Randomized, Controlled, Single-blinded Clinical Trial Comparing a Biosynthetic Cellulose Dressing with a Foam Dressing. *Adv. Skin Wound Care* **2013**, *26* (8), 352–359.

Dobos, A. M., et al. Rheological Properties and Microstructures of Cellulose Acetate Phthalate/hydroxypropyl Cellulose Blends. *Polym. Compos.* **2012**, *33* (11), 2072–2083.

Dobos, A. M., et al. Influence of Self-complementary Hydrogen Bonding on Solution Properties of Cellulose Acetate Phthalate in Solvent/non-solvent Mixtures. *Cell. Chem. Technol.* **2013**, *47* (1–2), 13–21.

Doelker, E. Cellulose Derivatives. In *Biopolymers I*; Langer, R. S., Peppas, N. A., Eds.; Springer: Berlin, Germany, 1993; Vol. 107, pp 199–265.

Duteille, F.; Jeffery, S. L. A Phase II Prospective, Non-comparative Assessment of a New Silver Sodium Carboxymethylcellulose (Aquacel® Ag Burn) Glove in the Management of Partial Thickness Hand Burns. *Burns* **2012**, *38* (7), 1041–1050.

Dugan, J. M., et al. Oriented Surfaces of Adsorbed Cellulose Nanowhiskers Promote Skeletal Muscle Myogenesis. *Acta Biomater.* **2013a**, *9* (1), 4707–4715.

Dugan, J. M.; Gough, J. E.; Eichhorn, S. J. Bacterial Cellulose Scaffolds and Cellulose Nanowhiskers for Tissue Engineering. *Nanomedicine* **2013b**, *8* (2), 287–298.

Duan, X., et al. Synthesis of a Novel Cellulose Physical Gel. *J. Nanomat.* **2014**, *2014*, 1–7.

Entcheva, E., et al. Functional Cardiac Cell Constructs on Cellulose-based Scaffolding. *Biomaterials* **2004**, *25* (26), 5753–5762.

Edgar, K. J. Cellulose Esters in Drug Delivery. *Cellulose* **2007**, *14* (1), 49–64.

Edgar, K. J. et al. Advances in Cellulose Ester Performance and Application. *Prog. Polym. Sci.* **2001**, *26* (9), 1605–1688.

Eldrige, J. H., et al. Controlled Vaccine Release in the Gut-associated Lymphoid Tissues. Part I. Orally Administered Biodegradable Micro-spheres Target the Peyer's Patches. *J. Control. Release* **1990**, *11* (1–3), 205–214.

Erceg, M.; Kovačić, T.; Perinović, S. Kinetic Analysis of the Nonisothermal Degradation of Poly(3-hydroxybutyrate) Nanocomposites. *Thermochim. Acta* **2008**, *476* (1–2), 44–50.

Ekici, S. Intelligent Poly(N-isopropylacrylamide)-carboxymethyl Cellulose Full Interpenetrating Polymeric Networks for Protein Adsorption Studies. *J. Mater. Sci.* **2011**, *46* (9), 2843–2850.

Fleige, E.; Quadir, M. A.; Haag, R. Stimuli-responsive Polymeric Nanocarriers for the Controlled Transport of Active Compounds: Concepts and Applications. *Adv. Drug. Deliv. Rev.* **2012**, *64* (9), 866–884.

Ford, J. L., et al. Importance of Drug Type, Tablet Shape and Added Diluents on Drug Release Kinetics from Hydroxypropylmethylcellulose Matrix Tablets. *Int. J. Pharm.* **1987**, *40* (3), 223–234.

Franz, G. Polysaccharides in Pharmacy: Current Applications and Future Concepts. *Planta Med.* **1989**, *55* (6), 493–497.

Fu, L., et al. Skin Tissue Repair Materials from Bacterial Cellulose by a Multilayer Fermentation Method. *J. Mater. Chem.* **2012**, *22*, 12349–1257.

Gallardo, D.; Skalsky, B.; Kleinebudde, P. Controlled Release Solid Dosage Forms Using Combinations of (Meth)Acrylate Copolymers. *Pharm. Dev. Technol.* **2008**, *13* (5), 413–423.

Grassmann, A., et al. ESRD Patients: Global Overview of Patient Numbers, Treatment Modalities and Associated Trends. *Nephrol. Dial. Transplant.* **2005**, *20* (12), 2587–2593.

Gong, C.; Wong, K.; Lam, M. H. Photoresponsive Molecularly Imprinted Hydrogels for the Photoregulated Release and Uptake of Pharmaceuticals in the Aqueous Media. *Chem. Mater.* **2008**, *20* (4), 1353–1358.

Gorrasi, G., et al. Vapor Barrier Properties of Polycaprolactone Montmorillonite Nanocomposites: Effect of Clay Dispersion. *Polymer* **2003**, *44* (8), 2271–2279.

Gu, H., et al. Control of in Vitro Neural Differentiation of Mesenchymal Stem Cells in 3D Macroporous, Cellulosic Hydrogels. *Regen. Med.* **2010**, *5* (2), 245–253.

Guo, J., et al. Pharmaceutical Applications of Naturally Occurring Water-soluble Polymers. *Pharm. Sci. Technol.* **1998**, *1* (6) 254–261.

Gupta, H.; Velpandian, T.; Jain, S. Ion-and pH-activated Novel In Situ Gel System for Sustained Ocular Drug Delivery. *J. Drug Targeting* **2010a**, *18* (7), 499–505.

Gupta, A., et al. Formulation and Evaluation of Topical Gel of Diclofenac Sodium Using Different Polymers. *Drug Invent. Today* **2010b**, *2* (5), 250–253.

Han, D.; Gouma, P. I. Electrospun Bioscaffolds that Mimic the Topology of Extracellular Matrix. *Nanomedicine* **2006**, *2* (1), 37–41.

Hoenich, N. A. Cellulose for Medical Applications: Past, Present, and Future. *BioResources* **2006**, *1* (2), 270–280.

Hosny, K. M., et al. Preparation and Evaluation of Ketorolac Tromethamine Hydrogel. *Int. J. Pharm. Sci. Rev. Res.* **2013**, *20* (2), 269–274.

Ikeda, K., et al. Enhancement of Bioavailability of Dopamine via Nasal Route in Beagle Dogs. *Chem. Pharm. Bull.* **1992**, *40* (8), 2155–2158.

Jain, S. P., et al. In Situ Ophthalmic Gel of Ciprofloxacin Hydrochloride for Once a Day Sustained Delivery. *Drug. Dev. Ind. Pharm.* **2008**, *34* (4), 445–452.

Jadhav, K. R.; Pawar, A. Y.; Talele G. S. Bioadhesive Drug Delivery System: An Overview. *J. Pharm. Clin. Res.* **2013**, *6* (2), 1–10.

Jaeger, T., et al. Acid-coated Textiles (pH 5.5–6.5)—A New Therapeutic Strategy for Atopic Eczema? *Acta Derm. Venereol.* **2015**, *95* (6), 659–663.

Jantharaprapap. R.; Stagni, G. Effects of Penetration Enhancers on In Vitro Permeability of Meloxicam Gels. *Int. J. Pharm.* **2007**, *343* (1–2), 26–33.

Jeong, S. I., et al. Toxicologic Evaluation of Bacterial Synthesized Cellulose in Endothelial Cells and Animals. *Mol. Cell. Toxicol.* **2010**, *6* (4), 373–380.

Jorfi, M.; Foster, E. J. Recent Advances in Nanocellulose for Biomedical Applications. *J. Appl. Polym. Sci.* **2015**, *132* (14), 41719.

Jones, D. *Pharmaceutical Applications of Polymers for Drug Delivery*; Smithers Rapra Publishing- Queen's University: Belfast, 2004; Vol. 15.

Josephine, L. J. J., et al. Formulation and Evaluation of Microparticles Containing Curcumin for Colorectal Cancer. *J. Drug Deliv. Ther.* **2012**, *2* (3), 125–128.

Kalia, S., et al. Cellulose-based Bio- and Nanocomposites: A Review. *Int. J. Polym. Sci.* **2011**, *2011*, 1–35.

Kalia, S.; Kaith, B. S.; Kaur, I. Pretreatments of Natural Fibers and Their Application as Reinforcing Material in Polymer Composites. *Polym. Eng. Sci.* **2009,** *49* (7), 1253–1272.

Kamel, S., et al. Pharmaceutical Significance of Cellulose: A Review. *eXPRESS Polym. Lett.* **2008,** *2* (11), 758–778.

Kapoor, D., et al. Site Specific Drug Delivery through Nasal Route Using Bioadhesive Polymers. *J. Drug Deliv. Ther.* **2015,** *5* (1), 1–9.

Khan, A., et al. Recent Progress on Cellulose-based Electro-active Paper, Its Hybrid Nanocomposites and Applications. *Sensors* **2016,** *16* (8), 1172.

Kim, H., et al. Swelling and Drug Release Behavior of Tablets Coated with Aqueous Hydroxypropyl Methylcellulose Phthalate (HPMCP) Nanoparticles. *J. Control Release* **2003,** *89* (2), 225–233.

Kim, J.; Yun, S.; Ounaies, Z. Discovery of Cellulose as a Smart Material. *Macromolecules* **2006,** *39* (12), 4202–4206.

Kingkaew, J., et al. Biocompatibility and Growth of Human Keratinocytes and Fibroblasts on Biosynthesized Cellulose-chitosan Film. *J. Biomater Sci. Polym. Ed.* **2010,** *21* (8–9), 1009–1021.

Klemm, D., et al. *Comprehensive Cellulose Chemistry: Fundamentals and Analytical Methods*; Wiley-VCH Verlag GmbH Weinheim: Germany, 1998(a); Vol. 1.

Klemm, D., et al. *Comprehensive Cellulose Chemistry: Functionalization of Cellulose*; Wiley-VCH Verlag GmbH Weinheim: Germany, 1998(b); Vol. 2.

Klemm, D., et al. Cellulose: Fascinating Biopolymer and Sustainable Raw Material. *Angew. Chem. Int. Ed.* **2005,** *44* (22), 3358–3393.

Klouda, L.; Mikos, A. G. Thermoresponsive Hydrogels in Biomedical Applications—A Review. *Eur. J. Pharm. Biopharm.* **2008,** *68* (1), 34–45.

Kontturi, E.; Tammelin, T.; Österberg, M. Cellulose – Model Films and the Fundamental Approach. *Chem. Soc. Rev.* **2006,** *35* (12), 1287–1304.

Kowalska-Ludwicka, K., et al. Modified Bacterial Cellulose Tubes for Regeneration of Damaged Peripheral Nerves. *Arch. Med. Sci.* **2013,** *9* (3), 527–534.

Kusumocahyo, S. P., et al. Pervaporative Separation of Organic Mixtures Using Dinitrophenyl Group-containing Cellulose Acetate Membrane. *J. Membr. Sci.* **2005,** *253* (1–2), 43–48.

Kouchak, M., Handali, S. Effects of Various Penetration Enhancers on Penetration of Aminophylline through Shed Snake Skin. *Jundishapur J. Nat. Pharm. Prod.* **2014,** *9* (1), 24–29.

Kumar, S.; Himmelstein, K. J. Modification of In Situ Gelling Behavior of Carbopol Solutions by Hydroxypropyl Methylcellulose. *J. Pharm. Sci.* **1995,** *84* (3), 344–348.

Kunioka, M.; Ninomiya, F.; Funabashi, M. Novel Evaluation Method of Biodegradabilities for Oil-based Polycaprolactone by Naturally Occurring Radiocarbon-14 Concentration Using Accelerator Mass Spectrometry Based on Iso 14855-2 in Controlled Compost. *Polym. Degrad. Stab.* **2007,** *92* (7), 1279–1288.

Kyyrönen, K.; Urtti, A. Improved Ocular: Systemic Absorption Ratio of Timolol by Viscous Vehicle and Phenylephrine. *Invest. Ophthalmol. Vis. Sci.* **1990,** *31* (9), 1827–1833.

Lagus, H., et al. Prospective Study on Burns Treated with Integra®, a Cellulose Sponge and Split Thickness Skin Graft: Comparative Clinical and Histological Study-randomized Controlled Trial. *Burns* **2013,** *39* (8), 1577–1587.

Lavanya, D., et al. Sources of Cellulose and Their Applications—A Review. *Int. J. Drug Formulation Res.(IJDFR)* **2011**, *2* (6), 19–38.

Lecomte, F., et al. pH-sensitive Polymer Blends Used as Coating Materials to Control Drug Release from Spherical Beads: Importance of the Type of Core. *Biomacromolecules* **2005**, *6* (4), 2074–2083.

Lin, H. R.; Sung, K. C. Carbopol/pluronic Phase Change Solutions for Ophthalmic Drug Delivery. *J. Control. Release* **2000**, *69* (3), 379–388.

Lin, N.; Dufresne, A. Nanocellulose in biomedicine: Current Status and Future Prospect. *Eur. Polym. J.* **2014**, *59*, 302–325.

Liu, F., et al. Evolution of a Physiological pH 6.8 Bicarbonate Buffer System: Application to the Dissolution Testing of Enteric Coated Products. *Eur. J. Pharm. Biopharm.* **2011**, *78* (1), 151–157.

Lowman, A. M.; Peppas, N. A. Hydrogels. In *Encyclopedia of Controlled Drug Delivery*; Mathiowitz, E., Ed.; Wiley: New York, NY, 2000; pp 397–417.

MacNeil, S. Progress and Opportunities for Tissue-engineered Skin. *Nature* **2007**, *445* (7130), 874–880.

Mahajan, A.; Aggarwal, G. Smart Polymers: Innovations in Novel Drug Delivery. *Int. J. Drug Dev. Res.* **2011**, *3* (3), 16–30.

Mahanta, R.; Mahanta, R. Sustainable Polymers and Applications, In *Handbook of Sustainable Polymers: Processing and Applications*; Thakur, V. K., Thakur, M. K., Eds.; Pan Stanford Publishing Pte. Ltd.: Boca Raton, FL, 2015; pp 1–56.

Malhotra, B.; Keshwani, A.; Kharkwal, H. Natural Polymer Based Cling Films for Food Packaging. *Int. J. Pharm. Pharm. Sci.* **2015**, *7* (4), 10–18.

Manchun, S.; Dass, C. R.; Sriamornsak, P. Targeted Therapy for Cancer Using pH-responsive Nanocarrier Systems. *Life Sci.* **2012**, *90* (11–12), 381–387.

Marques-Marinho, F. D.; Vianna-Soares, C. D. Cellulose and Its Derivatives Use in the Pharmaceutical Compounding Practice. In *Cellulose—Medical, Pharmaceutical and Electronic Applications*; van de Ven, Th., Godbout, L., Eds.; InTech: Croatia – European Union, 2013; Chapter 8, pp 141–162.

Mashkour, M., et al. Fabricating Unidirectional Magnetic Papers Using Permanent Magnets to Align Magnetic Nanoparticale Covers Natural Cellulose Fibers. *BioResources* **2011**, *6* (4), 4731–4738.

Messersmith, P. B.; Giannelis, E. P. Synthesis and Barrier Properties of Poly(e-caprolactone)-layered Silicate Nanocomposites. *J. Polym. Sci. Part A Polym. Chem.* **1995**, *33* (7), 1047–1057.

Miyamoto, et al. Tissue Compatibility to Cellulose and Its Derivatives. *J. Biomed. Mat. Res.* **1989**, *23* (1), 125–133.

Mohan, E. C.; Kandukuri; J. M.; Allenki, V. Preparation and Evaluation of In-situ-gels for Ocular Drug Delivery. *J. Pharm. Res.* **2009**, *2* (6), 1089–1094.

Mohamed, M. I. Optimization of Chlorphenesin Emulgel Formulation. *AAPS J.* **2004**, *6* (3), 81–87.

Mohammadi, H. Nanocomposite Biomaterial Mimicking Aortic Heart Valve Leaflet Mechanical Behaviour. *Proc. Inst. Mech. Eng. H.* **2011**, *225* (7), 718–722.

Mothe, A. J., et al. Repair of the Injured Spinal Cord by Transplantation of Neural Stem Cells in a Hyaluronan-based Hydrogel. *Biomaterials* **2013**, *34* (15), 3775–3783.

Murtaza, G. Ethylcellulose Microparticles. *Acta Poloniae Pharm.* **2012**, *69* (1), 11–22.

Neurath, A. R., et al. Cellulose Acetate Phthalate, a Common Pharmaceutical Excipient, Inactivates HIV-1 and Blocks the Coreceptor Binding Site on the Virus Envelope Glycoprotein gp120. *BMC Infect. Dis.* **2001,** *1,* 7.

Neurath, A. R.; Strick N.; Li, Y. Y. Water Dispersible Microbicidal Cellulose Acetate Phthalate Film. *BMC Infect. Dis.* **2003,** *3,* 27.

Nanjawade, B. K., et al. A Novel pH-triggered In Situ Gel for Sustained Ophthalmic Delivery of Ketorolac Tromethamine. *Asian J. Pharm. Sci.* **2009,** *4* (3), 189–199.

Ojantakanen, S. Effect of Viscosity Grade of Polymer Additive and Compression Force on Dissolution of Ibuprofen from Hard Gelatin Capsules. *Acta Pharm. Fennica* **1992,** *101* (33), 119–126.

Omidian, H.; Park, K.; Hydrogels. In *Fundamentals and Applications of Controlled Release Drug Delivery*; Siepmann, J., Siegel, R., Rathbone, M., Eds.; Springer: New York, NY, 2012; pp 75–106.

Onofrei, M. D.; Dobos, A. M.; Ioan, S. Processes in Cellulose Derivative Structures. In *Nanocellulose Polymer Nanocomposites: Fundamentals and Applications*; Thakur, V. K., Ed.; Wiley and Scrivener Publishing LLC: Beverly, MA, 2015; pp 355–391.

Pardeshi, S.; Dhodapkar, R.; Kumar, A. Molecularly Imprinted Polymers: A Versatile Tool in Pharmaceutical Applications. In *Handbook of Sustainable Polymers: Processing and Applications*; Thakur, V. K., Thakur, M. K., Eds.; Publishing Pte. Ltd.: Pan Stanford, 2015; Chapter 18, pp 611–653.

Paulsson, M. Acta Universitatis Upsaliensis. Comprehensive Summaries of Uppsala Dissertations from the Faculty of Pharmacy 259. In *Controlled Release Gel Formulation for Mucosal Drug Delivery*; Uppsala University: Sweden, 2001.

Petersen, N.; Gatenholm, P. Bacterial Cellulose-based Materials and Medical Devices: Current State and Perspectives. *Appl. Microbiol. Biotechnol.* **2011,** *91* (5), 1277–1286.

Peršin, Z., et al. Novel Cellulose Based Materials for Safe and Efficient Wound Treatment. *Carbohydr. Polym.* **2014,** *100* (16), 55–64.

Pooyan, P.; Tannenbaum, R.; Garmestani, H. Mechanical Behavior of a Cellulose-reinforced Scaffold in Vascular Tissue Engineering. *J. Mech. Behav. Biomed. Mater.* **2012,** *7,* 50–59.

Poersch-Parcke, H. G.; Kirchner, R. *Solutions,* 2nd ed.; Wolff Cellulosics GmbH: Walsrode, 2003.

Prakash, P. R.; Rao, N. G. R.; Soujanya, C. Formulation, Evaluation and Anti-inflammatory Activity of Topical Etoricoxib Gel. *Asian J. Pharm. Clin. Res.* **2010,** *3* (2), 126–129.

Qiu, X. Y.; Hu, S. W. Smart Materials Based on Cellulose: A Review of the Preparations, Properties, and Applications. *Materials* **2013,** *6* (3), 738–781.

Quadir, M.; Zia, H.; Needham, T. E. Toxicological Implications of Nasal Formulations. *Drug Del.* **1999,** *6* (4), 227–242.

Reid, M. L., et al. An Investigation into Solvent Membrane Interactions When Assessing Drug Release from Organic Vehicles Using Regenerated Cellulose Membranes. *J. Pharm. Pharmacol.* **2008,** *60* (9), 1139–1147.

Risbud, M. V.; Bhargava, S.; Bhonde, R. R. In Vivo Biocompatibility Evaluation of Cellulose Macrocapsules for Islet Immunoisolation: Implications of Low Molecular Weight Cut-off. *J. Biomed. Mater. Res A* **2003,** *66* (1), 86–92.

Ritger, P. L.; Peppas, N. A. A Simple Equation for Description of Solute Release. II. Fickian and Anomalous Release from Swellable Devices. *J. Control. Release* **1987,** *5* (1) 37–42.

Robertson, D., et Al. Adhesion Prevention in Gynaecological Surgery. *J. Obstet. Gynaecol. Can.* **2010,** *32* (6), 598–608.

Rogers, T. L.; Wallick, D. Reviewing the Use of Ethylcellulose, Methylcellulose and Hypromellose in Microencapsulation. Part 2: Techniques Used to Make Microcapsules. *Drug Dev. Ind. Pharm.* **2011,** *37* (11), 1259–1271.

Rogers, T. L.; Wallick, D. Reviewing the Use of Ethylcellulose, Methylcellulose and Hypromellose in Microencapsulation. Part 1: Materials Used to Formulate Microcapsules. *Drug Dev. Ind. Pharm.* **2012,** *38* (2), 129–157.

Rojas, J.; Bedoya, M.; Ciro, Y. Current Trends in the Production of Cellulose Nanoparticles and Nanocomposites for Biomedical Applications. In *Cellulose—Fundamental Aspects and Current Trends*; Poletto, M., Ornaghi, H. L. Jr., Eds.; Intech: Croatia - European Union, 2015; Chapter 8, pp 191–228.

Rowe, R. C. Molecular Weight Dependence of the Properties of Ethylcellulose and Hydroxypropyl Methylcellulose Films. *Int. J. Pharm.* **1992,** *88* (1–3), 405–408.

Spence, K. L., et al. Water Vapor Barrier Properties of Coated and Filled Microfibrillated Cellulose Composite Films. *BioResources* **2011,** *6* (4), 4370–4388.

Sabale, V.; Vora, S. Formulation and Evaluation of Microemulsion-based Hydrogel for Topical Delivery. *Int. J. Pharm. Invest.* **2012,** *2* (3), 140–149.

Sannino, A., et al. Biomedical Application of a Superabsorbent Hydrogel for Body Water Elimination in the Treatment of Edemas. *J. Biomed. Mater. Res. A* **2003,** *67* (3), 1016–1024.

Salerno, C.; Carlucci, A. M.; Bregni, C. Study of in Vitro Drug Release and Percutaneous Absorption of Fluconazole from Topical Dosage Forms. *AAPS Pharm. Sci. Tech.* **2010,** *11* (2), 986–993.

Sato, Y., et al. Evaluation of Proliferation and Functional Differentiation of LLC-PK1 Cells on Porous Polymer Membranes for the Development of a Bioartificial Renal Tubule Device. *Tissue Eng.* **2005,** *11* (9–10), 1506–1515.

Sawant, P. D., et al. Drug Release from Hydroethanolic Gels. Effect of Drug's Lipophilicity (Log P), Polymer-drug Interactions and Solvent Lipophilicity. *Int. J. Pharm.* **2010,** *396* (1–2), 45–52.

Siepmann, J., et al. HPMC-matrices for Controlled Drug Delivery: A New Model Combining Diffusion, Swelling, and Dissolution Mechanisms and Predicting the Release Kinetics. *Pharm. Res.* **1999,** *16* (11), 1748–1756.

Siepmann, J., et al. Calculation of the Required Size and Shape of Hydroxypropyl Methylcellulose Matrices to Achieve Desired Drug Release Profiles. *Int. J. Pharm.* **2000,** *201* (2), 151–164.

Siepmann F., et al. Blends of Aqueous Polymer Dispersions Used for Pellet Coating: Importance of the Particle Size. *J. Control. Release* **2005,** *105* (3) 226–239.

Singh, D. H., et al. A Review on Recent Advances of Enteric Coating. *IOSR J. Pharm.* **2012,** *2* (6), 5–11.

Shahin, M., et al. Novel Jojoba Oil-based Emulsion Gel Formulations for Clotrimazole Delivery. *AAPS Pharm. Sci. Tech.* **2011,** *12* (1), 239–247.

Shanmuganathan K., et al. Biomimetic Mechanically Adaptive Nanocomposites. *Prog. Polym. Sci.* **2010,** *35* (1–2), 212–222.

Shanbhag, A., et al. Application of Cellulose Acetate Butyrate-based Membrane for Osmotic Drug Delivery. *Cellulose* **2007,** *14* (1), 65–71.

Shi, Q. The Osteogenesis of Bacterial Cellulose Scaffold Loaded with Bone Morphogenetic Protein-2. *Biomaterials* **2012**, *33* (28), 6644–6649.

Shokri, I., et al. Comparison of HPMC Based Polymers Performance as Carriers for Manufacture of Solid Dispersions Using the Melt Extruder. *Int. J. Pharm.* **2011**, *419* (1–2), 12–19.

Shokri J., et al. Effects of Various Penetration Enhancers on Percutaneous Absorption of Piroxicam from Emulgels. *Res. Pharm. Sci.* **2012**, *7* (4), 225–234.

Siró, I.; Plackett, D. Microfibrillated Cellulose and New Nanocomposite Materials: A Review. *Cellulose* **2010**, *17* (3), 459–494.

Sklenář, Z., et al. Formulation and Release of Alaptide from Cellulose-based Hydrogels. *Acta Vet. Brno.* **2012**, *81* (3), 301–306.

Skoug, J. W., et al. In Vitro and in Vivo Evaluation of Whole and Half Tablets of Sustained-release Adinazolam Mesylate. *Pharm. Res.* **1991**, *8* (12), 1482–1488.

Tan, J., et al. Controllable Aggregation and Reversible pH Sensitivity of AuNPs Regulated by Carboxymethyl Cellulose. *Langmuir* **2010**, *26* (3), 2093–2098.

Thies, R.; Kleinebudde, P. Melt Pelletisation of a Hygroscopic Drug in a High–shear Mixer. Part 1. Influence of Process Variables. *Int. J. Pharm.* **1999**, *188* (2), 131–143.

Tous, S. S.; El Nasser, K. A. Acetazolamide Topical Formulation and Ocular Effect. *S. T. P. Pharm. Sci.* **1992**, *2* (1), 125–131.

Tripathi, G. K.; Singh, S. Formulation and in Vitro Evaluation of pH Trigger Polymeric Blended Buoyant Beads of Clarithromycin. *Int. J. Pharm. Tech. Res.* **2012**, *4* (1), 5–14.

Trojani, C., et al. Three-dimensional Culture and Differentiation of Human Osteogenic Cells in an Injectable Hydroxypropylmethylcellulose Hydrogel. *Biomaterials* **2005**, *26* (27), 5509–5517.

Trygg, J. *Functional Cellulose Microspheres for Pharmaceutical Applications*; Suomen Yliopistopaino Oy, Juvenes Print: Turku, 2015.

Ugwoke, M. I., et al. Bioavailability of Apomorphine Following Intranasal Administration of Mucoadhesive Drug Delivery Systems in Rabbits. *Eur. J. Pharm. Sci.* **1999**, *9* (2), 213–219.

Ugwoke, M. I., et al. Intranasal Bioavailability of Apomorphine from Carboxymethylcellulose Based Drug Delivery Systems. *Int. J. Pharm.* **2000**, *202* (1–2), 125–131.

Vlaia, L., et al. Percutaneous Penetration Enhancement of Propranolol Hydrochloride from HPMC-based Hydroethanolic Gels Containing Terpenes. *Farmacia* **2014**, *62* (5), 991–1008.

Vlaia, L., et al. Cellulose-derivatives-based Hydrogels as Vehicles for Dermal and Transdermal Drug Delivery (Chapter 7). In *Emerging Concepts in Analysis and Applications of Hydrogels*; Majee, S. B., Ed.; InTech: Croatia - European Union, 2016; pp 159–200.

Voinovich, D., et al. Preparation in High Shear Mixer of Sustained–Release Pellets by Melt Pelletisation. *Int. J. Pharm.* **2000**, *203* (1–2), 235–244.

Wander, A. H. Long-term Use of Hydroxypropyl Cellulose Ophthalmic Insert to Relieve Symptoms of Dry Eye in a Contact Lens Wearer: Case-based Experience. *Eye Contact Lens* **2011**, *37* (1), 39–44.

Wang, Y.; Heinze, T.; Zhang, K. Stimuli-responsive Nanoparticles from Ionic Cellulose Derivatives. *Nanoscale* **2016**, *8*, 648–657.

Wang, X.; Chi, N.; Tang X. Preparation of Estradiol Chitosan Nanoparticles for Improving Nasal Absorption and Brain Targeting. *Eur. J. Pharm. Biopharm.* **2008**, *70* (3), 735–740.

Waron, M. Sarcoid Nephrocalcinotic Renal Failure Reversed by Sodium Cellulose Phosphate. *Am. J. Nephrol.* **1986,** *6* (3), 220–223.

Williams III, R. O; Liu, J. Influence of Processing and Curing Conditions on Beads Coated with an Aqueous Dispersion of Cellulose Acetate Phthalate. *Eur. J. Pharm. Biopharm.* **2000,** *49* (3), 243–252.

Witt, C.; Mader, K.; Kissel, T. The Degradation, Swelling and Erosion Properties of Biodegradable Implants Prepared by Extrusion or Compression Moulding of Poly(Lactide-co-glycolide) and ABA Triblock Copolymers. *Biomaterials* **2000,** *21* (9), 931–938.

Wohl, B. M.; Engbersen, J. F. J. Responsive Layer-by-layer Materials for Drug Delivery. *J. Control. Release* **2012,** *158* (1), 2–14.

Xie, J.; Li, J. Smart Drug Delivery System Based on Nanocelluloses. *J. Bioresour.Bioprod.* **2017,** *2* (1), 1–3.

Yilmaz, N. D.; Çılgı, G. K.; Yilmaz K. Natural Polysaccharides as Pharmaceutical Excipients. In *Handbook of Polymers for Pharmaceutical Technologies*; Thakur, V. K., Thakur, M. K., Eds.; Wiley Scrivener Publishing LLC: Beverly, MA, 2015; Vol. 3, pp 483–516.

Zaki, N. M., et al. Enhanced Bioavailability of Metoclopramide HCl by Intranasal Administration of Mucoadhesive in Situ Gel with Modulated Rheological and Mucociliary Transport Properties. *Eur. J. Pharm. Sci.* **2007,** *32* (4), 296–307.

Zhang, Z., et al. Thermo-and pH-responsive HPC-*g*-AA/AA Hydrogels for Controlled Drug Delivery Applications. *Polymer* **2011,** *52* (3), 676–682.

CHAPTER 9

RECENT RESEARCHES ON PVA– CLAY NANOCOMPOSITES FOR TARGETED APPLICATIONS

SIMONA MORARIU*, MIRELA TEODORESCU, and MARIA BERCEA

Department of Electroactive Polymers and Plasmochemistry, "Petru Poni" Institute of Macromolecular Chemistry, 41A Grigore Ghica Voda Alley, 700487 Iasi, Romania

Corresponding author. E-mail: smorariu@icmpp.ro

CONTENTS

ABSTRACT

Clays have been used since the dawn of history for medicinal purposes. During the last decades, a particular interest has been shown by researchers on development of new materials containing clay minerals with promising properties, high availability, and low cost. Because of their unique layered structure, the clays have the capacity to interact with different nonionic/ionic polymers leading to high-performance nanocomposites for targeted applications. The main clays used for preparing of hybrid nanocomposites belong to the following classes: layered silicates (such as montmorillonite, laponite, and halloysite), chain silicates (sepiolite), and layered double hydroxide. One of the most used polymers in combination with various clays is poly(vinyl alcohol), resulting materials which cumulate the good mechanical, thermal, and gas barrier properties of the inorganic compound with the nontoxicity, noncarcinogenicity, biocompatibility, and biodegradability of poly(vinyl alcohol). Thus, by this work, we tried to offer an accurate overview on various poly(vinyl alcohol)/clay nanocomposites, their obtaining methods, properties, and applications.

9.1 INTRODUCTION

Clays have been used for a long time for medicinal purposes, being recorded in texts dating from Antiquity, as for example, the studies of Hippocrates and Aristotle, Pedanius Dioscorides, or Pliny the Elder (Carretero, 2002). Since those times, researches regarding the discovery of various clays led to tremendous advancement of their applications in different fields and still present interesting opportunities for development in polymer science and nanotechnology.

During the last years, nanocomposites based on polymer(s) matrix and an inorganic compound represented a research area that has attracted great interest due to their impressive properties resulted by combining two or more components. Because of their low cost, high availability, capacity to form interactions with different polymers, clays can provide, in combination with nonionic/ionic polymers, new materials with improved properties compared to pure polymers. The viscoelastic properties of the clay/polymer-based nanocomposites are influenced by composition, polymer molecular weight, components concentration, shear conditions, temperature, etc. (Morariu and Bercea, 2009, 2012, 2015a). For example, the increase of Laponite RD (LAP) concentration into a poly(ethylene oxide) (PEO)/LAP aqueous mixture determines the increasing of the viscoelastic parameters and the change of their rheological behavior from a preponderant viscous fluid to preponderant elastic one (Morariu and Bercea, 2012). Biocompatible and nontoxic injectable hydrogels were obtained by addition of a low amount of chitosan (CS) into PEO/LAP aqueous dispersion (Morariu et al., 2014).

Poly(vinyl alcohol) (PVA) gels registered wide applicability in various pharmaceutical and biomedical fields, as for example, in drug delivery systems, targeted and controlled release, tissue engineering, wound dressings, artificial organs, contact lenses, and many others (Hassan and Peppas, 2000). Due to the hydroxyl groups from its structure, PVA in aqueous solution is able to gelation by repeated freezing/thawing cycles (Bercea et al., 2013). Characterized by its solubility in water, easy processing, good chemical resistance, nontoxicity, noncarcinogenicity, biocompatibility, and biodegradability, PVA gives excellent properties to PVA/polymer blend gels obtained by physical methods (Bercea et al., 2014, 2015c, 2016; Morariu et al., 2015b; Teodorescu et al., 2016).

However, the hydrophilic polymers are recommended due to their ability to form complexes with different clay minerals. Among these polymers, PVA was used in combination with montmorillonite (MMT) since the early 1960s (Emerson and Raupach, 1964) and it was proved to be a valuable candidate to obtain hybrid materials containing clays. The nanomaterials based on the mixture of PVA and clay are characterized by a synergetic combination of properties of their individual components.

9.2   CLAYS STRUCTURE AND CLASSIFICATION

The terms "clay" and "clay mineral" refer to a class of fine-grained naturally occurring hydrophilic, water-swelling materials with particles smaller than 2 μm, found in large aggregates in nature, which become plastic when they contain water and harden when are dried or fired (Al-Ani and Sarapaa, 2008; Theng, 2012). Generally, clay minerals possess a unique crystalline structure of hydrous aluminum *phyllosilicates* (*layer* or *sheet silicates* from Greek word "*phyllon*" meaning "*leaf*"), which can contain different quantities of magnesium, iron, alkaline earth, alkaline metals, or various other cations (Kotal and Bhowmick, 2015).

Structural building components of phyllosilicates are represented by alumina-silicate layers formed from different arrangements of silica tetrahedral and alumina octahedral sheets. The two basic units are a silicon (Si^{4+}) tetrahedron and an aluminum (Al^{3+}), magnesium (Mg^{2+}), or iron (Fe^{2+}/Fe^{3+}) octahedron (Fig. 9.1a).

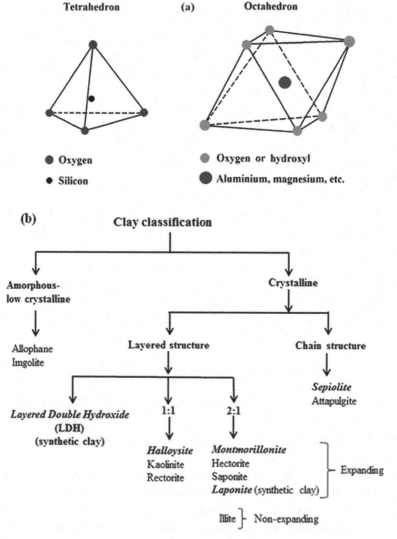

FIGURE 9.1   (a) Silica tetrahedron and alumina octahedron representation and (b) clay classification.

The silicon tetrahedron describes the arrangement of atoms of the crystalline structure, and thus it is formed by a central Si^{4+} cation (though the aluminum, ferric iron, or other elements cations can sometimes occur at this site) linked to four O^{2-} ions at the apices. Since Si has a 4+ charge and the four O have 2− charges for each of them, it results that an isolated tetrahedron possesses a net charge of 4−. Tetrahedral sheets are hexagonal networks built by the connection of adjacent SiO_4 tetrahedra. These tetrahedra share the three nearest-neighbor O^{2-} ions corners which form the basal plane (Barton and Karathanasis, 2002). The fourth O^{2-} ion (i.e., the apical oxygen) of the tetrahedron is left unbound and thus is free to connect with other polyhedra. Electrical neutrality of tetrahedral sheets is maintained by H^+ ions that with the apical oxygen ions form hydroxyl groups, considering that the basal O^{2-} ions confer a charge of 1− to each Si^{4+} ion (Schulze, 2005).

The second structural building block is called an octahedron and consists of six closely linked O^{2-} ions (or hydroxyl groups), three of which form a triangle in one plane intermeshed with another triangle rotated with 60°, relative to the first triangle, formed by the other three O^{2-} ions (or hydroxyl groups) (Schulze, 2005). The octahedral site is typically occupied by an Al^{3+}, Mg^{2+}, or Fe^{2+}/Fe^{3+} cation found in octahedral coordination with the six O^{2-} ions (or hydroxyl groups). Unlike tetrahedral sheets that share corners, octahedral sheets are composed by edge-sharing. The structures of octahedral sheets are similar to gibbsite $Al(OH)_3$ and brucite $Mg(OH)_2$ minerals, but they may contain hydroxyl groups as well as oxygen ions. Clay minerals with relatively high charged trivalent cations, such as Al^{3+} or Fe^{3+}, in the central site have only two of three octahedra occupied and a vacant octahedron. The resulted dioctahedral structure is called "gibbsite-like". Alternatively, when the central cation is divalent, such as Mg^{2+} or Fe^{2+}, then all three octahedral sites are filled and electrical neutrality is maintained, leading to a trioctahedral "brucite-like" structure. Dioctahedral "gibbsite-like" sheets and trioctahedral "brucite-like" sheets bound by shared oxygen atoms with tetrahedral sheets result in alumina-silicate layers that represent the structural building components of phyllosilicates (Barton and Karathanasis, 2002; Al-Ani and Sarapaa, 2008).

The main categories of clays that have been used for preparation of hybrid nanocomposites are the following: layered silicates (such as MMT, LAP, and halloysite (HL)), chain silicates (sepiolite (SEP)) and layered double hydroxide (LDH) (Fig. 9.1b).

The most commonly used clays for the preparation of PVA–clay nanocomposites are MMT and LAP which belong to 2:1 silicates. They are able to form stable suspensions in water due to their structure consisting of an alumina sheet (octahedral sheet) sandwiched between two silica sheets (tetrahedral sheets) (Fig. 9.2a). This structure permits also the diffusion of water-soluble polymers between their inorganic crystalline layers. HL, which belongs to 1:1 clay group, has a structure consisting of alternating layers of one aluminum octahedron sheet and one silicon tetrahedron sheet. HL has a hollow tubular structure with the tubules length from 500 to 1000 nm and the diameter between 25 and 100 nm, depending on its sources (Fig. 9.2b).

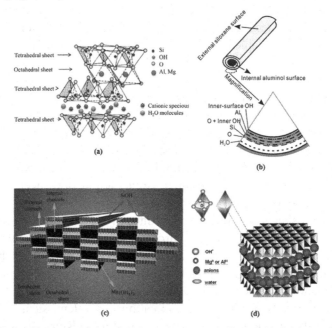

FIGURE 9.2   (See color insert.) Schematic representation of (a) 2:1 layered clay, (b) halloysite (1:1 layered clay), (c) sepiolite (chain silicate), and (d) layered double hydroxide structures.

a: Reprinted with permission from Ghadiri, M.; Chrzanowski, W.; Rohanizadeh, R. Biomedical Applications of Cationic Clay Minerals. *RSC Adv.* **2015**, *5* (37), 29467–29481. © 2015 Royal Chemical Society.

b: Reprinted with permission from Yuan, P.; Tan, D.; Annabi-Bergaya, F. Properties and Applications of Halloysite Nanotubes: Recent Research Advances and Future Prospects. *Appl. Clay Sci.* **2015**, *112–113*, 75–93. © 2015 Elsevier.

c: Reprinted with permission from Soheilmoghaddam, M., et al. Characterization of Bio Regenerated Cellulose/Sepiolite Nanocomposite Films Prepared via Ionic Liquid. *Polym. Test.* **2014**, *33*, 121–130. © 2014 Elsevier.

d: Reprinted with permission from Mohapatra, L.; Parida, K. A Review on the Recent Progress, Challenges and Perspective of Layered Double Hydroxides as Promising Photocatalysts. *J. Mater. Chem. A.* **2016**, *4* (28), 10744–10766. © 2016 Royal Chemical Society.

Compared with other 2:1 clay minerals, such as MMT and LAP, the structure of SEP can be considered as ribbons of 2:1 phyllosilicate structures which are linked to the next by inversion SiO_4 tetrahedra along a set of Si—O—Si bonds. SEP has a fibrous nature with a width of 10–30 nm and a thickness between 5 and 10 nm (Fig. 9.2c).

Other relevant clays used as component in polymer-based nanocomposites are LDHs which represent a class of ionic lamellar compounds consisted of positively charged brucite-like layers (Fig. 9.2d).

9.3 PVA–CLAY NANOCOMPOSITES

9.3.1 PVA—2:1 LAYERED CLAY NANOCOMPOSITES

Usually, the hydrogels based on PVA in the presence/absence of MMT are obtained by the freezing-thawing technique. During the freezing-thawing cycles, the hydrogel structure is formed due to the hydrogen bonds established between the hydroxyl groups of PVA chains and oxyanions from MMT surface (Fig. 9.3).

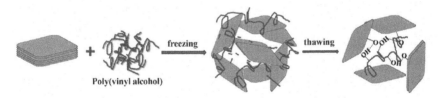

Poly(vinyl alcohol)

FIGURE 9.3   Preparation of PVA–clay hydrogels by freezing-thawing method. Reprinted with permission from Zhao, L. Z., et al. Recent Advances in Clay Mineral-containing Nanocomposite Hydrogels. *Soft Matter* **2015**, *11* (48), 9229–9246. © 2015 Royal Society of Chemistry.

Some nanocomposite hydrogels containing PVA and 10% organically modified-MMT clay were prepared by freezing-thawing method in order to obtain hydrogels as potential candidates for various drug delivery systems (Sirousazar et al., 2012a). The organically modified-MMT acts as a physical cross-linker in the network structure of PVA hydrogels affecting their structure and swelling properties. Scanning electron microscopy (SEM) micrographs of PVA hydrogels in absence/presence of organically modified-MMT clay have proven that the incorporation of clay creates a more densely cross-linked network into hydrogels and the pore sizes decrease leading to the decrease of the swelling ratios of the hydrogels (Sirousazar et al., 2012a; Paranhos et al., 2007).

Clay intercalated into PVA matrix interacts strongly by silanol group with the hydroxyl group of PVA chains. These interactions lead to a stable network with enhanced elastic modulus and a higher thermal stability (Paranhos et al., 2007).

The glass transition temperature of PVA hydrogel increases by addition of nanoclay. PVA/MMT nanocomposite hydrogels have also shown higher elastic moduli than those corresponding to PVA hydrogel without clay, in the temperature range of -100 to $100°C$, due to the reinforcing effect of MMT in the PVA matrix (Sirousazar et al., 2012b). The swelling kinetics of PVA hydrogels depend on the medium and can be expressed by diffusion-controlled mechanism. The swelling degree increases by increasing the medium temperature. The necessary time for reaching swelling equilibrium condition of hydrogels decreases with the increase of temperature. Both pure and nanocomposite PVA hydrogels reach faster the equilibrium swelling level at higher temperatures of the medium (Sirousazar et al., 2012a).

PVA/Na-MMT hybrid hydrogels with properties required for application as wound dressing (microbe penetration barrierity, high swelling degree, and good mechanical properties) were obtained by the freezing-thawing method (Kokabi et al., 2007). The swelling ratio of PVA/unmodified-MMT composite hydrogel is affected by clay amount due to the increase of the network rigidity and it increases by increasing the polymer content (Reguieg et al., 2017).

The structure of the polymer used influences the dispersing degree of the clay into polymer-clay nanocomposite films. For example, the silicate layers of MMT were better dispersed in the PVA/MMT mixture as compared to the PEO/MMT composite where the clay is in the form of a large tactoid. The silicate layers of MMT are parallel to the film surface in both mixtures but the orientation of polymer chains in PEO and PVA crystallites are parallel and perpendicular, respectively, to silicate layers (Ogata et al., 1997).

Recently, it was reported the preparation of PVA/Na-MMT aerogels by using the gamma irradiation method (Chen et al., 2014, 2015a). The aerogels obtained from PVA with higher molecular weights presented dense morphological structures and high values of viscoelastic parameters. The high molecular weight of PVA has the same effect as the irradiation-induced cross-linking for the aerogels preparation. The irra-diation dose used for obtaining a strong composite aerogel with a dense microstructure decreases by increasing the molecular weight of PVA. The aerogels with optimal mechanical, thermal, and flammability properties can be obtained by using PVA with very high molecular weight or with low molecular weight coupled with moderate gamma irradiation.

The gamma irradiation method was also used in order to obtain PVA/ Na-MMT membranes (low quantity of PVA) with extremely low flam-mability. The membranes have shown a layered microstructure in cross-section and an intercalated clay structure inside the composites. The low amount of PVA, on the one hand, and the low thermal conductivity of clay, on the other hand, contribute to reducing flammability of these composites (Chen et al., 2015b).

The freezing-thawing method and the electron beam irradiation were combined in order to develop PVA/clay hydrogels with application in treatment of wastewater with heavy and toxic metal ions and dye wastes (Ibrahim and El Naggar, 2013). The gel fraction increases by increasing the clay concentration at used irradiation doses (20, 25, and 30 kGy) and the X-ray diffraction and SEM methods revealed an intercalation and exfo-liation of clay into the PVA matrix. The investigation of PVA/MMT nano-composites in the form of film prepared under the effect of electron beam irradiation established that the appropriate dose of irradiation to obtain homogeneous nanocomposites is 20 kGy. The addition up to 4% MMT improves the tensile strength, elongation at break, and thermal stability of the PVA hydrogel. The melting transition of PVA was not affected by adding MMT and the glass transition temperature was suppressed (Abd Alla et al., 2006).

The addition of PVA into 2 wt% Turkey bentonite (MMT with minor amounts of illite, quartz, and feldspar) aqueous dispersion leads to the decrease of Bingham yield value from 0.170 Pa, for the clay-water system, to a minimum of about 0.033 Pa, for the polymer–clay–water system with the polymer concentration of 0.7 g/L PVA. All bentonite aqueous disper-sions in presence/absence of PVA showed a pseudoplastic flow with thixo-tropic behavior characterized by the existence of a hysteresis loop of the flow curves. The electrokinetic measurements revealed that, at low PVA concentration, the polymer chains are attached to the clay surface which is negatively charged and the screening of surface charge leads to the weak-ening of electrostatic interactions. The further addition of PVA causes the formation of bridges between the clay particles and the flocculation occurs (Isci et al., 2006).

κ-Carrageenan (κ-Carr)/PVA/MMT nanocomposite hydrogels obtained by freezing-thawing method have the capacity of adsorption of crystal violet from aqueous solution. The adsorption capacity depends on the MMT content, pH, added salt, as well as the solution temperature. The results suggested that the addition of MMT into κ-Carr/PVA composite enhances the adsorption capacity (Hosseinzadeh et al., 2015).

A new material with superabsorbent properties based on pullulan (PULL) was obtained by electrospinning technique followed by a heat treatment of aqueous solutions of PULL/PVA/MMT mixture (Islam et al., 2015). These materials have the capacity to retain 85% distilled water and 89% saline water after 1 day. The superabsorption property of PULL/PVA/ MMT nanofibers could be explained by their hydrophilic character. The water molecules are attracted into network structure of the polymers fibers over a diffusion gradient. Although the polymer chains tend to straighten, they cannot due to the cross-linking and the fibers expand as the water molecules enter into the network. The water is retained into the polymers network by hydrogen bonds formed between water and polymer chains or clay platelets (Fig. 9.4). By incorporation of poly(vinylpyrrolidone) (PVP) into PVA/MMT nanocomposite, it can be obtained films with good thermal stability and mechanical properties due to the strong hydrogen bonds established between the film components (Mondal et al., 2013).

Gaur et al. (2014) have prepared a proton-conducting polymer elec-trolyte nanocomposite membrane based on CS, PVA, MMT (Closide 30B), and poly(styrene sulfonic acid) by solution casting method. The membranes have shown improved thermal and mechanical properties and selectivity in the range of 10^4 Sscm^{-3}, and they can be a suitable candidate for direct methanol fuel cells. The major disadvantage of these membranes, consisting of their low stability in water at higher temperature, can be avoided by cross-linking of PVA and CS chains with chemical cross-linkers as glutaraldehyde (GL). The inclusion of low quantity of CS into nanocomposites based on PVA determines an appreciable increase of the antimicrobial activity (Sabaa et al., 2015). CS/PVA/MMT films present improved mechanical properties compared to the composite film which does not contain clay due to the hydrogen bonds formed between hydroxyl or amine groups belonging to the three components during manufacturing and drying processes (Mahdavi et al., 2013).

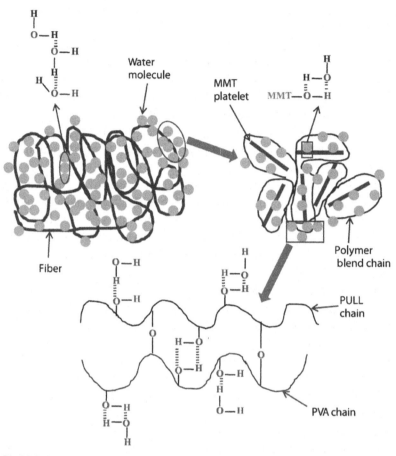

FIGURE 9.4   Schematic representation of the water absorption mechanism for the PULL/ PVA/MMT composite. Reprinted with permission from Islam, M. S.; Rahaman, M. S.; Yeum, J. H. Electrospun Novel Super-absorbent Based on Polysaccharide–Polyvinyl Alcohol–Montmorillonite Clay Nanocomposites. *Carbohydr. Polym.* **2015,** *115,* 69–77. 2015 Elsevier.

Taghizadeh et al. (2013) investigated the physical properties and the enzymatic degradation of PVA/starch/carboxymethyl cellulose nano-composites which contain low quantity of MMT. The film containing 5 wt% MMT exhibited the lowest water absorption capability due to the strong hydrogen bonds formed between hydroxyl groups of MMT and hydroxyl and carboxyl groups of PVA, starch, and carboxymethyl cellu-lose, respectively. These strong bonds improve the interactions between nanocomposite components, the cohesiveness of biopolymer matrix and decrease the water sensitivity.

The morphological investigation of PVA film containing ion-exchanged clays evidenced that the Na ion-exchanged clays are more easily dispersed in PVA matrix than alkyl ammonium ion-exchanged clays. PVA films which contain Na ion-exchanged clays have tensile strengths higher than those containing alkyl ammonium ion-exchanged clays. The presence of ion-exchanged clay into the PVA film affects the thermal stability and the tensile properties of PVA/clay nanocomposites (Chang et al., 2003).

Recently, Lai et al. (2015) investigated the effect of fumed silica and different clay types (modified MMT) on the thermal and mechanical properties of PVA/fumed silica/clay nanocomposite films obtained by solution intercalation method. The addition of modified MMT to PVA/fumed silica nanocomposites confers them the anti-wetting property and improved thermal stability.

Yu et al. (2003) have prepared a series of PVA/MMT nanocomposite materials with a morphology much smoother than pure PVA by in situ free radical polymerization. The orientation of crystalline polymer chains is destroyed and the polymer morphology approaches to amorphous state by addition of MMT. The incorporation of MMT into PVA matrix causes, on the one hand, an increase of the decomposition temperature, the glass transition temperature, and mechanical strength and, on the other hand, a reduction of transparency.

The addition of a low quantity of PVA (up to 16.7%) into nylon 6/ clay nanocomposites resins (with about 4.6 wt% exfoliated MMT) does not change the initial morphology of the nanocomposites and the crystallization of PVA is nearly completely inhibited due to strong hydrogen bonds between nylon 6 and PVA. By increasing the PVA content, the clay from the nylon 6 matrices migrates into the PVA phases leading to a γ-to-α crystal-phase transition of nylon 6 molecules (Cui and Yeh, 2010).

PVA/MMT nanocomposite films with different concentrations of clay were prepared by using a solution-intercalation casting method in order to improve the dispersing of MMT into PVA matrix (Allison et al., 2015). The individual clay stacks are much better dispersed in the polymer matrix for low MMT concentration (1 vol%) and, for high amount of MMT (25 vol%), in PVA matrix appear clay large agglomerates which determine the decrease of PVA/MMT nanocomposite films strength.

Alhassan et al. (2015) reported the preparation of directionally frozen ice-templated aerogels consisting of LAP and PVA. The aerogels are fragile at high LAP fractions while at higher polymer fraction more rigid aerogels are obtained.

Hydrogel membranes with no cytotoxic effect, based on PVA, CS, and LAP, were developed by Oliveira et al. (2012) using gamma radiation, which ensures the synthesis and the sterilization simultaneously. The hydrogel membranes of PVA or PVA/CS cross-linked by gamma irradiation in presence of LAP particles showed an increase of thermal stability with slower degradation of polymeric components as compared to the membranes based on PVA/CS obtained by chemical cross-linking (Oliveira et al., 2012; de Oliveira et al., 2013). The PVA/CS membranes become stronger and more opaque by adding LAP in their composition.

The thermosensitive nanocomposite hydrogel membranes based on PVA grafted with N-tert-butylacrylamide and LAP revealed the enhanced mechanical properties but low swelling degree in water (Nair et al., 2006). The multilayered composite films of PVA with LAP for potential applications as optical elements and humidity sensors were obtained by depositing self-assembled monolayers of a functional silane on substrates. Multilayered PVA/LAP films exhibited tensile strength, modulus, and toughness higher than those corresponding to the nanocomposites obtained by solvent casting method (Patro and Wagner, 2011). A nanocomposite aerogel based on PVA was obtained by freeze-drying of LAP/multigraphene platelets/PVA mixtures in water (Alhwaige et al., 2016). The addition of multigraphene platelets improves the mechanical properties of LAP/PVA aerogels. Thereby, a content of 3% multigraphene determines the increase of the compressive modulus of LAP/PVA nanocomposite aerogel by 180%. SEM micrographs evidenced that LAP/multigraphene platelets/PVA aerogels present a sheet-like morphology with ordered layers interconnected between them (Fig. 9.5). The crystallinity degree of nanocomposite aerogels increases by increasing the multigraphene platelets content.

PVA/sodium alginate/magnetic LAP hydrogel beads, cross-linked physically by Ca^{2+} and freezing-thawing method, were stable in different pH environments (Mahdavinia et al., 2016). Each component of this nanocomposite contributes to the improvement of the hydrogel properties: PVA increases the mechanical strength, sodium alginate induces pH response, and LAP nanoparticles confer magnetic properties. PVA/sodium alginate/ magnetic LAP hydrogel showed the maximum adsorption capacity of bovine serum albumin (BSA) at pH = 4.5 and its desorption on the beads occurred mainly at pH <4 and pH >5.

FIGURE 9.5 SEM micrographs of side direction of (a) neat PVA polymeric aerogel and (b) hybrid aerogels containing 5 wt% LAP/multigraphene platelets. Reprinted with permission from Alhwaige, A. A., et al. Laponite/Multigraphene Hybrid-reinforced Poly(vinyl alcohol) Aerogels. *Polymer* **2016**, *91*, 180–186. © 2016 Elsevier.

9.3.2 PVA—1:1 LAYERED CLAY NANOCOMPOSITES

The transparent films based on PVA/HL with potential biomedical applications in bone tissue engineering and drug delivery systems were obtained by using GL as the cross-linking agent. The elastic modulus of PVA films decreases by addition of HL up to 5% because the nanotubes cannot form a network and they are dispersed in PVA matrix. The increase of HL concentration above 5% determines the restriction of PVA chains movement and the clay acts as a network barrier by GL cross-linking. HL nanotube-dominant surface of the PVA/HL films increases the bone cell attachment ability (Zhou et al., 2010).

9.3.3 PVA–LAYERED DOUBLE HYDROXIDE NANOCOMPOSITES

The viscoelastic properties of CS/PVA/hydrotalcite-like anionic clay (MgAl-LDH) hybrid nanocomposites at 37°C depend on the pH conditions. The optimum pH value of network formation, characterized by a minimum loss tangent, was established around neutral pH (Bercea et al., 2015a). In weakly acid and basic conditions, the hydrotalcite-like anionic clay acts as a physical cross-linker for CS and PVA leading to a gel characterized by a high elasticity (Fig. 9.6). The efficiency of the clay to form a network at basic pH and the sensitivity of CS macromolecules to pH changes are responsible for the viscoelastic properties of CS/PVA/MgAl-LDH hybrid nanocomposites.

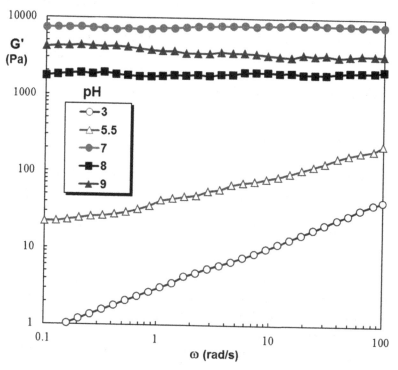

FIGURE 9.6 Elastic modulus as a function of oscillation frequency in different pH conditions for CS/PVA/MgAl-LDH hybrid nanocomposites at 37°C and 1 Pa. Reprinted with permission from Bercea, M., et al. pH Influence on Rheological and Structural Properties of Chitosan/poly(vinyl alcohol)/Layered Double Hydroxide Composites. *Eur. Polym. J.* **2015a,** *70,* 147–156. © 2015 Elsevier.

CS/PVA/MgAl-LDH chemical hydrogels were prepared in the presence of GL as cross-linker, starting from a CS/PVA mixture with 62.5% PVA, corresponding to composition for which the interactions between CS and PVA are the most favorable (Bercea et al., 2015b). These hydrogels revealed the highest elasticity of the network and solvent uptake at pH of 5.5 and 37°C.

The electrospinning method is a versatile method to process the polymer solutions into continuous fibers with diameters ranging from a few nanometers to a few micrometers. The nanofibers based on PVA/ hydrotalcite composite were obtained using this technique (Zhuo et al., 2011). The modification of the hydrotalcite surface with sodium dodec-ylbenzene sulfonate improves the distribution of clay particles in PVA matrix and its compatibility with PVA. The pH values of the medium influence the diameter distribution ranges of nanofibers. Thereby, the diameter of the PVA/hydrotalcite composite fibers is more uniformly distributed in weak acid medium than in alkaline or strong acidic conditions. This fact is due to the good spinnability of PVA in weak acidic condition.

9.3.4   PVA–CHAIN SILICATES NANOCOMPOSITES

The PVA/SEP nanocomposite containing up to 10 wt% SEP with potential uses as trabecular bone scaffolds was prepared by freezing-drying process and cross-linked with a small quantity of poly(acrylic acid) (PAA) with low molecular weight (Killeen et al., 2012). The addition of SEP decreases the anisotropy degree of nanocomposite due to its fibrous structure and increases the average pore size from 50 m for PVA hydrogel to 92 m for PVA hydrogel with 10 wt% SEP. Fourier Transform Infrared Spectroscopy analysis suggested the strong interfacial interactions by hydrogen bonds established between the silanol groups on SEP surface and the hydroxyl groups of PVA (Fig. 9.7).

The hydrogels containing 6 wt% SEP have shown the best mechanical properties. A higher quantity of SEP leads to a tendency of hydrogel deterioration explained by the aggregation of clay particles which act as the stress concentrators altering the mechanical properties. The heat treatment of SEP determines the decrease of the thermal stability and optical clarity of PVA/SEP nanocomposite films prepared by the solution dispersion method (Alkan and Benlikaya, 2009).

PVA-inorganic hybrid materials which represent a promising alternative to conventional polymer nanocomposites can be applied in electronics, food packaging, medicine, etc. The possible applications of polymer-inorganic nanocomposites (gel and film) based on PVA are summarized in Table 9.1.

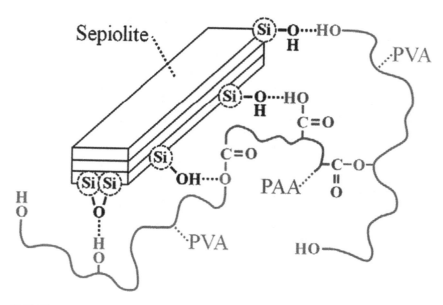

FIGURE 9.7 Interactions between sepiolite and thermally cross-linked PVA/PAA. Reprinted with permission from Killeen, D.; Frydrych, M.; Chen, B. Porous Poly(vinyl alcohol)/Sepiolite Bone Scaffolds: Preparation, Structure and Mechanical Properties. *Mater. Sci. Eng. C Mater. Biol. Appl.* **2012**, *32* (4), 749–757. © 2012 Elsevier.

9.4 CONCLUSIONS

Clays represent promising materials for obtaining organic–inorganic hybrid nanocomposites due to their layered structure, high availability, and low cost. The combination of clay with PVA provides a material which cumulates the good mechanical, thermal, and gas barrier properties of the inorganic compound with the nontoxicity, noncarcinogenicity, biocompatibility, and biodegradability of PVA.

This chapter reviews the main researches on the synthesis, properties, and applications of PVA–clay nanocomposites. Despite the important number of papers in the polymer–clay nanocomposites field, the interactions between the organic–inorganic components were not fully understood and elucidated until now. In this context, the future investigations should be focused on the designing of new PVA–clay nanocomposite for various applications as well as the identification of the interactions established between the composite components.

TABLE 9.1 Organic–Inorganic Nanocomposites Containing PVA and Various Clays and Their Possible Applications.

Form	Clay	Polymer A	Polymer B	Preparation method	Applications	References
Dried film/gel	MMT	PVA	PULL	Electrospinning	Superabsorbent material	Islam et al. (2015)
Dried film/gel	Rectorite	PVA	Sodium alginate	Electrospinning	Food packaging and wound dressing	Li et al. (2013)
Film	Na-MMT	PVA (small quantity)	–	Gamma irradiation	Fire retardant materials	Chen et al. (2015b)
Film	Organo-MMT	PVA	CS	Solvent casting	Carrier for controlled release of curcumin	Parida et al. (2011)
Film	LAP	PVA	–	Layer-by-layer deposition	Optical elements, humidity sensors	Patro and Wagner (2011)
	MMT	PVA	–	Freezing-thawing followed by electron beam irradiation	Treatment of wastewater	Ibrahim and El Naggar (2013)
	MMT	PVA	κ-Carr	Freezing-thawing	Adsorbent for crystal violet from wastewater	Hosseinzadeh et al. (2015)
	MMT	PVA		Freezing-thawing	Wound dressing	Kokabi et al. (2007)
	MMT	PVA	CS	Freezing-thawing	Wound dressing	Noori et al. (2015)
Porous scaffold	SEP	PVA	PAA (small quantity)	Freezing-drying	Trabecular bone scaffolds	Killeen et al. (2012)

KEYWORDS

- nanocomposite
- poly(vinyl alcohol)
- montmorillonite
- laponite
- sepiolite
- halloysite
- layered double hydroxide

REFERENCES

Abd Alla, S. G.; El-Din, H. M. N.; El-Naggar, A. W. M. Electron Beam Synthesis and Characterization of Poly(vinyl alcohol)/Montmorillonite Nanocomposites. *J. Appl. Polym. Sci.* **2006,** *102* (2), 1129–1138.

Al-Ani, T.; Sarapaa, O. *Clay and Clay Mineralogy; Report on Geological Survey of Finland;* Kaolinite Books: Litmanen, 2008.

Alhassan, S. M.; Qutubuddin, S.; Schiraldi D. A. Mechanically Strong Ice Templated Laponite/Poly(vinyl alcohol) Aerogels. *Mater. Lett.* **2015,** *157,* 155–157.

Alhwaige, A. A., et al. Laponite/Multigraphene Hybrid-reinforced Poly(vinyl alcohol) Aerogels. *Polymer* **2016,** *91,* 180–186.

Alkan, M.; Benlikaya, R. Poly(vinyl alcohol) Nanocomposites with Sepiolite and Heat-treated Sepiolites. *J. Appl. Polym. Sci.* **2009,** *112* (6), 3764–3774.

Allison, P. G., et al. Mechanical, Thermal, and Microstructural Analysis of Polyvinyl Alcohol/montmorillonite Nanocomposites. *J. Nanomater.* **2015,** *2015,* 9 (Article ID 291248). DOI:10.1155/2015/291248.

Barton, C. D.; Karathanasis, A. D. Clay Minerals. In *Encyclopedia of Soil Science;* Lal, R., Ed.; Marcel Dekker: New York, 2002; pp 187–192.

Bercea, M.; Morariu, S.; Rusu, D. In-situ Gelation of Aqueous Solutions of Entangled Poly(vinyl alcohol). *Soft Matter* **2013,** *9* (4), 1244–1253.

Bercea, M., et al. Investigation of Poly(vinyl alcohol)/Pluronic F127 Physical Gel. *Polym. Plast. Technol. Eng.* **2014,** *53* (13), 1354–1361.

Bercea, M., et al. pH Influence on Rheological and Structural Properties of Chitosan/poly(vinyl alcohol)/Layered Double Hydroxide Composites. *Eur. Polym. J.* **2015a,** *70,* 147–156.

Bercea, M., et al. Chitosan/Poly(vinyl alcohol)/LDH Biocomposites with pH-sensitive Properties. *Int. J. Polym. Mater. Polym. Biomater.* **2015b,** *64* (12), 628–636.

Bercea, M., et al. In-situ Gelling System Based on Pluronic F127 and Poly(vinyl alcohol) for the Smart Biomaterials. *Rev. Roum. Chim.* **2015c,** *60* (7–8), 787–795.

Bercea, M.; Morariu, S.; Teodorescu, M. Rheological Investigation of Poly(vinyl alcohol)/ Poly(N-vinylpyrrolidone) Mixtures in Aqueous Solution and Hydrogel State. *J. Polym. Res.* **2016**, *23* (142), 1–9.

Carretero, M. I. Clay Minerals and Their Beneficial Effects upon Human Health. A Review. *Appl. Clay Sci.* **2002**, *21* (3–4), 155–163.

Chang, J. H., et al. Poly(vinyl alcohol) Nanocomposites with Different Clays: Pristine Clays and Organoclays. *J. Appl. Polym. Sci.* **2003**, *90* (12), 3208–3214.

Chen, H.-B., et al. Fabrication and Properties of Irradiation-cross-linked Poly(vinyl alcohol)/clay Aerogel Composites. *ACS Appl. Mater. Interfaces* **2014**, *6* (18), 16227–16236.

Chen, H. B., et al. Effects of Molecular Weight upon Irradiation-cross-linked Poly(vinyl alcohol)/Clay Aerogel Properties. *ACS Appl. Mater. Interfaces* **2015a**, *7* (36), 20208–20214.

Chen, H. B., et al. Effects of Gamma Irradiation on Clay Membrane with Poly(vinyl alcohol) for Fire Retardancy. *Ind. Eng. Chem. Res.* **2015b**, *54* (43), 10740–10746.

Cui, L.; Yeh, J. T. Nylon 6 Crystal-Phase Transition in Nylon 6/clay/poly(vinyl alcohol) Nanocomposites. *J. Appl. Polym. Sci.* **2010**, *118* (3), 1683–1690.

de Oliveira, M. J. A., et al. Hydrogel Membranes of PVAl/clay by Gamma Radiation. *Radiat. Phys. Chem.* **2013**, *84*, 111–114.

Emerson, W. W.; Raupach, M. The Reaction of Polyvinyl Alcohol with Montmorillonite. *Aust. J. Soil. Res.* **1964**, *2* (1), 46–55.

Gaur, S. S., et al. Prospects of Poly(vinyl alcohol)/Chitosan/Poly(styrene sulfonic acid) and Montmorillonite Cloisite 30B Clay Composite Membrane for Direct Methanol Fuel Cells. *J. Renew. Sustain. Energ.* **2014**, *6* (5), 53135.

Ghadiri, M.; Chrzanowski, W.; Rohanizadeh, R. Biomedical Applications of Cationic Clay Minerals. *RSC Adv.* **2015**, *5* (37), 29467–29481.

Hassan, C. M.; Peppas, N. A. Structure and Applications of Poly(vinyl alcohol) Hydrogels Produced by Conventional Crosslinking or by Freezing/Thawing Methods. *Adv. Polym. Sci.* **2000**, *153*, 37–65.

Hosseinzadeh, H.; Zoroufi, S.; Mahdavinia, G. R. Study on Adsorption of Cationic Dye on Novel κ-Carrageenan/Poly(vinyl alcohol)/Montmorillonite Nanocomposite Hydrogels. *Polym. Bull.* **2015**, *72* (6), 1339–1363.

Ibrahim, S. M.; El Naggar, A. A. Preparation of Poly(vinyl alcohol)/clay Hydrogel Through Freezing and Thawing Followed by Electron Beam Irradiation for the Treatment of Wastewater. *J. Thermoplast. Compos.* **2013**, *26* (10), 1332–1348.

Isci, S., et al. Rheology and Structure of Aqueous Bentonite – Polyvinyl Alcohol Dispersions. *Bull. Mater. Sci.* **2006**, *29* (5), 449–456.

Islam, M. S.; Rahaman, M. S.; Yeum, J. H. Electrospun Novel Super-absorbent Based on Polysaccharide–Polyvinyl Alcohol–Montmorillonite Clay Nanocomposites. *Carbohydr. Polym.* **2015**, *115*, 69–77.

Killeen, D.; Frydrych, M.; Chen, B. Porous Poly(vinyl alcohol)/Sepiolite Bone Scaffolds: Preparation, Structure and Mechanical Properties. *Mater. Sci. Eng. C Mater. Biol. Appl.* **2012**, *32* (4), 749–757.

Kokabi, M.; Sirousazar, M.; Hassan, Z. M. PVA-Clay Nanocomposite Hydrogels for Wound Dressing. *Eur. Polym. J.* **2007**, *43* (3), 773–781.

Kotal, M.; Bhowmick, A. K. Polymer Nanocomposites from Modified Clays: Recent Advances and Challenges. *Prog. Polym. Sci.* **2015**, *51*, 127–187.

Lai, J. C. H., et al. Impact of Nanoclay on Physicomechanical and Thermal Analysis of Polyvinyl Alcohol/Fumed Silica/Clay Nanocomposites. *J. Appl. Polym. Sci.* **2015,** *132* (15), 41843.

Li, W., et al. Poly(vinyl alcohol)/Sodium Alginate/Layered Silicate Based Nanofibrous Mats for Bacterial Inhibition. *Carbohydr. Polym.* **2013,** *92* (2), 2232–2238.

Mahdavi, H., et al. Poly(vinyl alcohol)/Chitosan/Clay Nano-composite Films. *J. Am. Sci.* **2013,** *9* (8), 203–214.

Mahdavinia, G. R., et al. Magnetic Hydrogel Beads Based on PVA/Sodium Alginate/Laponite RD and Studying Their BSA Adsorption. *Carbohydr. Polym.* **2016,** *147,* 379–391.

Mohapatra, L.; Parida, K. A Review on the Recent Progress, Challenges and Perspective of Layered Double Hydroxides as Promising Photocatalysts. *J. Mater. Chem. A.* **2016,** *4* (28), 10744–10766.

Mondal, D., et al. Effect of Poly(vinylpyrrolidone) on the Morphology and Physical Properties of Poly(vinyl alcohol)/Sodium Montmorillonite Nanocomposite Films. *Prog. Nat. Sci. Mater. Int.* **2013,** *23* (6), 579–587.

Morariu, S.; Bercea, M. Effect of Addition of Polymer on the Rheology and Electrokinetic Features of Laponite RD Aqueous Dispersions. *J. Chem. Eng. Data.* **2009,** *54* (1), 54–59.

Morariu, S.; Bercea, M. Effect of Temperature and Aging Time on the Rheological Behavior of Aqueous Poly(ethylene glycol)—Laponite RD Dispersions. *J. Phys. Chem. B.* **2012,** *116* (1), 48–54.

Morariu, S.; Bercea, M. Viscoelastic Properties of Laponite RD Dispersions Containing PEO with Different Molecular Weights. *Rev. Roum. Chim.* **2015a,** *60* (7–8), 777–785.

Morariu, S.; Bercea, M.; Brunchi C. E. Effect of Cryogenic Treatment on the Rheological Properties of Chitosan/Poly(vinyl alcohol) Hydrogels. *Ind. Eng. Chem. Res.* **2015b,** *54* (45), 11475–11482.

Morariu, S.; Bercea, M.; Sacarescu, L. Tailoring of Clay/Poly(ethylene oxide) Hydrogel Properties by Chitosan Incorporation. *Ind. Eng. Chem. Res.* **2014,** *53* (35), 13690–13698.

Nair, S. H., et al. Swelling and Mechanical Behavior of Modified Poly(vinyl alcohol)/ Laponite Nanocomposite Membranes. *J. Appl. Polym. Sci.* **2006,** *103* (5), 2896–2903.

Noori, S.; Kokabi, M.; Hassan, Z. M. Nanoclay Enhanced the Mechanical Properties of Poly(vinyl alcohol)/Chitosan/Montmorillonite Nanocomposite Hydrogel as Wound Dressing. *Procedia Mater. Sci.* **2015,** *11,* 152–156.

Ogata, N.; Kawakage, S.; Ogihara, T. Poly(vinyl alcohol)-Clay and Poly(ethylene oxide)-Clay Blends Prepared Using Water as Solvent. *J. Appl. Polym. Sci.* **1997,** *66* (3), 573–581.

Oliveira, M. J. A., et al. Hybrid Hydrogels Produced by Ionizing Radiation Technique. *Rad. Phys. Chem.* **2012,** *81* (9), 1471–1474.

Paranhos, C. M., et al. Poly(vinyl alcohol)/Clay-based Nanocomposite Hydrogels: Swelling Behavior and Characterization. *Macromol. Mater. Eng.* **2007,** *292* (5), 620–626.

Parida, U. K., et al. Synthesis and Characterization of Chitosan-Polyvinyl Alcohol Blended with Cloisite 30B for Controlled Release of the Anticancer Drug Curcumin. *J. Biomater. Nanobiotechnol.* **2011,** *2* (4), 414–425.

Patro, T. U.; Wagner, H. D. Layer-by-layer Assembled PVA/Laponite Multilayer Freestanding Films and Their Mechanical and Thermal Properties. *Nanotechnology* **2011,** *22* (45), 455706.

Reguieg, F.; Nabahat Sahli, N.; Belbachir, M. Hydrogel Composite of Poly(vinyl alcohol) with Unmodified Montmorillonite. *Curr. Chem. Lett.* **2017,** *6,* 69–76.

Sabaa, M. W., et al. Synthesis, Characterization and Application of Biodegradable Cross-linked Carboxymethyl Chitosan/Poly(vinyl alcohol) Clay Nanocomposites. *Mater. Sci. Eng. C* **2015**, *56*, 363–373.

Schulze, D. G. Clay Minerals. In *Encyclopedia of Soils in the Environment*; Hillel, D., Ed.; Elsevier/Academic Press: Boston, 2005; Vol. 1, pp 246–254.

Sirousazar, M.; Kokabi, M.; Hassan, Z. M. Swelling Behavior and Structural Characteristics of Polyvinyl Alcohol/Montmorillonite Nanocomposite Hydrogels. *J. Appl. Polym. Sci.* **2012a**, *123* (1), 50–58.

Sirousazar, M., et al. Polyvinyl Alcohol/Na-montmorillonite Nanocomposite Hydrogels Prepared by Freezing–Thawing Method: Structural, Mechanical, Thermal, and Swelling Properties. *J. Macromol. Sci. B* **2012b**, *51* (7), 1335–1350.

Soheilmoghaddam, M., et al. Characterization of Bio Regenerated Cellulose/Sepiolite Nanocomposite Films Prepared via Ionic Liquid. *Polym. Test.* **2014**, *33*, 121–130.

Taghizadeh, M.; Sabouri, N.; Ghanbarzadeh, B. Polyvinyl Alcohol: Starch: Carboxymethyl Cellulose Containing Sodium Montmorillonite Clay Blends; Mechanical Properties and Biodegradation Behavior. *SpringerPlus* **2013**, *2*, 376.

Teodorescu, M., et al. Viscoelastic and Structural Properties of Poly(vinyl alcohol)/Poly(vinylpyrrolidone) Hydrogels. *RSC Adv.* **2016**, *6* (46), 39718–39727.

Theng, B. K. G. *Formation and Properties of Clay-polymer Complexes*, 2nd ed.; Elsevier Science: Amsterdam, Netherlands, 2012; Vol. 4.

Yuan, P.; Tan, D.; Annabi-Bergaya, F. Properties and Applications of Halloysite Nanotubes: Recent Research Advances and Future Prospects. *Appl. Clay Sci.* **2015**, *112–113*, 75–93.

Yu, Y. H., et al. Preparation and Properties of Poly(vinyl alcohol)-Clay Nanocomposite Materials. *Polymer* **2003**, *44* (12), 3553–3560.

Zhao, L. Z., et al. Recent Advances in Clay Mineral-containing Nanocomposite Hydrogels. *Soft Matter* **2015**, *11* (48), 9229–9246.

Zhou, W. Y., et al. Poly(vinyl alcohol)/Halloysite Nanotubes Bionanocomposite Films: Properties and In Vitro Osteoblasts and Fibroblasts Response. *J. Biomed. Mater. Res. A* **2010**, *93* (4), 1574–1587.

Zhuo, Q., et al. Poly(vinyl alcohol)/Hydrotalcite Composite Nanofibre: Preparation and Characterization. *Iran. Polym. J.* **2011**, *20* (5), 357–365.

CHAPTER 10

SURFACE ENGINEERING OF POLYMERIC MEMBRANES

ADINA MARIA DOBOS*

Department of Physical Chemistry of Polymers, "Petru Poni" Institute of Macromolecular Chemistry, 41A Grigore Ghica Voda Alley, 700487 Iasi, Romania

E-mail: necula_adina@yahoo.com

CONTENTS

ABSTRACT

Membrane technology is an early concern; however, due to the scientific evolution and also to the growth of daily life requirements, this domain represents a continuous challenge for researchers. The surface properties of the polymeric membranes play an important role in various applications like adhesion, surface immobilization, and separation of gases or liquids mixtures. Thus, the shaping of the hydrophilicity/hydrophobicity, enhancing of the biocompatibility for membrane target applications involves the using of some surfaces manufacturing techniques. In this context, the chapter is a brief presentation of the most used processing methods of membrane surfaces in order to improve their properties to be used in different areas of activity and especially in the biomedical field. Chemical treatments such as wet chemical treatment, surface grafting, chemical modification, as well as the physical methods for polymeric membrane modification by plasma exposure, flame treatment, bombardment with charged particle beams (ions or electrons) can be used for membrane materials obtaining with biomedical applications as support for cell growth as well as implanted materials.

10.1 INTRODUCTION

Increasing of needs from health and implicitly biomedical domain led to an increase of techniques for the obtaining and characterization of new materials with properties that meet these requirements. Recently, studies on surface properties of polymers have become important in biomedicine, because the polymers are able to form biomaterials that may come into contact with the physiological medium (i.e., blood or living tissues) (Pignatello, 2011; Hench, 1998; Leeuwenburgh et al., 2008). By surface modification, the polymer properties can be improved without altering the bulk mechanical and physical properties. Metals were used as first biomaterials but, in recent years, they have been replaced by polymeric materials to which, researchers have focused their studies. Thus, this chapter is revealed the importance of the surface changes and surface analytical tools for polymer biomaterials field.

The materials used in contact with the biological environment, so-called *biomaterials*, must satisfy two basic conditions:

- to possess physicochemical properties that give them the possibility to replace the defective functions of the body;
- it must be inherent (not have undesirable reactions with the body) and must to induce the desired answers. In this context, the answers are governed by the interface interactions, being dependent on the material surface properties, such as roughness and chemical composition which directly influences the amount of surface energy.

The most common polymers present advantages as low cost and ease processability, but also disadvantages like adhesion, ability to be printed, barrier properties, and low chemical resistance. From this reason, in order to improve the wettability, adhesivity, and ability to be printed a number of techniques to modify the polymeric surfaces, by introduction of functional groups, have been developed. When the surface modification represents an earlier stage for bioactive compound attachment, these techniques must be applied, so that, on the surface to be introduced, specific functional groups. Modification of the surface properties has an important role in various domains (i.e., biosensors, biomaterials, and biopackages) and involves evaluation of the existing surface treatment methods and development of new other methods and techniques. For example, it is known that biosensors consist of an element that shows biological affinity, connected by an interface to a designated transducer. Consequently, for a material to be used in the biosensors field is necessary to apply treatments that improve the binding capacity of the biological component to its surface, property responsible for its bio-sensitivity.

Before applying any modification method, the surfaces must be cleaned by removing the impurities. Surface treatment facilitates the adhesion, making possible the effective wetting of the current surface of the substrate, compared to the apparent surface. In many cases, what seems to be a surface, in reality is a layer of grease, oil, or other contaminants. The methods used to prepare such surfaces for an adequate wetting will depend on type of contaminant and adhesive. In conclusion for surface treatment should be followed the next steps:

- surface preparation—by removing of oil, grease, or other contaminants from the polymeric surface;
- surface pre-treatment—using physical and chemical methods to remove highly absorbent surfaces in order to activate the surfaces;

• effective treatment of polymeric surfaces—by using physical and chemical methods to obtain surfaces with new properties by compounds immobilization.

Applying these procedures, polymeric materials with enhanced properties can be obtained and used in applications from various fields. Figure 10.1 schematically presents the main techniques for modifying the polymeric surfaces (Chen et al., 2008).

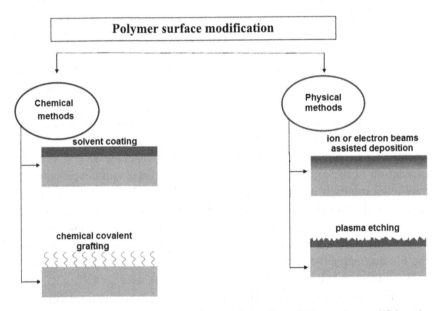

FIGURE 10.1 Schematic representation of the main techniques for modifying the polymeric surfaces. Adapted from Chen et al. (2008).

As was mentioned, these processed polymeric materials can be used in many areas, but in terms of their applicability in the medical domain, the introduction into the human body, sometimes, can generate a number of acute and chronic responses. Some of these reactions have been found in the case of bone implants where the mechanical strength, high compressive strength, and fine fatigue strength of the used substrate represent necessary properties (Galetz et al., 2010; Guo et al., 2013; Shikinami and Okuno, 2003). For fulfillment of these conditions, the metallic biomaterials, bioceramics, and polymers with good mechanical properties were chosen.

Accordingly, in order to increase the potential applications in the biomedical field, numerous polymeric surface modification techniques have been designed and implemented. In this context, this chapter cumulates a series of these kinds of techniques, specifying for each, polymers types, and target applications for which these treatments have been applied. Thus, through this nature, this work provides a useful database for scientists, providing information necessary to obtain polymeric materials for medical applications.

10.2 CHEMICAL TREATMENTS FOR POLYMERIC SURFACE MODIFICATIONS

An important role in evolutions concerning the chemical modifications of polymeric surfaces, it has the understanding of structure-property relationships and aspects on surfaces performance. This fact makes possible the rational obtaining of polymeric surfaces with target applications. On the other hand, a development of the processing techniques and characterizing methods was observed, thus achieving a correlation between the experimental protocol and resulted in physicochemical structure. There are many surface modification techniques mentioned in literature, but every time, the scientists have tried something new by combining the already known techniques (Penn and Wang, 1994).

10.2.1 WET CHEMICAL TREATMENT

Wet chemical treatments represent a way to modify the chemical composition of the polymeric surface by direct chemical reaction with a certain solution, being among the first types of methods used for the surface modification (Garbassi et al., 1994). The most important of techniques that this method covers are: hydrolysis, corrosion, and functionalization with sodium fluoropolymer. In the case of wet chemical modification, the surfaces are treated with a liquid which allows the generation of reactive functional groups on its surface. This classical method of surface modification does not require special equipment and can be done in most laboratories. Compared to the plasma treatment or other techniques that require energy consumption, wet treatment present the advantage to penetrate the porous three-dimensional substrates, being able to perform the in situ functionalization of surfaces.

10.2.2 SURFACE GRAFTING

Unlike the wet chemical treatment, where the surface modification is achieved by modification of existing chains, in the case of *surface grafting* the goal is achieved by covalent binding of new macromolecules on the surface. The first step that must be performed for surface grafting is to create reactive groups on the substrate surface. This can be accomplished by chemical means or, most often, through radiation using ionizing radiation, UV light, or plasma exposure (Bhattacharyaa and Misrab, 2004).

10.2.3 CHEMICAL TREATMENT APPLIED TO BIOPOLYMERS

By *chemical modification* of polymers has been trying the introduction of some reactive groups through processes like oxidation, hydrolysis, sulfonation, cauterization, polymerization, or other processes. These procedures can be materialized by adding some ionic functional groups at polystyrene (PS) resins, dextran (Mauzac et al., 1982), polyurethanes, or polymer chains grafting. Also, the alkylation of polymers that come into contact with blood has been suggested as a solution to reduce the deposit of thrombi at blood–polymer interface through preferential adsorption of albumin. This procedure can be achieved in solution or by surface grafting. It was established that the highly hydrophilic surfaces, namely, polyurethanes (Szleifer, 1997) or cellulose (Sirolli et al., 2000) grafted with polyethylene oxide or polyethylene glycol shows a reduced tendency for adsorption of proteins or platelet adhesion.

10.3 PHYSICAL METHODS FOR POLYMERIC SURFACE MODIFICATIONS

10.3.1 MODIFICATION OF POLYMERIC SURFACE BY PLASMA EXPOSURE

Most industrial applications that require printing or coating of polymers involves their surface treatment. The most common treatment techniques are based on the use of chemical solvents or reagents in proper amounts. As result of adverse effect of solvents, the development of this technique is limited in favor of those which use the gaseous phase and that are already applied in areas that require a clean process (microelectronic and biomedical filed). Some of these techniques relate to the plasma or corona treatment. Plasma represents a complex energy source that can adapt the materials properties in order to use them in various applications, being an active medium consisting in a variety of components (excited and ionized particles) which have significant energy to induce chemical reactions, both in the plasma volume, and, at its interface with solid surfaces (Borcia et al., 2011). Thus, by plasma treatment, the bulk properties of the basic material are not affected being favored the changing of physical and chemical features only at its surface. There are two types of plasma treatments: treatments in hot and in cold plasma (Won et al., 2000; Lee et al., 2000). Hot plasma is represented by plasma generated at atmospheric pressure which attains very high temperatures (up to 10,000K), the electrons and gas temperatures being equal. This plasma is mainly used in metallurgy for hard coatings and less for the polymers and biological tissues due to their low resistance at high temperatures to which they should be exposed (Dumitrascu & Borcia, 2006; Borcia et al., 2005). As opposed to the hot plasma, in cold plasma, the electrons have a different temperature than gas, but identical with the environment, and has the advantage that not cause the damage of surfaces with which they come into contact. However, they are capable of cleavage of some molecular bonds, being used for in situ polymer synthesis and modification of synthetic/natural polymers (Roth, 2001).

10.3.2 MODIFICATION OF POLYMERIC SURFACES BY FLAME TREATMENT

Flame treatment is a method of chemical modification of the surfaces, especially of polymeric ones, giving them improved polar properties, energetic efficiency and preparing them for coating, or lamination processes.

It is well known that the flame treatment along with the corona discharge represents the most used method for polyolefin surface activating (Farris et al., 2010). Combustion is a process that involves chemical reactions occurring between a hydrocarbon and oxygen from the air, resulting heat, and light in the form of flame. Water and carbon dioxide are the final products of these reactions, but other radicals, such as OH, HO_2, H_2O_2, and O_2, as well as ions and electrons, can be formed and can react with the polymeric surface leading to obtain of functional groups (hydroxyl and carbonyl) that improve the surface adhesion.

Generally, this process supports the migration of chemical species in the flame through subsonic wave, leading to obtaining of active radical species (Glassman and Yetter, 2008). Briefly, the combustion process is represented as a succession of reactions (see Fig. 10.2).

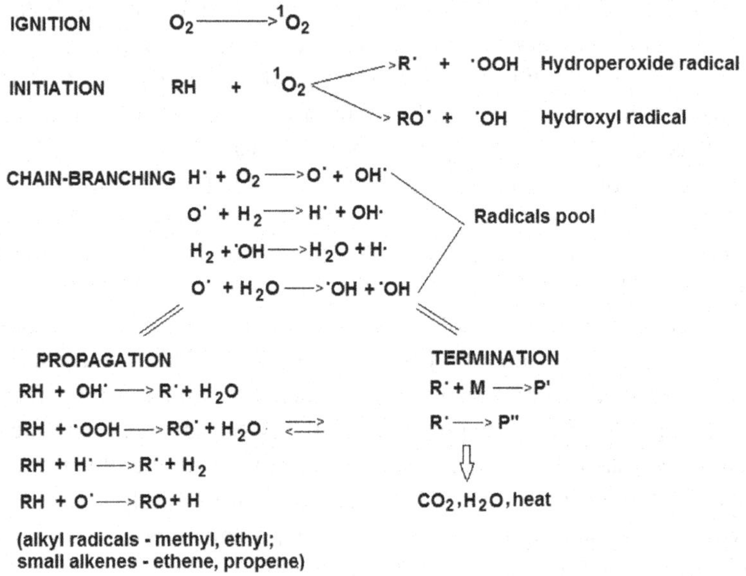

IGNITION $\quad O_2 \longrightarrow {}^1O_2$

INITIATION $\quad RH + {}^1O_2 \Big\langle {\overset{\longrightarrow R^\cdot \ + \ {}^\cdot OOH \quad \text{Hydroperoxide radical}}{\searrow RO^\cdot + \ {}^\cdot OH \quad \text{Hydroxyl radical}}}$

CHAIN-BRANCHING $\quad H^\cdot + O_2 \longrightarrow O^\cdot + OH^\cdot$
$\qquad\qquad\qquad\quad O^\cdot + H_2 \longrightarrow H^\cdot + OH^\cdot \qquad$ Radicals pool
$\qquad\qquad\qquad\quad H_2 + {}^\cdot OH \longrightarrow H_2O + H^\cdot$
$\qquad\qquad\qquad\quad O^\cdot + H_2O \longrightarrow {}^\cdot OH + {}^\cdot OH$

PROPAGATION $\qquad\qquad\qquad$ TERMINATION

$RH + OH^\cdot \longrightarrow R^\cdot + H_2O \qquad\qquad R^\cdot + M \longrightarrow P'$

$RH + {}^\cdot OOH \longrightarrow RO^\cdot + H_2O \rightleftharpoons \qquad R^\cdot \longrightarrow P''$

$RH + H^\cdot \longrightarrow R^\cdot + H_2 \qquad\qquad\qquad \Downarrow$

$RH + O^\cdot \longrightarrow RO + H \qquad\qquad\qquad CO_2, H_2O, \text{heat}$

(alkyl radicals - methyl, ethyl;
small alkenes - ethene, propene)

FIGURE 10.2 Schematic representation of the combustion process. Reprinted with permission from Farris, S., et al. The Fundamentals of Flame Treatment for the Surface Activation of Polyolefin Polymers—A Review. *Polymer*. **2010**, *51* (16), 3591–3605. © 2010 Elsevier.

10.3.3 SURFACE CHANGES BY BOMBARDMENT WITH CHARGED PARTICLE BEAMS

Other techniques often used for modification of the polymeric surface properties are bombardment with loaded beams, the most common species of used particles being electrons, and ions.

10.3.3.1 ION BEAM TREATMENT OF THE POLYMER SURFACE

Literature mention (Qiu et al., 2014) that a distinction between ion-beam-assisted deposition (IBAD)—which combine physical vapor deposition (PVD) with ion implantation—and other surface modification techniques (including ion beam deposition (IBD), ion-beam-induced deposition (IBID), and ion beam-sputtering deposition (IBSD)) must be done. IBAD is a process of continuous ion bombardment accomplished in order to clean polymeric surfaces and to prepare them for deposition process. Applying IBAD method, a gradual transition from the deposited material to substrate is formed, favoring thus strong adhesion to the substrate. Also, between the IBD and ion implantation are a difference concerning the particle beam energy which is higher in the latter case, particles penetrating deeper into the polymer substrate (Oechsner, 1989). IBID represents a chemical vapor deposition which uses a focused Ga^+ ion beam that decompose the gaseous molecules and deposits them at polymeric surface (Reyntjens and Puers, 2000), while ion beam sputtering deposition is a process by which an ion beam bombards a target and releases particles at the atomic level from this target in order to form a thin layer on the polymeric surface (Ong et al., 1992). Among all the ion bombardment methods above mentioned, IBAD is the most effective, the other technique being unable to create a gradual transition between the substrate and deposited film, often used for modification of biomaterials.

Ion beam treatments are very promising in terms of fulfilling of all requirements of advanced surface modification technique which can cross-link the polymer chains at significant depths. Moreover, this technique can be easily combined with traditional processes—that is, wet chemical treatment—proving the ability to modify more complex surfaces. Applying this method, the adhesion between polymeric surfaces and a material deposited on this surface improves significantly. The ability to obtain ion beams of different energies and fluxes involve a wide range of changes in surface properties, depending on the further applications of treated materials. Ion beam sources with large aperture, which have the ability to operate in reactive gases, make possible physical and chemical modification of the polymeric surfaces. Thus, an alteration in these surface properties leads to an improvement of the adhesion or metallization processes. As the researchers sustain in their studies (Sioshansi, 1984), ion beams can be used both for direct modification of the materials surfaces and for the pulverization of target species that will be deposited on the surface. In this context, the surface morphology, adherence, bio-adhesiveness, the attachment, and proliferation of endothelial cells are the main properties of the material surfaces that can be improved under the action of ionic bombardment (Hong et al., 1997; Qiu et al., 2014; Griffith et al., 1982; Song et al., 2005; Bilek, 2014).

10.3.3.2   ELECTRON BEAM TREATMENT

It is well known that after the electron bombardment a deterioration of the polymeric surfaces occurs as result of the electrons diffusion, interactions, or desorption of adsorbed species, these effects leading to obtaining of a superficial layer of deteriorated material (Licari and Hughes, 1990). Generally, the effects of polymers radiation with high-energy radiation can lead to significant changes both in the surface properties and in volume, high-energy electron irradiation being a complex, and random process. In this context, the changes that occur after electronic radiation are a consequence of two main phenomena (Pruitt, 2003): (1) *electron absorption*—accompanied by the chemical bonds cleavage and obtaining of radicals that would lead to the formation of crosslinks and (2) *disproportionation* which will generate the gas release by chain cleavage. Following the polymers exposure to the action of ionizing radiation are generated new type of chromophore groups which efficiently absorb light from UV and IR. The main effects of irradiation with electron beams are: crosslinking, degradation (generated by the cleavage of bonds from main and side chain), gas forming (H, CH_4, and CO), oxidation, obtaining of double bonds (C=C), and cyclization by forming intermolecular bonds depending on the oxygen content from environmental exposure. As a result of these processes, the surfaces and interfaces properties of synthetic polymers and biopolymers can be purposefully modified. Among these features can be mentioned: mechanical strength, swelling and dissolution, surface topography (smooth, rough), wetting, surface reactivity and adhesion of organic coatings—in the latter case, a major role having the use of low-energy electrons (<25 keV). The polymeric materials thus obtained can be utilized as coatings, food packaging, and protection for certain surfaces in medical technology as well as composite materials (Ghoranneviss et al., 2007).

10.3.4   MODIFICATION OF POLYMERIC SURFACES UNDER UV RADIATION ACTION

Besides the above-listed techniques, another easy and economical method for modifying the polymeric surfaces is the exposure to UV radiation, the material processed in this way finding applications both in industry (for disinfection of surfaces and packaging) and in medical field (Weibel et al.,

2009). Usually, ultraviolet radiation is high-energy radiation with photons whose energy varies between 10 and 124 eV (corresponding to wavelengths of 124–10 nm, respectively). Thus, in order to prevent the bulk degradation of materials, a short wavelength radiation from the extreme ultraviolet range which is absorbed in a very thin layer of the polymer surface (<100 nm) is used (Ahad et al., 2014). The most important criterion that must be satisfied in order to use photochemical method for the surface properties improving is that the polymer substrate to become photoreactive during the irradiation process (Wells et al., 1993). After UV irradiation, at polymeric surface, reactive functional groups are generated. These groups exposed to different gases can become reactive or can participate at UV-induced graft polymerization. This technique differs from other treatments that use ionized gas, through ability to obtain different depths (that can be modified by varying the wavelength of the radiation and adsorption coefficient, implicitly) on the irradiated surface. UV irradiation was used for the introduction of carboxyl groups on the poly(methyl methacrylate) (PMMA) surface (Situma et al., 2007) and for activates the PS surface in view of the enzymes immobilization (Nahar et al., 2001). Moreover, in order to improve wettability, adhesion, and antistatic properties of the polymeric surfaces UV irradiation in the presence of oxygen or ozone was used (Hedenberg and Gatenholm, 1996). There are many literature studies performed on various classes of polymers which indicated that exposure to UV radiation improves the wettability, compatibility, adhesion, and cell proliferation of the polymers especially of thermoplastic ones (Le et al., 2013; Rajajeyaganthan et al., 2011; Olbricha et al., 2007; O'Connell et al., 2009; Heitz et al., 2004; Ku et al., 2007).

10.4 PRODUCING OF NOVEL MATERIALS WITH BIOMEDICAL APPLICATIONS BY CHEMICAL MODIFICATION OF THE POLYMERIC SURFACES

As previously was mentioned, the *chemical modification* of polymeric surfaces has an important role in obtaining new materials with improved properties that can meet certain medical requirements. There are a lot of literature data which present different polymeric surfaces subjected to chemical treatments and used especially for active species immobilization (Costa et al., 2005; Gorecka and Jastrzebska, 2011; Vesel et al., 2012). In this sense can be mentioned studies conducted on polyacrylonitrile (PAN) membranes, which are of great interest due to their excellent properties namely: thermal stability and tolerance to most solvents, atmosphere, and bacteria as well as to photo-irradiation (Gorecka and Jastrzebska, 2011). However, they present some inconvenience as low hydrophilicity and biocompatibility, which does not make them suitable for protein adsorption or cell adhesion, determining the biofouling processes, and immunoreactions when they are used in red blood system. In this context, Wang et al. (2007) have used the *grafting polymerization* (known as "grafting-from" method), partially hydrolysis and macromolecule immobilization (also known as "grafting-to" method), to improve surface properties and enzyme immobilization. The advantages of this technique are: on the one side can lead to easily obtaining of polymeric surfaces with desired properties by simply amending of monomers, and, on the other hand, the stability of these surfaces is greater than those resulting from physical treatments. Following these procedures, the PAN membranes will present the ability to attach proteins or blood components, their hemocompatibility being improved. The major role, in this sense, it has the presence of reactive groups, at PAN membrane surface, which favors the attachment of other biomacromolecules. The improved properties will make them suitable for application, besides dialysis and filtration, in enzyme attachment. Compared with this, other scientists have used the grafting methods to reduce the attachment of blood components. Thus, Govindarajan and Shandas (2014) have grafted the polyethylene glycol (PEG) monoacrylates at biomaterial surface in order to prevent the erythrocyte attachment and consequently the thrombosis risk (Feng et al., 2013; Liang et al., 2009). High hydrophilicity and non-toxicity confer to PEG biocompatible properties, so that when is grafted to a hydrophobic surface a decrease of the hydrophobicity can occur, making the substrate proper for cell adhesion (Ma et al., 2007).

In order to increase the biocompatibility, another author's studies were focused on polymer surfaces changes by using *wet chemical treatment*. This method does not involve the attachment of chemical groups or chemical alteration of the surface but can cause a sufficient change for biocompatibility increasing. In this sense, dimethyl sulfoxide (DMSO), known as nontoxic for epithelial cells, was used to prevent the vascular smooth muscle cell activity on the stents surface and restenosis and thrombosis implicitly (Luscher et al., 2007). Also, the polyvinyl chloride (PVC), which is known to be widely used for the manufacture of plastic materials used in biomedical field, presents some inconvenience (lower hydrophilicity and surface free energy), so that, in pure state, is not biocompatible favoring the attachment of microbes and some unwanted proteins. Thereby, to solve these problems, it was imposed the PVC surface treatment (Asadinezhad et al., 2012; Wilkes and Summers, 2005; Williams, 1982). Wet chemical treatment method has often been used because is not very expensive, but also has the disadvantage of toxic solvents utilization. In literature, studies have been presented where PVC polymeric surfaces were activated using various solvents. Kurian and Sharma have subjected the PVC surfaces to treatment with ionic polyelectrolyte using dilute zephiran chloride and then immersion in polyelectrolyte solutions in order to improve the surface energy and platelet adhesion (Kurian and Sharma, 1984). This method is also effective for the obtaining of PVC materials that find applications in the medical field as oxygenators or blood bags. Reyes-Labarta et al. (2003) have immersed the PVC films in a 0.5 M solution of amino thiophenol, and then, for analyzing the reaction kinetics, they were removed at various time intervals, washed with water, kept for 24 h in air and then dried (Juan et al., 2003). Surface free energy and hydrophilicity of the PVC films have been also improved through physisorption of the azobisisobutyronitrile onto surface, followed by radical graft polymerization of hydrophilic monomers (Reyes-Labarta et al., 2003).

10.5 ADVANCED BIOMATERIALS OBTAINED BY PHYSICAL MODIFICATION OF THE POLYMERIC SURFACES

As was mentioned, the *flame treatment* is a technique to improve the surface properties of polymers, especially of polyolefins. To meet the performance criteria and to become high-quality products, these polymers are subjected to flame treatment—a process that determines an increase of wettability and adhesion properties. Thus, they will form the basis for development of new compounds with specific properties, able to replace PVC, and successfully used to obtain material—with impact resistance and ductility at low temperatures—for medical applications (Goodman, 1994). Literature shows that the polypropylene (PP) is one of the polyolefins trying to overcome the limits in order to find applications in medical field as less expensive and more environmentally friendly material (Laurence Mckeen, 2014). In this context was obtained the medical device that present:

- sterilization resistance—materials based on PP/metallocene cata-lyzed blends and ethylene-based plastomers are proper for the production of highly radiation-resistant thin gauge medical device;
- gamma radiation resistance—polypropylenic materials presenting sufficiently antioxidant can be sufficiently stabilized, thus resisting to irradiation process;
- autoclave sterilization—degradation temperature of PP materials is high enough to be used in autoclaving applications.

Regarding the *plasma treatment* of the polymeric surfaces, it is well known, that, so called the fourth state of matter consists in negatively charged electrons, positively charged ions, and neutral atoms or molecules, being obtained when gases are excited in energetic states through radio frequency or electrons from a hot filament discharge (Li et al., 1997). The modification technique occurs fastly and cleanly, with consequent forma-tion of functional groups, being very much applied in the biomedical field because determines an improvement of the polymeric substrate biocom-patibility. On these assumptions are based the studies of Siow et al. (2006) which selected an appropriate plasma in order to create, at the polymeric surfaces, covalent interfacial bonds—necessary for permanent immobili-zation of biologically active molecules. Among the functional groups that can be formed after plasma treatment can be mentioned: aldehyde, carboxy, amine, or hydroxyl groups. The oxygen from carboxyl or hydroxyl groups plays an important role in the cell adhesion (Ertel et al., 1991a), but this content cannot be correlated with the surfaces affinity for the cell attach-ment that inhibits plasma proteins (Ertel et al., 1991b). However, as the oxygen amount resulting from plasma treatment is greater, the polymeric substrate is more effective in terms of cell growth and favors the albumin and immunoglobulin binding at the surface. In the same context has been studied the importance of hydroxyl or carboxyl groups from the plasma copolymers treated surface, in culture of the keratinocytes and osteoblasts (France et al., 1998). The researchers have compared the results with those obtained from experimental investigations where collagen I was used as plasma treated support, and have observed that, for an oxygen content of 25%, the ability of plasma treated surfaces to fix the keratinocytes is the same. Studies concerning the increasing of polymeric membranes biocompatibility were also realized by Ulbricht and Belfort (1995). They subjected the PAN membranes to plasma treatment (helium or helium/ water plasma), and then, as results of the peroxide species which were formed at the polymeric surface they have grafted the membrane using 2-hydroxy-ethylmethacrylate (HEMA) and acrylic or methacrylic acid (Ulbricht and Belfort, 1996). At the end of these operations was obtained membranes for which the protein fouling was reduced because of low static protein adsorption, while the permeability was improved as result of the same protein retention. The hydrophilic PAN membranes with improved biocompatibility properties were also achieved by introduction of the acrylic acid (hydrophilic monomer) in membrane, using argon (Ar) plasma treatment and appropriate grafting reactions. Following the Ar plasma treatment, the C—N bonds did not break, grafting reaction being generated by scission of these bonds. All these processes have generated changes in membrane topography and also in pore sizes which become smaller as a consequence of the surface graft (Wang et al., 2007).

Usually, for polymeric surfaces modification is used the oxygen or argon plasma treatment, because both produce changes in roughness and increase hydrophilicity, favoring thus the cell adhesion (Pavithra and Doble, 2008; Ormiston et al., 2008). The advanced materials obtained in this way have found application in medical field especially for stents processing (Di Mario and Ferrante, 2008). Some literature studies have shown the way in which the cells adhesion process on the polymeric surface can be improved by promoting in situ roughening. In this regard, Shadpoura and Allbritton (2010) have used an aluminum suspension in order to enhance the cell attachment. Increasing of the surface roughness will facilitate both cell and biomolecules adhesion, making possible the use of obtained materials in manufacturing of the most advanced types of stents. This method is provided because, besides increasing of roughness generates an increase of wettability and facilitates the adhesion of another substrate at the treated surface (Govindarajan and Shandas, 2014). Jaganathan et al. (2015) have mentioned in their paper that argon plasma treatment of the polymeric surfaces promotes the cell attachment and proliferation. Also, the treated substrate shows the ability to inhibit the inflammation and support fibroblast adhesion. In this sense, Melnig et al. have tested in vivo biocompatibility of the polyurethane substrates (Melnig et al., 2005). They have observed that, unlike the untreated surfaces, the treated ones show no inflammation and tissue develops and adheres much better. Additionally, a large number of fibroblasts and collagen fibers are found in areas with high density of argon ions, while in the vicinity of the implanted material an inflamed area and a slight thrombosis appears (Fig. 10.3). Thus, the researchers have concluded that the poly(lactaturethane) surfaces treated with argon plasma ions lead to the improvement of polymeric material biocompatibility, these being used in biomedical field for tissue regeneration.

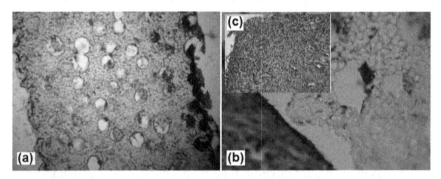

FIGURE 10.3 Viewing of collagen adherence on the poly(lactaturethane) surfaces subjected to argon ions plasma treatment (a), untreated surfaces of poly(lactaturethane) (b), and areas that highlight the presence of tissue inflammation and thrombosis (c). Adapted from Melnig et al. (2005).

Studies conducted in order to improve properties like adhesion, wettability, permeability, and biocompatibility were realized by Arefi et al. (1992) on PP surfaces treated with nitrogen plasma. Also, the polyethylene terephthalate surfaces (PTFE) were modified by exposure to argon plasma. It was observed that, after plasma treatment, the morphology and roughness have been completely changed. In addition, by varying plasma exposure time the values of the contact angle decrease and, simultaneously with that, an increase of the oxygen content and hydrophilicity occurs (Kolska et al., 2012).

As previously was noted, most used plasmas for surface modifications were those of oxygen and argon. Oxygen plasma was applied for modification of poly(L-lactic acid) (PLLA) and poly(d,l-lactic acid-co-glycolic acid) (PLGA) surfaces (Khorasani et al., 2008). The researchers have reported that the chemical modification of the polymeric surfaces has led to an improvement in wettability, the two polymers being designated as suitable for attachment, growth, and proliferation of nervous cell. Antibacterial activity of PMMA against to *Staphylococcus aureus* (*S. aureus*) and *Escherichia coli* (*E. coli*) bacteria was enhanced firstly, by plasma treatment of surface and secondly by TiO_2 deposition (Wenyue et al., 2010). It was observed that both samples show an antimicrobial activity of 100% after they have been subjected to plasma treatment for 2 h. Also, the immersion in a colloidal solution of TiO_2 led to superior anti-adhesion capability of the surfaces. Slepicka et al. (2013) have modified polymeric surfaces of poly(ethylene terephthalate) (PET), high-density polyethylene (HDPE), PTFE, and PLLA by applying plasma treatment technique. The values of the contact angles, immediately measured after the surface treatment, have shown a sudden decrease comparatively with values of the untreated samples. PLLA is the first polymer reaching saturation of wettability after 100 h of plasma treatment followed by the other two polymers, at a difference of 226 h. The studies have concluded that the PTFE is the most suitable for bioapplications as results of its highly improved biocompatibility. Effect of the plasma treatment and grafting with Au nanoparticles of polyethylene surfaces (PE) was studied by Kasalkova et al. (2012). They have observed that improving the morphology and roughness of the surfaces, by applying this method, has facilitated the cell attachment (Kasalkova et al., 2012). Comparatively, Junkar et al. have demonstrated that oxygen and nitrogen from radio frequency (RF), plasma treatments have a negative effect on wettability and morphology, leading to alteration of PET surfaces. As result of this effect, on the one hand, the platelet adhesion was altered, and on the other hand, has been improved the adhesion and proliferation of fibroblast and endothelial cells (Junkar et al., 2011). Comparing the results of the studies, the researchers have concluded that, the use of oxygen plasma is the most appropriate to enhance the hemocompatibility properties of PET surfaces than the nitrogen plasma. In the latter case, the adhesion at the polymeric surface was the same with those for untreated samples. A more complex study refers to the effect of plasma treatment on the metallic stents at the surface of which an ultrathin polymeric layer of [poly (2-chloroparaxylylene)] was added (Lahann et al., 1999). The scientists have used in their work the sulfur dioxide plasma, and the results have shown an increase in surface hydrophilicity. Moreover, the hemocompatibility was improved and the platelet adhesion has decreased. In context of the biocompatibility improvement, cell growth, and proliferation, Sharma et al. (2004) have used the helium plasma treatment on the polyurethane surfaces (PU). They have observed that this process increases the wettability facilitating the endothelial cell adhesion at the polymeric surface. These properties get better as exposure time at helium plasma is higher. In literature are mentioned various polymeric surfaces subjected to plasma treatment. Among these, the low-density polyethylene surfaces were tested in terms of their antibacterial activity after application of this surface modification technique and grafting with polyallyamine. The polyallyamine is designed to facilitate the immobilization of some antibacterial agent as benzalkonium chloride, bronopol, chlorhexidine, and triclosan at the tested surfaces. The results of biocidal activity analysis against to *S. aureus* and *E. coli* have shown that triclosan presents the highest antimicrobial activity, while bronopol the smallest. Subsequently, the triclosan was chosen for enhancing the antimicrobial properties of plasma-treated LDPE surface grafted with other compounds like allylamine (AA), Nallylmethylamine (AMA), and N,N-dimethylallylamine (DMAA). It has been shown that, in this case, a good answer it has the samples containing AMA and DMAA (Bilek et al., 2011; Bilek et al., 2013).

Considering those abovementioned it means that, there are numerous classes of synthetic, aromatic, plastic, thermoplastic, fluoropolymer, and polyester polymers, that subjected to plasma treatment and grafting processes led to the improvement of wettability, adhesion, properties, and consequently of biocompatibility. As was demonstrated, these aspects are of a real interest in order to use the listed compounds in obtaining of advanced biomaterials for cell growth and proliferation and, therefore, in tissue repair.

Because the polymers manifest some drawbacks as low mechanical strength and poor resistance, the ion implantation represents the way to solve these problems. The most important classes of compounds for which this method can be used are polyurethane and silicone rubber. Applying this method the polyurethane will enhance their wettability, anticoagulant, and anticalcific properties, while the critical surface tension of silicon rubber, that dictated the biofouling effect, also increases (Li and Zhao, 1994; Li and Zhao, 1993). Often, the polyurethane was chosen for bioapplications because of its durability, resistance to continuous stress, flexibility, and especially for tolerability when is used as implanted material. Li and Zhao have applied the silicon ion implantation in order to improve the blood compatibility and anticalcific behavior of polyurethane—features inferior to those of body organs when polymer is used in pure and unprocessed state (Li and Zhao, 1993). Also, the silicone rubber was subjected to ion beam bombardment to obtain materials suitable for catheter processing. Following this method of polymeric surfaces modification, the sticky and fouling features are drastically reduced, and biocompatibility and flexibility are increased (Sioshansi and Tobin et al., 1996). Other studies realized in the same sense are those of Cui et al. (1997). They have conducted a deposition of hydroxyapatite (HA) on alpha–beta titanium alloy (Ti–6Al–4V) substrate with an interface realized by IBAD. Initially, they have cleaned the substrate surface using Ar^+ ion beam bombardment, and then, have splashed a composite target containing HA and tricalcium phosphate with the same type of ion beam to form the coating on the substrate (simultaneous subjected to another energetic Ar^+ ion beam). At first, the ion beam energy was high and aimed the obtaining of gradual transition between the substrate and deposited film, and then descended leading to a thickening of the deposited film. In order to not affect the substrate, all these procedures have occurred at temperatures below 100°C. Polymeric materials, based on calcium phosphate and titanium, were usually used in medical field especially for implants, favoring the bone growth around them without being necessary a cemented fixation (Narayanan et al., 2008).

A similar study, but based on the *electron beam bombardment*, is that of Chen et al. (2010). They have obtained, by applying this method, a thin layer of calcium phosphate (CaP) on pure titanium support surface. Calcium phosphate was achieved through the mixing of calcium oxide with hydroxyapatite for 2 h, at 1200°C, in air. For monitoring the training and development of hydroxyapatite, the system was immersed in Dulbecco's phosphate-buffered saline solutions. This solution is designed to improve the biological activity of supports due to the content of calcium chloride and biomolecules, making them suitable for osteoblast adhesion (Qiu et al., 2014).

Among the polymeric surface modification techniques involving radiation and discharge, *UV treatment* method is the most easily applied, both as processing technique, but also from the economical point of view. The materials obtained in this way have found application in industry (as disinfecting agents of the surfaces and packaging) and, particularly, in the biomedical field (Jaganathan et al., 2015). In literature, it has been shown that depending on their nature, the materials can absorb certain wavelengths of radiation. For example, the organic materials absorb the radiation between 126 and 222 nm wavelengths, while the materials with energy enough to break the bonds from the surfaces are able to absorb the UV laser generated by the excimer lamp. Under the action of this type of radiation, both the surface properties and the mass characteristics are modified (Kuper and Stuke, 1988). The polycarbonate (PC) and polyether-ether ketone (PEEK) surfaces were modified by irradiation with different UV wavelengths in order to increase their wettability and biocompatibility (Laurens et al., 2000). Introduction of reactive functional groups at the polymeric surfaces optimizes various surface properties and creates suitable sites for different cell attachments. Thus, biomaterials with desired properties of biocompatibility and adhesiveness can be designed and successfully used in tissue engineering, the occurrence of infections being excluded. In literature are mentioned functional groups (e.g., $X(CH_2)_{15}SH$ where X can be $-CH_3$, $-CH_2OH$, $-CO_2CH_3$, or $-CO_2H$.) that are capable to facilitate the growth and proliferation of bovine aortic endothelial cells (Tidwell et al., 1997). It was found that the two processes are strikingly evident, when R = $-CO_2H$, and smaller when R = $-CH_2OH$. Moreover, at all monolayer surfaces, the cell growth was not very much favored as in the case of multiple chemical multilayers (i.e., PS), which means that the multiple functionalities were designated the most suitable for cell growth. Also, the plasma proteins adsorption, especially of the albumin, and fibro-nectin, at self-assembled monolayer was not greatly influenced by the type of functional groups. Fictionalization of the surfaces after UV treatment is a pretty delicate process, considering that they must remain non-toxic and friendly for being used in tissues regeneration. In this context, is attempted, as much as possible, the using of non-toxic and compatible compounds. UV treatment itself represents a way to control the biocompatibility of the samples. Thus, for increasing the wettability and biocompatibility, Heitz et al. have treated the polytetrafluoroethylene (PTFE) with Xe_2-excimer lamp at 172 nm wavelength for 30 min, and then, they grafted him with amino acid alanine. It was found that besides the improvement of mentioned properties an increase in optical absorbance for the treated samples occurs. In vivo tests have demonstrated that, on such surfaces, the rat aortic smooth muscle cells, endothelial fibroplastele, and human umbilical vein endothelial adhere much better than at the untreated surfaces (Heitz et al., 2003; Inam et al., 2014). Another group of scientist has studied the possibility of the PU to be used as vascular prosthesis (Doi et al., 1996). In this respect, they created a microporous polymer surface by using excimer laser ablation technique. The same authors have demonstrated that these kinds of materials are suitable for in vivo development of tissues. Zhihong has demonstrated in his work that the interactions between DNA molecules and polymer chains are favored by UV treatment of the surfaces (Zhihong, 2003). The behavior resulted before UV action against polysulfone (PSU), PP, or PS in presence of acrylic vapor was analyzed and compared with that for the same polymers subjected to the same radiation, but in presence of trimethoxy propyl silane (TMPSi). It was observed that, after 65 days of UV treatment, in presence of acrylic vapor, all samples present a high hydrophilic character, while, in presence of TMPSi, PSU shows a strong hydrophobic character (Rajajeyaganthan et al., 2011). This study is a proof that the same polymeric materials may become hydrophilic/hydrophobic and compatible/incompatible implicitly, depending on the conditions of synthesis. In order to improve the cytocompatibility was also created new nanocomposite that, under UV action, improves their hydrophilicity by creation of new hydrophilic group of N and O. In vivo test has shown that these materials represent favorable environments for development of endothelial cell in a very small period of time after they have been seeded (Olbricha et al., 2007). As was mentioned, the wavelength is an important factor influencing the wetting properties. In this context, Subedi et al. have studied the wetting properties of polycarbonate subjected to UV radiation action with two different wavelengths. Thus was found that a shorter wavelength of 254 nm was sufficient to enhance the wettability features of the polycarbonate comparatively with a longer wavelength of 375 nm (Subedi et al., 2009).

As a result of those noted in this section, it can be concluded that UV treatment can be used for a wide range of polymers (synthetic, thermoplastic polyesters), best results in terms of the wettability and cell compatibility being achieved for thermoplastic compounds.

10.6 CONCLUSIONS

This chapter provides an overview of approaches from recent years on polymeric surface engineering. It was established that the hydrophilic properties, surface free energy, binding properties of different chemical/biological compounds, can be enhanced without affecting mass properties of polymeric support. Also, the biocompatibility and cell adhesion can be greatly improved by using these processability techniques. Least-applied methods are chemical ones, based on the use of different solvents or flame treatment, which are recommended especially for polyolefins in order to improve the active species immobilization. Ion/electron beam bombardment, plasma treatment, or ultraviolet radiation represents the main polymer surface modification techniques which led to the improvement of abovementioned properties. The materials obtained in this way can be successfully used in medical field for enhancing the cell growth and proliferation as well as implanted materials.

Consequently, this chapter has been designed as a database—that offers information on polymeric surface technology, that is, modeling, processing, and medical application—useful both for academic and for research sectors.

KEYWORDS

- wet chemical treatment
- flame treatment
- surface grafting
- plasma exposure
- ion/electron beam bombardment
- biocompatibility
- cell proliferation

REFERENCES

Ahad, I. U., et al. Surface Modification of Polymers for Biocompatibility via Exposure to Extreme Ultraviolet Radiation. *J. Biomed. Mater. Res. A.* **2014**, *102* (9), 3298–3310.

Arefi, F., et al. Plasma Polymerization and Surface Treatment of Polymers. *Pure Appl. Chem.* **1992**, *64* (5), 715–723.

Asadinezhad, A., et al. Recent Progress in Surface Modification of Polyvinyl Chloride. *Materials* **2012**, *5* (12), 2937–2959.

Bhattacharyaa, A.; Misrab, B. N. Grafting: A Versatile Means to Modify Polymers Techniques, Factors and Applications. *Prog. Polym. Sci.* **2004**, *29* (8), 767–814.

Bilek, F., et al. Preparation of Active Antibacterial LDPE Surface through Multistep Physicochemical Approach II: Graft Type Effect on Antibacterial Properties. *Coll. Surf. B. Biointerfaces* **2013**, *102*, 842–848.

Bilek, F.; Krízova, T.; Lehocky M. Preparation of Active Antibacterial LDPE Surface through Multistep Physicochemical Approach: I Allylamine Grafting, Attachment of Antibacterial Agent and Antibacterial Activity Assessment. *Coll. Surf. B. Biointerfaces* **2011,** *88* (1), 440–447.

Bilek, M. M. M. Biofunctionalization of Surfaces by Energetic Ion Implantation: Review of Progress on Applications in Implantable Biomedical Devices and Antibody Microarrays. *Appl. Surf. Sci.* **2014,** *310,* 3–10.

Borcia, C.; Borcia, G.; Dumitrascu, N. Surface Treatment of Polymers by Plasma and UV Radiation. *Rom. J. Phys.* **2011,** *56* (1–2), 224–232.

Borcia, G.; Dumitrascu, N.; Popa, G. Influence of Dielectric Barrier Discharge Treatment on the Surface Properties of Polyamide-6 Films. *J. Optoelectron. Adv. Mater.* **2005,** *7* (5), 2535–2538.

Chen, C., et al. Biomimetic Apatite Formation on Calcium Phosphate-coated Titanium in Dulbecco's Phosphate-buffered Saline Solution Containing CaCl$_2$ with and Without Fibronectin. *Acta Biomater.* **2010,** *6* (6), 2274–2281.

Chen, J., et al. Preparation of Blood Compatible Hydrogels by Preirradiation Grafting Techniques. In *Topics in Tissue Engineering*; Ashammakhi, N., Reis R., Chiellini, F., Eds.; Expertissues e-Book: 2008; Vol. 4, pp 1–40.

Costa, S. A.; Azevedo, H. S.; Reis, R. L. Enzyme Immobilization in Biodegradable Polymers for Biomedical Applications. In *Biodegradable Systems in Tissue Engineering and Regenerative Medicine*; Reis, R. L., Román, J. S., Eds.; CRC Press LLC: Boca Raton, FL, 2005; pp 109–112.

Cui, F. Z.; Luo, Z. S.; Feng, Q. L. Highly Adhesive Hydroxyapatite Coatings on Titanium Alloy Formed by Ion Beam Assisted Deposition. *J. Mater. Sci. Mater. Med.* **1997,** *8* (7), 403–405.

Di Mario, C.; Ferrante, G. Biodegradable Drug-eluting Stents: Promises and Pitfalls. *Lancet* **2008,** *371* (9616), 873–874.

Doi, K.; Nakayama, Y.; Matsuda, T. Novel Compliant and Tissue permeable Microporous Polyurethane Vascular Prosthesis Fabricated Using an Excimer Laser Ablation Technique. *J. Biomed. Mater. Res.* **1996,** *31* (1), 27–33.

Dumitrascu, N.; Borcia, C. Adhesion Properties of Polyamide-6 Fibres Treated by Dielectric Barrier Discharge. *Surf. Coat. Technol.* **2006,** *201* (3–4), 1117–1123.

Ertel, S. I., et al. Endothelial Cell Growth on Oxygen-containing Films Deposited by Radio-frequency Plasmas: The Role of Surface Carbonyl Groups. *J. Biomater. Sci. Polym. Ed.* **1991a,** *3* (2), 163–183.

Ertel, S. I.; Ratner, B. D.; Horbett, T. A. The Adsorption and Elutability of Albumin, IgG, and Fibronectin on Radiofrequency Plasma Deposited Polystyrene. *J. Colloid Interface Sci.* **1991b,** *147* (2), 433–442.

Farris, S., et al. The Fundamentals of Flame Treatment for the Surface Activation of Polyolefin Polymers—A Review. *Polymer.* **2010,** *51* (16), 3591–3605.

Feng, Y., et al. Grafting of Poly(Ethylene Glycol) Monoacrylates on Polycarbonateurethane by UV Initiated Polymerization for Improving Hemocompatibility. *J. Mater. Sci. Mater. Med.* **2013,** *24* (1), 61–70.

France, R. M., et al. Plasma Copolymerization of Allyl Alcohol/1,7-octadiene: Surface Characterization and Attachment of Human Keratinocytes. *Chem. Mater.* **1998,** *10* (4), 1176–1183.

Galetz, M. C.; Fleischmann, E. W.; Konrad, C. H. Abrasion Resistance of Oxidized Zirconium in Comparison with CoCrMo and Titanium Nitride Coatings for Artificial Knee Joints. *J. Biomed. Mater. Res. B Appl. Biomater.* **2010,** *93* (1), 244–251.

Garbassi, F.; Morra, M.; Occhiello, E. Modification Techniques. In *Polymer Surfaces: From Physics to Technology*; John Wiley & Sons: West Sussex, England, 1994; pp 243–273.

Ghoranneviss, M., et al. In *Electron Beam Modification of Polypropylene Fabrics*, Proceedings of the 3rd International Conference on the Frontiers of Plasma Physics and Technology (PC/5099), P-15, Tehran, Iran, 2007.

Glassman, I.; Yetter, R. *Combustion*, 4th ed.; Academic Press: Cambridge, MA, 2008.

Goodman, D. Global Marketsf Chlorine and PVC: Potential Impacts of Greenpeace Attacks. *J. Vinyl Addit. Technol.* **1994,** *16* (3), 156–161.

Gorecka, E.; Jastrzebska, M. Immobilization Techniques and Biopolymer Carriers. *Biotechnol. Food Sci.* **2011,** *75* (1–2), 65–86.

Govindarajan, T.; Shandas R. A Survey of Surface Modification Techniques for Next-generation Shape Memory Polymer Stent Devices. *Polymers* **2014,** *6* (9), 2309–2331.

Griffith, J. E.; Qiu, Y.; Tombrello. T. A. Ion-beam-enhanced Adhesion in the Electronic Stopping Region. *Nucl. Instrum. Methods Phys. Res.* **1982,** *198* (2–3), 607–609.

Guo, W. G., et al. Strength and Fatigue Properties of Three-step Sintered Dense Nanocrystal Hydroxyapatite Bioceramics. *Front. Mater. Sci.* **2013,** *7* (2) 190–195.

Hedenberg, P.; Gatenholm, P. Conversion of Plastic/cellulose Waste into Composites. II. Improving Adhesion between Polyethylene and Cellulose Using Ozone. *J. Appl. Polym. Sci.* **1996,** *60* (13), 2377–2385.

Heitz, J., et al. Cell Adhesion on Polytetrafluoroethylene Modified by UV-irradiation in an Ammonia Atmosphere. *J. Biomed. Mater. Res.* **2003,** *67* (1)130–137.

Heitz, J., et al. Adhesion and Proliferation of Human Vascular Cells on UV-light-modified Polymers. *Pubmed.* **2004,** *39*, 59–69.

Hench, L. L. Biomaterials: A Forecast for the Future. *Biomaterials* **1998,** *19* (16), 1419–1423.

Hong, J., et al. Effect of Ion Bombardment on In-plane Texture, Surface Morphology, and Microstructure of Vapor Deposited Nb Thin Films. *J. Appl. Phys.* **1997,** *81* (10), 6754–6761.

Inam, U., et al. Surface Modification of Polymers for Biocompatibility Via Exposure to Extreme Ultraviolet Radiation. *J. Biomed. Mater. Res. A.* **2014,** *102* (9), 3298–3310.

Jaganathan, S. K., et al. Review: Radiation-induced Surface Modification of Polymers for Biomaterial Application. *J. Mater. Sci.* **2015,** *50* (5), 2007–2018.

Juan, A., et al. Wetchemical Surface Modification of Plasticized PVC. Characterization by FTIR-ATR and Raman Microscopy. *Polymer* **2003,** *44* (8), 2263–2269.

Junkar, I.; Cvelbar, U.; Lehocky, M. Plasma Treatment of Biomedical Materials. *Mater. Technol.* **2011,** *45* (3), 221–228.

Khorasani, M. T.; Mirzadeh, H.; Irani, S. Plasma Surface Modification of Poly (L-lactic Acid) and Poly (Lactic-co-glycolic Acid) Films for Improvement of Nerve Cells Adhesion. *Radiat. Phys. Chem.* **2008,** *77* (3), 280–287.

Kolska, Z., et al. PTFE Surface Modification by Ar Plasma and its Characterization. *Vacuum* **2012,** *86* (6), 643–647.

Kuper, M.; Stuke, S. Ablation of Polytetrafluoroethylene (Teflon) with Femtosecond UV Excimer Laser Pulses. *Appl. Phys. Lett.* **1988,** *54* (1), 4–6.

Kurian, G.; Sharma, C. P. Surface Modification of Polyvinyl Chloride towards Blood Compatibility. *Bull. Mater. Sci.* **1984,** *6* (6), 1087–1091.

Lahann, J., et al. Improvement of Haemocompatibility of Metallic Stents by Polymer Coating. *J. Mater. Sci Mater. Med.* **1999**, *10* (7), 443–448.

Laurence, W. McKeen, Plastics Used in Medical Devices. In *Handbook of Polymer Applications in Medicine and Medical Devices*, 1st ed.; Modjarrad, K., Ebnesajjad, S., Eds.; Elsevier: Oxford, 2014; Chapter 3, pp 21–53.

Laurens, P., et al. Characterization of Modifications of Polymer Surfaces After Excimer Laser Treatments Below the Ablation Threshold. *Appl. Surf. Sci.* **2000**, *154–155*, 211–216.

Le, Q. T., et al. Mechanism of Modification of Fluorocarbon Polymer by Ultraviolet Irradiation in Oxygen Atmosphere. *ECS J. Solid State Sci. Technol.* **2013**, *2* (5), N93–N98.

Lee, K. R., et al. Dehydration of Ethanol/water Mixtures by Pervaporation with Composite Membranes of Polyacrylic Acid and Plasma-treated Polycarbonate. *J. Membr. Sci.* **2000**, *164* (1–2), 13–23.

Leeuwenburgh, S. C. G., et al. Trends in Biomaterials Research: An Analysis of the Scientific Programme of the World Biomaterials Congress. *Biomaterials* **2008**, *29* (21), 3047–3052.

Li, D. J.; Zhao J. Surface Modification of Medical Polyurethane by Silicon Ion Bombardment. *Nucl. Instrum. Meth. Phys. Res. Sect. B Beam Interact. Mater. Atoms.* **1993**, *82* (1), 57–62.

Li, D. J.; Zhao, J. The Structure and Biomedical Behavior of Ion Bombarded and Plasma Polymerized Segmented Polyurethane. *Appl. Surf. Sci.* **1994**, *78* (2), 195–201.

Li, R. Z.; Ye, L.; Mai, Y. W. Application of Plasma Technologies in Fibre-reinforced Polymer Composites: A Review of Recent Developments. *Compos. Part A Appl. Sci.* **1997**, *28* (1), 73–86.

Liang, C., et al. In *Synthesis and Characterization of Shape-memory Polyurethane Films with Blood Compatibility*, Proceedings of the Second International Conference on Smart Materials and Nanotechnology in Engineering, Weihai, China, 2009.

Licari, J. J.; Hughes, A. L. *Handbook of Polymer Coatings for Electronics: Chemistry, Technology and Applications*, 2nd ed.; William Andrew: Norwich, NY, 1990.

Luscher, T. F., et al. Drug-eluting Stent and Coronary Thrombosis: Biological Mechanisms and Clinical Implications. *Circulation.* **2007**, *115* (8), 1051–1058.

Ma, Z.; Mao, Z.; Gao, C. Surface Modification and Property Analysis of Biomedical Polymers Used for Tissue Engineering. *Colloid Surf. B* **2007**, *60* (2), 137–157.

Mauzac, M.; Aubert, N.; Jozefonvicz, J. Antithrombic Activity of Some Polysaccharide Resins. *Biomaterials* **1982**, *3* (4), 221–224.

Melnig, V., et al. Improvement of Polyurethane Surface Biocompatibility by Plasma and Ion Beam Techniques. *J. Optoelectron. Adv. Mater.* **2005**, *7* (5), 2521–2528.

Nahar, P.; Wali, N. M.; Gandhi, R. P. Light-induced Activation of an Inert Surface for Covalent Immobilization of a Protein Ligand. *Anal. Biochem.* **2001**, *294* (2), 148–153.

Narayanan, R., et al. Calcium Phosphate-based Coatings on Titanium and Its Alloys. *J. Biomed. Mater. Res. Part. B. Appl. Biomater.* **2008**, *85B* (1), 279–299.

O'Connell, C., et al. Investigation of the Hydrophobic Recovery of Various Polymeric Biomaterials after 172 nm UV Treatment Using Contact Angle, Surface Free Energy and XPS Measurements. *Appl. Surf. Sci.* **2009**, *255* (8), 4405–4413.

Oechsner, H. Ion and Plasma Beam Assisted Thin Film Deposition. *Thin Solid Films* **1989**, *175* (L153–L156), 119–127.

Olbricha, M., et al. UV Surface Modification of a New Nanocomposite Polymer to Improve Cytocompatibility. *J. Biomater. Sci.* **2007**, *18* (4), 453–468.

Ong, J. L., et al. Structure, Solubility and Bond Strength of Thin Calcium Phosphate Coatings Produced by Ion Beam Sputter Deposition. *Biomaterials* **1992**, *13* (4), 249–254.

Ormiston, J. A., et al. A Bioabsorbable Everolimus-eluting Coronary Stent System for Patients with Single De-novo Coronary Artery Lesions (ABSORB): A Prospective Open-label Trial. *Lancet* **2008**, *371* (9616), 899–907.

Pavithra, D.; Doble, M. Biofilm Formation, Bacterial Adhesion and Host Response on Polymeric Implants-issues and Prevention. *Biomed. Mater.* **2008**, *3* (3), 1–11.

Penn, L. S.; Wang, H. Chemical Modification of Polymer Surfaces: A Review. *Polym. Adv. Technol.* **1994**, *5* (12), 809–817.

Pignatello, R. *Biomaterials Science and Engineering*; InTech: London, 2011.

Pruitt, L. A. The Effects of Radiation on the Structural and Mechanical Properties of Medical Polymers. In *Radiation Effects on Polymers for Biological Use*; Henning Kausch, N., Anjum, Y., Chevolot, B., Gupta, D., Léonard, H. J., Mathieu, L. A., Pruitt, L., Ruiz-Taylor, Schol, M., Eds.; Springer: Berlin, Germany, 2003; Vol. 162, pp 63–93.

Qiu, Z. Y., et al. Advances in the Surface Modification Techniques of Bone-related Implants for Last 10 Years. *Regen. Biomater.* **2014**, *1* (1), 67–79.

Rajajeyaganthan, R., et al. Surface Modification of Synthetic Polymers Using UV Photochemistry in the Presence of Reactive Vapours. *Macromol. Symp.* **2011**, *299/300*, 175–182.

Reyes-Labarta, J., et al. Wetchemical Surface Modification of Plasticized PVC. Characterization by FTIR-ATR and Raman Microscopy. *Polymer* **2003**, *44* (8), 2263–2269.

Reyntjens, S.; Puers, R. Focused Ion Beam Induced Deposition: Fabrication of Three-dimensional Microstructures and Young's Modulus of the Deposited Material. *J. Micromech. Microeng.* **2000**, *10* (2), 181–188.

Roth J. R. *Industrial Plasma Engineering: Applications to Nonthermal Plasma Processing*; CRC Press: Boca Raton, FL, 2001; Vol. 2.

Sioshansi, P.; Tobin, E. J. Surface Treatment of Biomaterials by Ion Beam Processes. *Surf. Coat. Technol.* **1996**, *83* (1–3), 175–182.

Sirolli, V., et al. Biocompatibility and Functional Performance of a Polyethylene Glycol Acid-grafted Cellulosic Membrane for Hemodialysis. *Int. J. Artif. Organs* **2000**, *23* (6), 356–364.

Slepicka, P., et al. Surface Characterization of Plasma Treated Polymers for Applications as Biocompatible Carriers. *eXPRESS Polym. Lett.* **2013**, *7* (6), 535–545.

Kasalkova, N. S., et al. Cell Adhesion and Proliferation on Polyethylene Grafted with Au Nanoparticles. *Nucl. Instrum. Methods Phys. Res. Sect. B* **2012**, *272*, 391–395.

Shadpoura, H.; Allbritton, N. L. In-Situ Roughening of Polymeric Microstructures. *ACS Appl. Mater. Interfaces* **2010**, *2* (4), 1086–1093.

Sharma, D. S.; Ali, R.; Mazumender, N. K. Enhancement and Blood Compatibility of Implants by Helium Plasma Treatment. *Ind. Appl. Conf.* **2004**, *4* (2), 932–936.

Shikinami, Y.; Okuno, M. Mechanical Evaluation of Novel Spinal Interbody Fusion Cages Made of Bioactive, Resorbable Composites. *Biomaterials* **2003**, *24* (18), 3161–3170.

Sioshansi, P. Ion Beam Modification of Materials for Industry. *Thin Solid Films* **1984**, *118* (1, 3), 61–72.

Siow, K. S., et al. Plasma Methods for the Generation of Chemically Reactive Surfaces for Biomolecule Immobilization and Cell Colonization—A Review. *Plasma Process. Polym.* **2006**, *3* (6–7), 392–418.

Situma, C., et al. Immobilized Molecular Beacons: A New Strategy Using UV-activated PMMA Surfaces to Provide Large Fluorescence Sensitivities for Reporting on Molecular Association Events. *Anal. Biochem.* **2007,** *363* (1), 35–45.

Song, J. S., et al. Surface Modification of Silicone Rubber by Ion Beam Assisted Deposition (IBAD) for Improved Biocompatibility. *J. Appl. Polym. Sci.* **2005,** *96* (4), 1095–1101.

Subedi, D. P.; Tyata, R. B.; Rimal, D. Effect of UV-treatment on the Wettability of Polycarbonate. *Kathmandu Univ. J. Sci.* **2009,** *5* (2), 37–41.

Szleifer, I. Protein Adsorption on Surfaces with Grafted Polymers: A Theoretical Approach. *Biophys. J.* **1997,** *72* (2 Pt 1), 595–612.

Tidwell, C. D.; Ertel, S. I.; Ratner, B. D. Endothelial Cell Growth and Protein Adsorption on Terminally Functionalized, Self-assembled Monolayers of Alkanethiolates on Gold. *Langmuir* **1997,** *13* (13), 3404–3413.

Ulbricht, M.; Belfort G. Surface Modification of Ultrafiltration Membranes by Low Temperature Plasma I. Treatment of Polyacrylonitrile. *J. Appl. Polym. Sci.* **1995,** *56* (3), 325–343.

Ulbricht, M.; Belfort, G. Surface Modification of Ultrafiltration Membranes by Low Temperature Plasma II. Graft Polymerization onto Polyacrylonitrile and Polysulfone. *J. Membr. Sci.* **1996,** *111* (2) 193–215.

Vesel, A.; Elersic, K.; Mozetic M. Immobilization of Protein Streptavidin to the Surface of PMMA Polymer. *Vacuum* **2012,** *86* (6), 773–775.

Wang, Z. G.; Wan, L. S.; Xu, Z. K. Review: Surface Engineerings of Polyacrylonitrile-based Asymmetric Membranes towards Biomedical Applications: An Overview. *J. Membr. Sci.* **2007,** *304* (1–2), 8–23.

Weibel, D. E., et al. Ultraviolet-induced Surface Modification of Polyurethane Films in the Presence of Oxygen or Acrylic Acid Vapours. *Thin Solid Films* **2009,** *517* (18), 5489–5495.

Wells, R. K., et al. A Comparison of Plasma-oxidized and Photo-oxidized Polystyrene Surface. *Polymer* **1993,** *34* (17), 3611–3613.

Wenyue, S., et al. Plasma Pre-treatment and TiO2 Coating of PMMA for the Improvement of Antibacterial Properties. *Surf. Coat. Technol.* **2010,** *205* (2), 465–469.

Williams, D. F. *Biocompatibility in Clinical Practice*; CRC Press: Boca Raton, FL, 1982.

Wilkes, C.; Summers, J.; Daniels, C. *PVC Handbook*; Hanser: Munich, Germany, 2005.

Won, J., et al. Surface Modification of Polyimide and Polysulfone Membranes by Ion Beam for Gas Separation. *J. Appl. Polym. Sci.* **2000,** *75* (12) 1554–1560.

Zhihong, Z. Surface Modification by Plasma Polymerization and Application of Plasma Polymers as Biomaterials. Doctoral Dissertation. Johannes Gutenberg-Universitate Mainz: Mainz, Germany, 2003.

CHAPTER 11

DESIGN OF BIOLOGICALLY ACTIVE POLYMER SURFACES: CATIONIC POLYELECTROLYTES AS MULTIFUNCTIONAL PLATFORM TO PREVENT BACTERIAL ATTACHMENT

SIMONA DUNCA[1]* and ANCA FILIMON[2]

[1]*Microbiology Department, Faculty of Biology, "Alexandru Ioan Cuza" University of Iasi, 20A Carol I Bvd., 700505 Iasi, Romania*

[2]*Physical Chemistry of Polymers Department, "Petru Poni" Institute of Macromolecular Chemistry, 41A Gr. Ghica Voda Alley, 700487 Iasi, Romania*

Corresponding author. E-mail: sdunca@uaic.ro

CONTENTS

ABSTRACT

Modern medicine has been challenged with complex problems, which have led to technological advances, so that the application of antimicrobial polymer-based surfaces on medical devices and implants seems to be a necessary tool for fighting microbial infections, particularly in hospitals, by preventing the spread of microbial cells. Therefore, the design and synthesis of antimicrobial polymers have gained increasing attention by the scientific community as a safe and effective strategy to combat multidrug-resistant.

The development of the antimicrobial materials, as multifunctional platform to prevent bacterial attachment, has been the focus of present research, with numerous methods being applied to prevent and control the occurrence of these complications. As shown in this chapter, the performance of these biologically active polymer surfaces demonstrates their great potential and the significance of bolstering the research regarding the improvement of their design, activity, and mechanisms of action, as well as on their potential applications.

The use of these antimicrobial materials is a promising approach to lower the propensity of pathogen-resistant development. Therefore, the study of antimicrobial polymers is necessary to gain insight into many biomedical and industrial processes such as the biomedical field, water treatment, food packaging and storage, and textile product.

11.1 INTRODUCTION

It is generally accepted that the human society is dominated by microorganisms, such as bacteria, yeast, fungi, and algae, and the microbial infections treatment became increasingly difficult. For this reason, modern medicine has been challenged with complex problems, which have led to technological advancements, so that the application of antimicrobial surface on medical devices and implants seems to be a necessary tool for fighting microbial infections, particularly in hospitals, by preventing the spread of microbial cells (Zilberman and Elsner, 2008). Another less life-threatening but economically important, area that requires the control of microbial populations is the formation of biofilms on manmade materials, which corrode and deteriorate, and are rendered dysfunctional by these biofilms (Meyer, 2003).

Polymers are part of everyday life, being used in various medical devices, including implants, drug carriers, protective packaging materials, and healthcare items, textile fibers, and construction materials, as well as in electronics field. Antimicrobial agents are natural or synthetic compounds that inhibit microbial growth. Different classes of antimicrobial agents, most of which biocides, can be used in various industries. From the inorganic materials, there has been an increasing interest in the use of a number of metals and metal oxides as antimicrobial agents due to their durability, heat resistance, resistance to intensive processing conditions, selective toxicity to bacteria, and minimal effects on human cells (Turos et al., 2007). Zinc oxide (ZnO), copper oxide (CuO), magnesium oxide (MgO), titan dioxide (TiO_2), and silver (Ag) are some of the inorganic compounds most frequently used in antimicrobial coatings. Such compounds act against bacteria by binding intracellular proteins and inactivating them, generating oxygen-reacting species, and destroying cell walls (Gao and Cranston, 2008).

During the last years, researchers have focused on creating microbiocidal polymeric materials containing metal ions to be used in different medical fields such as implantation of medical devices and artificial organs, prosthetics, bone regeneration, dentistry, ophthalmology, as well as controlled release drug delivery systems. Antimicrobial polymers, the polymeric materials with resistance to microbial colonization and pathogenic microorganism spreading, represent one of the examples of the active material functionality. These polymers protect against negative impact of the pathogenic microorganisms which can seriously affect the society from the viewpoint of both health damages and unwanted economical loads connected with that.

Developing new biomedical polymers involves thorough knowledge of the interaction between the polymer structure/surface and blood components, of how such components are absorbed onto the polymer surface, the influence of the polymer on biological responses such as homeostasis, platelet adhesion, or complement activation (Siedenbiedel and Tiller, 2012). One of the main challenges in developing materials that come into contact with blood and tissues is the hemocompatibility of the biomaterial, considered only when the interaction with blood does not result in the modification of the structure of plasma proteins or deterioration of the blood cells (Lufrano et al., 2000).

Polymer-based surfaces are often vulnerable to bacterial attachment, which can give rise to serious complications in the form of so-called infections. Therefore, development of the biologically active polymer surfaces has been the focus of extensive researches, with numerous methods being applied to prevent and control the occurrence of these complications.

11.2   BIOLOGICALLY ACTIVE POLYMER SURFACES

11.2.1   ANTIMICROBIAL SOURCE AND MODIFICATION STRATEGIES OF POLYMER SURFACES

The design and synthesis of antimicrobial polymers have gained increasing attention by the scientific community as a safe and effective strategy to combat multidrug-resistant. Due to their properties, such as chemically stable, limited residual toxicity, long-term activity, and do not permeate through the skin, the use of polymers as antimicrobial agents presents several advantages. Depending on their action, the antimicrobial polymers fall into three categories, namely, *polymeric biocides*, *biocidal polymers*, and *biocide-releasing polymers* (Fig. 11.1).

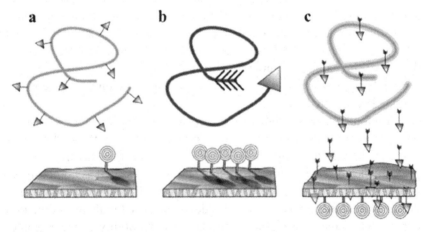

FIGURE 11.1   Types of antimicrobial polymers: (a) polymeric biocides; (b) biocidal polymers; and (c) biocide-releasing polymers. Reprinted from Siedenbiedel, F.; Tiller, J. C. Antimicrobial Polymers in Solution and on Surfaces: Overview and Functional Principles. *Polymers* **2012,** *4* (1), 46–71. Open access: http://www.mdpi.com/2073-4360/4/1/46. https://creativecommons.org/license

Polymeric biocides can be designed by synthetic routes, generating the macromolecules with antimicrobial activity. They are characterized by the fact that the biocidal groups attached to the polymer act similarly to the analogs with low molecular weight, which means that the repetitive unit is a biocide. The active principle of biocidal polymers (polymers with intrinsic antimicrobial activity) is represented by the entire macromolecule, which does not necessarily require antimicrobial repeating units. Polymers with intrinsic antimicrobial activity are based on polycations. Usually, the cationic groups found in these polymers are quaternary ammonium, quaternary phosphonium, guanidinium, or tertiary (Filimon and Ioan, 2015). The biocide-releasing polymers do not act through the actual polymeric part. Instead, they function as carriers for biocidal products which are transferred somehow to the attacked microbial cells (Siedenbiedel and Tiller, 2012).

As abovementioned, considering the working principles of antimicrobial systems, the design of polymer systems with antibacterial action involves special polymerization reactions or chemical functionalization. Thus, an antimicrobial polymer system consisting of a polymer matrix and an antimicrobial agent is considered a material chemically or physically modified in order to prevent bacterial attachment and/or colonization. In this context, the optimization of material properties can be achieved through different antimicrobial modification techniques, taking into account the polymer properties (chemical and physical) and characteristics of the antimicrobial agent (toxicity, thermal stability, and affinity with a certain component).

(1) *Chemical deposition of an antimicrobial agent on a polymer surface* involves first synthesizing the polymer, followed by modification with an active species, either directly attached after prior activation, or through of a "mediator" (e.g., polyacrylic acid) which is grafted on the polymer surface and determines the immobilization of antimicrobial component. The backbone of the homopolymers/copolymers (such as methyl methacrylate, vinyl alcohol, vinyl benzyl chloride, 2-chloroethyl vinyl ether and maleic anhydride) is formed by used monomers and subsequently, the polymers are activated by anchoring antimicrobial species (e.g., phenol groups, phosphonium, and ammonium salts). In this stage, the antimicrobial agent is released from the surface and deactivates any present bacteria.

Conventional antimicrobial agents prepared based on natural or low-molecular-weight compounds, are easily susceptible to resistance and can result in environmental contamination and toxicity to the human body due to biocidal diffusion (Fuchs and Tiller, 2006; Thomassin et al., 2007). In contrast to the antimicrobial polymeric materials which are achieved by physically entrapping or coating with organic and/or inorganic active agents (during or after processing), polymers containing covalently linked antimicrobial fragments avoid the problem of the permeation of low-molecular-weight biocides from the polymer matrices. Among them, the antimicrobial polymeric materials containing quaternary ammonium and/ or phosphonium salts are probably most widely used and studied antimicrobial polymers (Filimon and Ioan, 2015).

For determining the antimicrobial efficiency of water-soluble quaternary pyridinium polymers Eren et al. have synthesized a series of amphiphilic polymers with various quaternary alkyl pyridinium side chains, using two different methods, namely, direct-polymerization and post quaternization (Eren et al., 2008). Following the evaluation of their antibacterial activity in terms of minimum inhibitory concentration against *Escherichia coli* and *Bacillus subtilis* and hemolytic activity against fresh human red blood cells, it can be concluded that, the synthetic route of polymeric biocides with pendant quaternary ammonium salt has impact on their biological activity due to the quaternization degree and consequently, on the hydrophobic/hydrophilic balance of the polymer surfaces.

Literature (Abel et al., 2002) indicates that antimicrobial activity of surface is strongly correlated with the length of the adjacent alkyl chain and the size and number of cationic ammonium groups in the molecule. For example, a long alkyl chain substituent, that is, at least eight carbons renders polymers with pendant quaternary ammonium salts very antimicrobial. Considering this, Dizman et al. (2004) have synthesized a methacrylate monomer containing pendant quaternary ammonium salts (either a butyl or a hexyl group) based on 1,4-di-azabicyclo-[2.2.2]-octane which did not show any antimicrobial properties. Nevertheless, the corresponding homopolymers exhibit biocidal activity against *Staphylococcus aureus* and *E. coli*, and their activity was found to be dependent on the length of the hydrophobic segment, that is, the polymer with hexyl groups was more effective than the one with butyl groups. Therefore, it has been shown that an increase in the alkyl chain length of compound is followed by an increase in the hydrophobic interaction with the lipid bilayer of the cell wall, which in turn increases the antimicrobial activity of the compound. By copolymerization of N-isopropylacrylamide with methacryloyloxyethyl trialkyl phosphonium chloride, Nonaka et al. (2002) were achieved antimicrobial polymers containing quaternary phosphonium salts with thermosensitivity properties. The copolymers with octyl groups showed a lower critical solution temperature (LCST) and higher antimicrobial activity compared to those with either ethyl or butyl groups.

Phosphorous derivatives with antimicrobial activity have been investigated as medicals and antiseptics (Basri et al., 2012; Gorbunova, 2013). Moreover, phosphorous compounds have been designed to prevent virus-cell fusion/attachment, principally in sexually transmitted diseases (Muñoz-Bonilla and Fernández-García, 2012).

(2) *Polymer surface modification technique without an antimicrobial compound* is based on the assumption that, the bacterial adhesion during the initial stage of the biofilm formation process can be diminished by changing the surface properties of material, namely, the topography, polarity, surface free energy, etc. By this method, antimicrobial modification can be obtained in various ways, including the chemical reaction with different monomers, by cold plasma technique (polymer surface modification by ionized gas, usually N_2, O_2, or Ar, and CF_4) or by applying of a high energy electromagnetic radiation (laser, ultraviolet radiation, and gamma rays).

The synthetic method involves covalently linking of antimicrobial agents that contain functional groups with high antimicrobial activity, such as amino, hydroxyl, or carboxyl groups to variety of polymerizable derivatives or monomers before polymerization. It is known that, by polymerization, the antimicrobial activity of the active agent can be enhanced or reduced, this depends both on the type of monomer used and also how the agent kills bacteria. In this context, Ren et al. were synthesized the monomer 3-(4'-vinyl benzyl)-5,5-dimethylhydantoin and subsequently, by polymerization with a cationic surfactant, using it to coat cotton fibers. After chlorination with dilute sodium hypochlorite, the polymeric-coated cotton fabrics inhibit the growth both *S. aureus* and *E. coli* microorganisms in relatively brief contact times (Ren et al., 2008).

The polymerization of the synthesized vinyl monomers with phenol and benzoic acid as pendent groups was carried out by Park et al. (2001). The antimicrobial activity of the polymers was evaluated in terms of the diameter of the inhibition zone after a contact time of 72 h at 28°C. It was found out that the polymerization of the monomers decreased their antimicrobial activities. However, they could be coated on glassy polymers, even though the antimicrobial activity of the polymers is much lower than that of the corresponding monomers.

It is important to note that antibacterial surfaces created by plasma are characterized by weak stability over time and tend to return to their original chemical state (Jampala et al., 2008; Asadinezhad et al., 2010; Asanovic et al., 2010; North et al., 2010; Jacobs et al., 2012). Additionally, the surface activation induced by the interactions between polymer surface and electromagnetic radiation generates the breakage of accessible polymer bonds, permitting subsequent chemical modification (Piccririllo et al., 2009; Alvarez-Lorenzo et al., 2010; Gozzelino et al., 2010).

Han et al. have reported data on antimicrobial activity and surface properties of low-density polyethylene (LDPE)/polyamide obtained by the electron beam irradiation (Han et al., 2007). The antimicrobial polymer effectiveness of several coating of FDA-approved antimicrobial compounds including sorbic acid, carvacrol, trans-cinnamaldehyde, thymol, and rosemary oleoresin using selected food pathogen surrogates have been established. Films were irradiated using a 10 MeV linear electron beam accelerator at room temperature. The analysis of results shows that all films possess inhibition zones against *Listeria innocua* ATCC 33090 and *E. coli* ATCC 884; the antimicrobial agent reduced the specific growth rate of *L. innocua* by 3.8–8.5% and decreased final cell concentration of both strains by 5.7–14.6 and 7.2–16.8%, respectively. All tested active compounds retained the antimicrobial activity when exposed to 1–3 kGy. Moreover, the tensile strength and toughness of the films were influenced neither by the presence of active compound nor by the dose used, so it become more ductile and has an improved functionality to moisture barrier. These results are a first step toward the development of self-sterile active packaging materials for use in combination with irradiation treatment of food.

(3) *Direct deposition of an antimicrobial agent on a polymer surface* represents the simplest technique. This is widely used for practical applications in medicine and in this case, the antimicrobial agent (in the form of a solution or ointments) is applied to the surface of polymer-based medical device, just before use. Nevertheless, method presents a low efficiency due to rapid resorption of the active component (Laporte, 1997).

(4) To constitute the active systems with biocide characteristics, another method is used and is based on *direct incorporation of the antimicrobial agent in a polymer matrix*. Polymer solution is suitable for preparing

special coatings and films by cast subsequently. The model was based on the idea the development of a "polymeric spacer" capable of penetrating the bacterial cell wall, so reaching its cell membrane, and finally, at killing the microorganism. The proposed mechanism for developments of the concept of "polymeric spacers" is yet discussed and difficult to imagine, due to the active lengths of most polymer chains being required high stretching of the macromolecules to reach the inner cell membrane of attached microbes. Indeed, their cationic/hydrophobic balance, that is, a sufficient charge density, is not all that is required.

Various antimicrobial surfaces have been developed by combining polymeric materials with releasable nanoparticles (Shi et al., 2004), metal ions and/or clays (Podsiadlo et al., 2005), which results in dual-functionalized antimicrobial properties. In this context, by employing the layer-by-layer deposition method, Grunlan et al. (2005) and Li et al. (2006) developed antimicrobial multilayer films containing quaternary ammonium salts and silver ions. Grunlan et al. have reported that the polyelectrolyte multilayer films, prepared by alternately dipping a poly(ethylene terephthalate) substrate into solutions of biocidal agents (i.e., cetyltrimethylammonium bromide (CTAB)), possess inhibition zones against Gram-negative bacteria, *E. coli* and Gram-positive bacteria, *S. aureus* higher for films made with CTAB comparatively with the films containing either silver alone or both CTAB and silver. Instead, Li et al. (2006) have designed the antimicrobial thin film coatings using two distinct functional layers, namely, a reservoir for loading and releasing of silver ions and a nanoparticle surface cap immobilized with [3-(trimethoxysilyl)propyl]octadecyldimethylammonium chloride. In this case, the dual-functional coatings, which possess both biocidal-releasing and contact bacterial killing properties, exhibited high bactericidal efficiency, maintaining at the same time the antimicrobial activity, even after the silver depletion.

Literature presents researches on the analysis of antibacterial activity of metal-doped sol–gel surfaces. Jaiswal et al. have evaluated antibacterial activity of metal nitrate (silver, zinc, and copper) doped methyltriethoxysilane coating, using polypropylene microtiter plates coated with different volumes of liquid sol–gel and cured under various conditions (Jaiswal et al., 2012). It was found out that, the bioactivity of metal-doped sol–gel at equivalent concentrations against *E. coli* and *S. aureus* microorganisms increase as follows: copper>zinc>silver. Analysis of antimicrobial activity against Gram-positive and Gram-negative bacteria indicates the influence of several factors, including the increased presence of silver nanoparticles at the sol–gel coating surface which leads to higher elution rates (evaluated by inductively coupled plasma atomic emission spectroscopy). Therefore, it is demonstrated that the antibacterial activity of the silver-doped sol–gel offer potential applicability as broad-spectrum biocidal coatings toward smart biomaterials. This fact is especially true in biomedicine, where many applications benefit of the inorganic materials presence, for example, for control over cell proliferation and differentiation.

Today, medicinal inorganic/organic chemistry remains a field of great promise with many challenges. Despite the existence of several research already conducted on polymeric materials with biocidal activity, the potential for a major expansion of chemical diversity into new structural and reactivity materials with high therapeutic impact is unquestionable. Therefore, polymeric materials may be introduced into a biological system for therapeutic or diagnostic purposes.

11.2.2 ACTION MECHANISM OF ANTIMICROBIAL POLYMERS: ELECTROSTATIC INTERACTIONS BETWEEN THE CHARGED SUBSTRATE AND BACTERIA

Antimicrobial polymers are a group of biocidal substances that has become more and more important as an alternative to the existing biocides and, in some cases, even to antibiotics. The mechanism of action of a large number of polymers with different structure has not been fully elucidated, some of which showed limited action even against resistant bacterial strains (Milovic et al., 2005).

Designing synthetic antimicrobial polymers takes into account the structural particularities of the outside layer of different bacterial cells. The antibacterial activity of some polymers is correlated to a great extent to the surface of the bacterial cell, which is very complex in terms of structure and chemically heterogeneous. Thus, in Gram-positive bacteria, peptidoglycan accounts for 80–90% of the dry weight of the cell wall. Their cell wall also contains teichoic acids and teichuronic acids which are covalently linked to peptidoglycan. It can be digested by lysozyme which hydrolyzes the bonds between the N-acetylmuramic acid and N-acetyl-glucosamine or by endopeptidases which cleave the bonds among the tetrapeptides in peptidoglycan. In Gram-negative bacteria, the cell wall does not contain teichoic acids and teichuronic acids. In turn, they have an external membrane consisting of phospholipids, globular proteins, lipoproteins, membrane transport protein (porins), and a periplasmic space delineated on the outside by the external membrane and on the inside by the cytoplasmic membrane, where enzymes such as alkaline phosphatase, cyclic phosphodiesterase, DNase, RNase, or other nonenzymatic proteins binding or transporting different substances (i.e., carbohydrates, amino acids, inorganic ions, etc.) are eliminated. The peptidoglycan layer of the Gram-negative bacteria is much thinner than that of the Gram-positive ones (Brown et al., 2015) (Fig. 11.2).

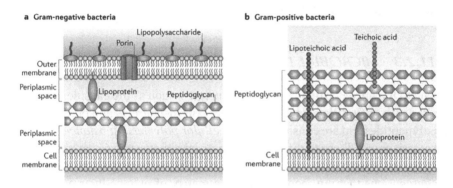

FIGURE 11.2 **(See color insert.)** Cell wall structures of Gram-negative bacteria (a) and Gram-positive bacteria (b). Reprinted with permission from Brown, L., et al. Through the Wall: Extracellular Vesicles in Gram-positive Bacteria, *Mycobacteria* and Fungi. *Nat. Rev. Microbiol.* **2015,** *13* (10), 620–630. © 2015 Nature Publishing Group.

Another important structural element of the bacterial cells is the cytoplasmic membrane; its phospholipid bilayer has a hydrophobic end oriented toward the inside of the membrane and a hydrophobic end oriented toward the outside of the membrane. Thus, the microbial cells have a positive charge on the surface due to the proteins inside their membrane and the teichoic acids (in Gram-positive bacteria) or the phospholipids present in the external membrane (in Gram-negative bacteria). Therefore, polycations are attracted and if they have commensurate amphiphile properties, they are capable of destroying the cell wall and the cytoplasmic membrane, which ultimately results in the cell lysis (Timofeeva and Kleshcheva, 2011).

It is generally accepted the fact that the biocidal action of the cationic polymeric biocides triggers destructive interactions with the cell wall and cytoplasmic membranes. The Gram-negative bacteria cell has an additional bilayer, a phospholipid membrane, which provides higher protection to the cytoplasmic membrane on the inside against the deleterious action of the polymeric biocide. For example, the antimicrobial function of polysulfones (PSFs) can be seen in the attractive interactions between the ammonium cation group in the quaternary amines (positive charge) and the negatively charged bacterial cell membranes. Consequently, such interactions result in the formation of a surface complex which determines a discontinuation of all the essential functions of the cell membrane and thus, the disruption of the protein activity.

11.2.3 MICROBIAL LIFE ON SURFACES: DEVELOP BIOFILMS

Due to the presence of a number of structures, such as the capsule, glycocalyx, pili, or fimbriae, which are usually polysaccharides and occasionally polypeptides and are considered *extracellular polymeric substances* (EPS), bacteria may adhere to different living or inanimate surfaces or substrates in the environment and develop biofilms. These are defined as microbial aggregations that are incorporated in a matrix of extracellular polymeric substances produced by the cells. The bacterial adherence to different substrates, either natural (e.g., teguments and mucous membranes) or artificial (e.g., catheters) are very important in the pathogenesis of infectious diseases, protein-containing materials used in medicine being seen as risk factors for infections (Novel, 1999).

The substrate has a series of physical and chemical properties that influence the initial absorption of bacteria and successive accumulation of biofilms, such as hydrophobicity, surface available, surface energy, etc. Amorphous materials, like most of the biopolymers, have a very rugged surface, the irregularities of which foster bacterial adhesion. These materials are hydrophobic, which positively influence bacterial adhesion since the cell surface of bacteria has a hydrophobic nature. Thus, cellular hydrophobicity promotes the adherence to an inert substrate due to the hydrophobic forces which draw bacteria very close to the substrate and thus allowing the creation of hydrophobic interactions; as a result, bondings are created between both the surface of the bacterial cells and the substrate among them, contributing to the formation of bacterial biofilms (Lazar, 2003). Studies have shown that the quaternary ammonium groups can reduce the multiplication rate of microorganisms. When in the quaternary ammonium salts the long hydrocarbon chain is linked by ammonium cations, two types of interaction can take place between such compounds and the microbial agent, namely, polar interactions, with the ammonium nitrogen cation, and nonpolar ones, with the hydrophobic chain (Siedenbiedel and Tiller, 2012).

The formation of a fully developed biofilm has two stages (Lichter et al., 2009). The first stage, which is rapid and reversible, consists in the interaction between the bacterial cell wall and the surface of the material, while the second, which is considered irreversible, includes specific and nonspecific interactions between the proteins of surface structures, such as fimbriae or pili, and the binding molecules on the surface of the material (Fig. 11.3).

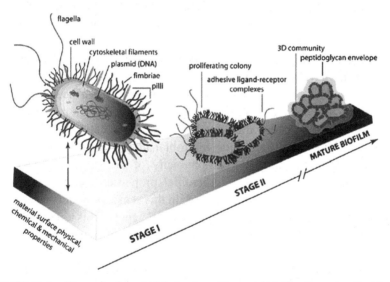

FIGURE 11.3 (See color insert.) Mechanism of bacterial biofilm formation. Reprinted with permission from Lichter, A. J., et al. Design of Antibacterial Surfaces and Interfaces: Polyelectrolyte Multilayers as a Multifunctional Platform. *Macromolecules* **2009,** *42* (22), 8573–8586. © 2009 American Chemical Society.

Biofilm formation gives numerous advantages to bacteria, among which protection against antibiotics (Goldberg, 2002) or disinfectants (Peng et al., 2002), etc. The use of implantable medical devices (IMD)

has become essential in all medical fields, be it for diagnostic or therapeutic purposes (von Eiff et al., 2005). The medical devices, depending
on their purpose, are made of polymeric, metal, or ceramic biomaterials.
According to certain authors (Darouiche, 2001), polymeric biomaterials
are prone to microbial contamination, representing a risk factor for IMD-
associated infections. Biofilm-associated infections require a different
approach, both clinically and paraclinically (Lazar and Chifiriuc, 2010).
The IMD most commonly compromised by infections due to microbial
biofilms on their surfaces are catheters, joint prosthesis, devices for joint
fixation, reconstructive orthopedic implants, etc. (von Eiff et al., 2005).

The adherence of microorganisms to the surface of medical devices is
considered a crucial stage in the pathogenesis of the infection. Adherence
is influenced by the characteristics of the material surface; rugged and
hydrophobic materials promote the microbial adherence capacity (Lazar,
2003), and the hydrophobic nature of polymeric biomaterials encourage
protein adsorption from the body fluids thus creating the so-called condi-
tioning film, the components of which serve as receptors for the microbial
adhesions involved in adherence (Arciola et al., 2005).

The biofilm generated by the microorganisms are extremely hard to
remove and have an increased resistance to all types of biocides. Therefore, preventing biofilm formation on antimicrobial surfaces is the best
way to prevent disease spreading and material deterioration. For this, the
material should not allow the adhesion of microbial cells.

Taking into consideration that microbial adherence and biofilm formation are facilitated by the hydrophobicity of the materials, designing
hydrophilic materials with antimicrobial properties could represent an
alternative in fighting IMD-associated. In this sense, polymers can act as a
matrix for materials containing antimicrobial agents. This alternative does
not have the drawback of the diffusion of low-molecular-weight biocides
through the polymeric matrix, which can often cause toxicity in humans.
Moreover, antimicrobial polymers have usually a long-term action.

11.2.4 CONTACT-ACTIVE SURFACES VERSUS RELEASING SYSTEMS

There is a high need for an effective antimicrobial technology on designing
polymer surfaces in a way, which largely prevents microorganisms from

growing upon them. The focus is laid on coating different surfaces, films, fibers, or three-dimensional substrates with materials exhibiting antimicrobial properties and which function via release mechanism or contact mechanism. In this context, it is supported idea to develop contact-active surfaces for killing microorganisms or delaying their growth. A surface that kills by contact it works by a different mechanism compared to one that releases a biocide and consequently, understanding and knowledge of the action mode is important for the differentiation those two processes.

Microbial infections and toxins are spread by biofilm, which generates further infections and inflammations. Adhesion of planktonic cells or small-dispersed biofilm fragments generates biofilm formation virtually on every surface (Landini et al., 2010). Once formed biofilm, it is difficult to eliminate this contamination and consequently, the antimicrobial surfaces should prevent the primary attack. It is known that there is a category of antimicrobial compounds which prevent the primary attack by creating surfaces do not allow adhesion of microbial cells and also another major category capable of the killing of approaching microbes, as illustrated in Figure 11.4. Both approaches can be carried out either by surface modifications or by releasing bioactive compounds.

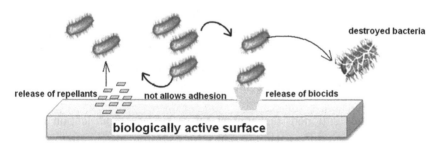

FIGURE 11.4   Schematic representation of the action mode of biologically active polymer surfaces.

When antimicrobial agents are fixed to the polymeric network by covalent bonding, microbes which have direct contact with the surface cannot grow and are subsequently effectively inhibited. The release mechanism works by incorporating active agents in such a way as to allow its diffusion into the surroundings and thus retards microbial attack. A large number of combinations and variations of inorganic and organic components allow adaptation to various surfaces and active agent systems. The establishment and growth of microbes on material surfaces can be minimized or even completely prevented (Fig. 11.5).

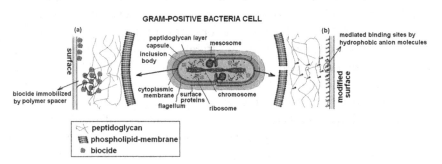

FIGURE 11.5   (See color insert.) Bactericidal action mode for (a) Active-surface with immobilized biocides via polymeric spacer and (b) active-modified surface inducing the removal of hydrophobic anion from microbial cell wall.

The low surface energy chemistry minimizes microbial attachment, while antimicrobial agents inhibit growth of microorganisms and finally kill them. Additionally, antimicrobial coating can be engineered in order to release the active agents over a prolonged time period. Furthermore, the action mechanism of antimicrobial system is based on either the slow release of toxic antibacterial agents or contact killing, no release of the active agent (Tiller et al., 2001; Klibanov, 2007). A modern approach toward permanently sterile materials can be represented by surface modi-fication that effectively kills microbes on contact without releasing a biocide.

As abovementioned, the contact-active surface modification can be achieved through different approaches, for example, the chemical grafting of N-alkylated poly(4-vinylpyridine) quaternized polyethyleneimine (Lin et al., 2002) and quaternary derivatives of acrylic acid (Lee et al., 2004) to numerous common materials such as glass, cellulose, and plastics (Tiller et al., 2002), that prevent the adhesion of bacteria onto the surface (Martin et al., 2007; Vasilev et al., 2009). First studies on the explain of contact activity mechanisms at polymer surfaces were described since 1972s by Isquith et al. using glass substrates were modified with silane 3-(trimethoxysilyl)-propyl dimethyloctadecyl ammonium chloride (Isquith et al., 1972) and subsequently, were expanded to several plastics (Tiller et al., 2001; Tiller et al., 2002).

It is generally accepted that the surface modification with a hydrophobic and/or a hydrophilic antimicrobial compound is often essential. In this sense, the polymers are positively charged, for example, polymers with quaternary ammonium groups, are the most explored kind of polymeric biocides (Filimon, et al., 2009). When the bacterial cells are in a contact with surfaces which present sufficiently long hydrophobic tails and long hydrophobic chains penetrate and disrupt the cell membranes. Whereas the antimicrobial polymers which contain the long chains seem to present the polymeric spacer effect, the attached molecules directly onto surface, act as a good surfactant (negatively charged phospholipids) seem to attract hydrophobic anionic molecules from the attached microbial cells, having the high binding capability for bacterial cells (Denyer and Stewart, 1998; Munch and Sahl, 2015; Shaikh et al., 2015).

Antimicrobial efficacy is affected by some factors, namely, molecular weight of the polymer, spacer length between active site, and the backbone hydrophobic tail length attached to the active site, hydrophilic/hydrophobic balance of the material and nature of counterions (Tashiro, 2001; Dizman et al., 2004; Klibanov, 2007). Moreover, Dunne asserts that the adhesion process is dictated by a number of variables, including bacterial species, surface composition, and environmental factors (Dunne, 2002). The adhesion mechanism is reversible; bacteria can attach onto the surface due to the existence of adhesion and the absence of physical and chemical interventions and subsequently, this will lead to biofilm formation and will cause biofouling onto membrane surfaces. Polymeric chains may interact more effectively with the cells of Gram-positive bacteria, as their polyglycane outer layer is weak enough to facilitate the deep penetration of the polymer chain inside the cell, to interact with the cytoplasmic membrane. Instead, Gram-negative bacterial cell presents an additional membrane with a bilayer phospholipid structure, which protects the inner cytoplasmic membrane to a higher extent against the adverse action of the polymeric biocide.

Consequently, the ways for an antimicrobial surface to function are repelling microbes or killing them on contact. The oldest approach for keeping surfaces free of biofilms is the controlled release of biocides which form an outer inhibition zone and an inner kill zone that destroys microbes in the proximity of the contact surface. Design of antimicrobial surfaces by tailored variations of the components and synthetic processes, using modern biotechnology is an increasing research area for many fields

(Rajendra et al., 2010; Moustafa and Fouda, 2012; Rathnayake et al., 2014). So, modern coatings, contact-active surfaces, tend toward microbe-repelling solutions and other marine organisms by replacing the existing toxin-releasing products. Considering such a speculation and the urgent need in the control of infectious diseases, in recent decades, is an extensive search for natural or at least environmentally benign biocides that can development of new microbicidal polymers, which are both bactericidal and biocompatible, as well as nanotechnological materials.

11.3 POLYMER-BASED SURFACES AS SMART BIOMATERIALS FOR MEDICINE, HEALTHCARE PRODUCT, AND ENVIRONMENT

In recent decades, researches concerning antimicrobial polymers have gained a great interest, since microbial contamination represents a serious issue with impact on public health. Polymeric materials with intrinsic antimicrobial properties have become very important as coatings and adhesives, being used in numerous fields, such as the disinfection of healthcare facilities and surface of the equipment. The performance of these antimicrobial materials demonstrates their great potential and the significance of bolstering the research regarding the improvement of their design, activity, and mechanisms of action, as well as on their potential applications. It is crucial to achieve innocuous materials, noncytotoxic to the human organisms, with potent and a broad range of antimicrobial activity, long-lasting response, and even reusable to maintain the activity, while never neglecting the environmental and recycling aspects. Therefore, the study of antimicrobial polymers is necessary to gain insight into many biological and industrial processes such as the biomedical field, water treatment, food packaging and storage, and textile product.

Intensive research into antibacterial agents continues to be a necessity, due to microbial resistance and infectious diseases. In the domain of tissue engineering, a multidisciplinary field combining the principles of biology, medicine, and engineering, which aims at replacing damaged, injured, or missing organs and tissues with a functional artificial substitute, the spread of bacteria represents a common problem and can be transferred from patient-to-patient, or can contaminated the surfaces of medical devices and hospital equipment. Thus, was revealed that the antimicrobial polymers are powerful candidates for polymeric drugs with high activities, due to their characteristics of carrying the high local charge density of the active groups in the vicinity of the polymer chains. Additionally, the development of new polymers, which are both bactericidal and biocompatible, represents a substantial research interest for many applications in medicine.

In this context, the most diverse group of biomaterials with medical applications is that of the chemically modified PSFs (Filimon et al., 2015). Due to a number of distinctive qualities, such as high mechanical strength, high transparency, resistance to acids, chemical and thermal stability, and their film-forming property and biocompatibility, PSFs can resist to different sterilization techniques (Noshay and Robeson, 1997), maintaining their properties within a broad temperature range, that is, from $-100°C$ to $150°C$ (Siedenbiedel and Tiller, 2012). Their rigidity derives from the inflexibility and relative immobility of the phenyl and SO_2 groups, while their strength comes from the etheric oxygen bond (Filimon et al., 2015).

The relevant literature contains also information on the use of PSFs in producing films for hemodialysis, wastewater recovery, gas separation (Siedenbiedel and Tiller, 2012). Taking into account that the inherent hydrophobic nature of such materials can sometimes limit their use as membranes, which require hydrophilic properties, researchers need to modify the structure of the conventional materials, to molecularly design variants of PSF and improve their qualities for specific applications in order to obtain nontoxic compounds with stable mechanical and morphological properties and adherent while used. Considering the broad range of industrial and environmental applications of these compounds, the chemical modification of PSFs by chloromethylation and quaternization are two main objectives of the present-day research (Filimon and Ioan, 2015). In this sense, the quaternization with ammonium groups has proved effective in developing polymers to be used as biomaterials and semipermeable membranes. The polymers containing quaternary ammonium groups are the most widely used type of biocidal polymers (Filimon et al., 2009), as such groups can modify hydrophilicity and increase water permeability (which is of special interest for biomedical applications), have antimicrobial action as well as solubility characteristics. Their antimicrobial effectiveness is tightly correlated with the length of the adjacent alkyl chain, the size, and number of cationic ammonium groups in the molecule. However, irritating and cytotoxic effects of these compounds on the human cells/tissues have been reported; therefore, improving the quaternary ammonium compounds becomes necessary and should have in view both the antimicrobial action and safety of the human cells. To overcome this challenge, attaching the quaternary ammonium compounds to the polymer chain seems promising in the attempt of developing materials with antimicrobial properties (Siedenbiedel and Tiller, 2012). Thus, the quaternary ammonium salts are known as antimicrobial compounds which act effectively against Gram-negative and Gram-positive bacteria, fungi and some types of viruses, and therefore, they can be included in the composition of antimicrobial polymers (Ioan and Filimon, 2012).

The biological activity of the quaternary ammonium compounds depends on the nature of organic groups attached to nitrogen and the number of existing nitrogen atoms. They usually contain four nitrogen-bound organic groups with similar or different chemistry and structure. The organic substituents are alkyl, aryl, or heterocyclic compounds. Studies have shown that increasing the alkyl chain length in an amphiphile compound triggers the enhancement of the hydrophobic interaction with the lipid bilayer of the cell wall, which in its turn, determines the increase of the antimicrobial action of the compound. The quaternary ammonium compounds containing an alkyl substituent of at least eight carbon atoms in length proved to have high biocidal performance in water (Noshay and Robeson, 1997). Two main hypotheses on how the antimicrobial substances containing quaternary ammonium salts act can be found in the literature. It is generally known, in this sense that most bacterial cell walls are negatively charged as they contain phosphatidylethanolamine (70%) (Ioan and Filimon, 2012).

The polymers that contain phosphorous derivatives have been found to exhibit bactericidal action. They were initially used as antiseptics and later on, as disinfectants. Despite the numerous positive characteristics, the antimicrobial action of the polymers containing quaternary ammonium salts is not as efficient against a broad spectrum of bacteria as that of other polymeric compounds. Quaternary ammonium-functionalized polysulfones (PSFQs) have less powerful antimicrobial effects generated by the quaternary nitrogen, which is far from the backbone chain. Therefore, an improved biocidal effect requires the introduction of hydrophobic groups with greater charge density. To achieve this, another successful antimicrobial system was made of polymers containing insoluble quaternary phosphonium salts transplanted onto the main chain (Ioan and Filimon, 2012). This was supported by the hydrophobic nature of the polymer, expansion of the polymer coil and enhancement of the charge density, which is generated by the active cationic centers found on a single polyelectrolyte molecule. As a result, such polymers exhibit a relatively high biocidal activity as long as the polymeric chain retains the positive charge density. In addition to the electrostatic interactions, such interactions involve also hydrogen and hydrophobic bonds (Siedenbiedel and Tiller, 2012). The mechanism of action of the antimicrobial PSF compounds containing quaternary phosphonium against bacteria is similar to that of the quaternary ammonium-containing compounds.

In conclusion, the traditional antimicrobial drugs focus on just a few cellular processes and derive from distinctive chemical classes. A promising strategy—the development of new biomaterials as antibacterial agents (for instance, PSFs containing ammonium and phosphorous compounds)—provides the opportunity to consolidate the broad-spectrum therapy against infections with multiple drug-resistant pathogens, infections in which many successful compounds have failed. This advantage is unique as it allows the selective destruction of the target pathogens. Exploring the ways to adapt the antimicrobial spectrum to the use of conventional antibiotics by enhancing their action and reversed resistance is also a topic of particular interest (Ioan and Filimon, 2012).

In the last decades, the mixtures of PSF or modified PSFs with other synthetic polymers have been a subject of great interest for the researchers, considering the efficacy of the mixture of two different polymers to create new materials. For example, PSFs containing ammonium (PSFQ) was combined with polyvinyl alcohol (PVA) and/or cellulose acetate phthalate (CAP) into its matrix, yielding flexible, biocompatible, and biodegradable composites with improved bioactivity compared to the PSFQ (Dobos et al., 2016; Filimon et al., 2016). Researchers have shown that interactions between blood and polymer surface depend on blood composition, blood flow, and physicochemical properties of the polymer surface, for example, flexibility, hydrophobicity/hydrophilicity, and roughness, or on the toxicological and antimicrobial properties. These results seem to be applicable for evaluating bacterial cell adhesion on polymer surfaces and could be employed for studying possible induced infections or for obtaining biomembranes.

Hart et al. have evaluated the efficacy of polymeric surface coated with antimicrobial compound (polyethylene terephthalate coated with polymer/levofloxacin in different compositions) against of bacterial infection generated by foreign body implant in an animal (Hart et al., 2010). The analysis of results states, beyond any doubt, that use of a specific polymer coating incorporating levofloxacin act as means of reducing device-related infections.

Studies of the antimicrobial polymers for applications in biomedical field were expanded. In this context, solutions processable of cationic polyelectrolytes have received widespread attention for their promising roles as exchange membranes and antibacterial coatings. In particular, the solutions of the quaternized PSF were processed by electrospinning to create new fibrous materials that can modulate biomembrane properties (Filimon et al., 2015). Their biological activity (investigated against Gram-positive *S. aureus* and the Gram-negative *E. coli* bacteria) depends on their structure and physicochemical properties which affects the interaction with the cytoplasmic membrane of bacteria and influences their cell metabolism. Consequently, the electrospinning proved to extend the possible applications of quaternized PSFs by preparing continuous fine fibers with characteristics which recommend them for applications as biomaterials and membranes in various fields.

Cellulose and cellulose derivatives (CAP) are commonly used in cosmetics, as skin and hair conditioners and also their excellent biocompatibility determine extent of use as pharmaceutical compounded and industrialized products. Thus, cellulose compounds are polymers with good film-forming characteristics which find a variety of uses, as classical material-coatings and controlled-release systems, hydrophilic matrices, and semipermeable membranes for applications in pharmacy, hospital products, hygiene products, agriculture, and cosmetics. In this context, the CAP added in different amounts in PSFQ matrix represents an important tool in adjusting the properties of the membranes, for example, to control the porosity, hydrophilicity, and permeability, as well as to minimize the adhesion of organic foulants/bacteria and maximize the inactivation of bacterial cells. Thus, due to its features, CAP has contributed to the successful development of environmentally friendly blends and can be considered as possible candidates for industrial use (Onofrei and Filimon, 2016).

Currently, development of antimicrobial paper (i.e., medical chart paper, paper towel, toilet paper, wallpaper, and bank notes) presents an opportunity to reduce the bacteria proliferation from different environments (hospitals, office, vehicles, and many other places). Literature (Rajendra, 2010; Zhang, 2010; Rizzotto, 2012; Rathnayake, 2014) discusses different possible approaches that prevent the adhesion of bacteria (e.g., coating with ZnO nanoparticles onto paper), giving an antibacterial surface suitable for use as wallpaper in hospitals.

Environmental issues are known to play an important role in human progress, so the application of polymeric disinfectants, as the coated surface, membrane, hollow-fiber membranes, in the treatment, and disinfection of water, it is ideal. Li and Shen have evaluated the antibacterial activity of insoluble pyridinium-type polymers with different structures against *E. coli* bacteria suspended in sterilized and distilled water (Li and Shen, 2000). The results on antibacterial activity showed that the insoluble pyridinium-type polymers present ability to capture bacterial cells by adsorption/adhesion process, partially irreversible. Instead, the antibacterial activity of the corresponding soluble polymers is characterized by the ability to kill bacterial cells from water. Finally, the general analysis of the relationship between the chemical structure and the activity of insoluble/soluble pyridinium-type polymers against different strain types indicates that they possess broad prospects for development in new water treatment techniques by whole-cell immobilization.

Polyacrylonitrile has been employed as the antimicrobial polymer and has been proposed as biomaterial for environmental applications. Alamri et al. (2012) have reported a simple method for achieving the amine-terminated polyacrylonitrile by the immobilization of benzaldehyde derivatives, generating materials with antimicrobial properties for biomedical applications and water treatment. Biocidal efficacy of the prepared polymers were explored by viable cell counting methods against of different type of microorganism (e.g., Gram-positive bacteria—*S. aureus* and Gram-negative bacteria—*E. coli, Pseudomonas aeruginosa*, and *Salmonella typhi*), as well as fungi (*Aspergillus flavus, Aspergillus niger, Candida albicans*, and *Cryptococcus neoformans*). It was found that the antimicrobial activity depends on the nature of the functionalized group, generating different interactions of the bioactive groups with the bacterial cell membrane and consequently, increases with increasing number of phenolic hydroxyl group in the bioactive group.

11.4   CONCLUSIONS AND OUTLOOK

Microbial infections have a dramatic effect worldwide and affect a broad range of human activities. This very serious problem is owing to the inappropriate prescription of antibiotics for viral diseases and use of antibiotics in livestock feedstuff, and also incomplete patient treatment regimens. Undoubtedly, the antimicrobial polymers could have a decisive role in this global effort to find effective solutions for a problem that affects everybody due to their intrinsic features. Therefore, research into polymeric systems, which maximize the efficiency of active bactericidal agents, continues to be a necessity, being taken into account as a feasible alternative for bactericidal applications (such as artificial organs, drugs, healthcare products, implants, bone replacement, water treatment, food and textile industry, etc.). Majority of antimicrobial polymer materials produced by either the combination of organic and inorganic antimicrobial agents or by coating polymer surfaces with biocides including chemical binding represent a class of materials with unique biochemical properties, which can be in cure effective antimicrobial therapy. The use of these antimicrobial materials is a promising approach to lower the propensity of pathogen-resistant development.

As shown throughout this chapter, there are many strategies to design synthetic antimicrobial polymers with diverse manners of action over microorganisms. These developments have opened new ways for the synthesis of polymers with stringent control over the structure, morphology, and topology by using radical methods. Their activity may be inherent in their original structure, but the chemical modification or the introduction of organic or inorganic antimicrobial agents confers the biocidal behavior. Another strategy deals with the use of molecules with biocide characteristics that can be released from the polymer matrix. However, constraints associated with the potential resistance development when antibiotic molecules are involved or safety issues associated with the use of releasing nanoparticles could limit the potential of these strategies in the future.

As a result, high demand for intensifying efforts in the research and development of antimicrobial polymers has placed heavy reliance on both academia and industry to find viable solution for producing safer materials in medical and healthcare industry, food packaging, textiles, coating of catheter tubes, necessarily sterile surfaces, etc. The greater need for materials that fight infection will give incentive for discovery and use of antimicrobial polymer.

One important issue for new developments is to find the true mechanism of existing and new antimicrobial surfaces because only that knowledge allows useful predictions for their optimization in terms of reactivity and long-term activity. Efforts will be also focused to reduce costs, by investigating inexpensive systems. In short, the future of these materials will go through a combination of different approaches, along with a continuous investigation of the killing action of these promising systems.

KEYWORDS

- polymer-based surfaces
- multifunctional platform
- releasing systems
- biomaterials for everyday life
- environment protection

REFERENCES

Abel, T., et al. Preparation and Investigation of Antibacterial Carbohydrate-based Surfaces. *Carbohydr. Res.* **2002,** *337* (24), 2495–2499.

Alamri, A.; El-Newehy, M. H.; Al-Deyab, S. Biocidal Polymers: Synthesis and Antimicrobial Properties of Benzaldehyde Derivatives Immobilized onto Amine-terminated Polyacrylonitrile. *Chem. Cent. J.* **2012,** *6* (1), 111–124.

Alvarez-Lorenzo, C., et al. Medical Devices Modified at the Surface by Gamma-ray Grafting for Drug Loading and Delivery. *Expert Opin. Drug Deliv.* **2010,** *7* (2), 173–185.

Arciola, C. R., et al. Prevalence of cna, fnbA and fnbB Adhesin Genes among *Staphylococcus aureus* Isolates from Orthopedic Infections Associated to Different Types of Implant. *FEMS Microbiol. Lett.* **2005,** *246* (1), 81–86.

Asadinezhad, A., et al. An In Vitro Bacterial Adhesion Assessment of Surface-modified Medical-grade PVC. *Colloid Surf. B* **2010,** *77* (2), 246–256.

Asanovic, K., et al. Some Properties of Antimicrobial Coated Knitted Textile Material Evaluation. *Text. Res. J.* **2010,** *80* (16), 1665–1674.

Basri, H.; Ismail, A. F.; Aziz, M. Microstructure and Anti-adhesion Properties of PES/TAP/Ag Hybrid Ultrafiltration Membrane. *Desalination.* **2012,** *287* (1), 71–77.

Brown, L., et al. Through the Wall: Extracellular Vesicles in Gram-positive Bacteria, *Mycobacteria* and Fungi. *Nat. Rev. Microbiol.* **2015,** *13* (10), 620–630.

Darouiche, R. O. Device-associated Infections: A Macroproblem that Starts with Microadherence. *Clin. Infect. Dis.* **2001,** *33* (9), 1567–1572.

Denyer, S. P.; Stewart, G. S. A. B. Mechanisms of Action of Disinfectants. *Intern. Biodeterior. Biodegrad.* **1998,** *41* (3–4), 261–268.

Dizman, B.; Elasri, M. O.; Mathias, L. J. Synthesis and Antimicrobial Activities of New Water-soluble Bis-quaternary Ammonium Methacrylate Polymers. *J. Appl. Polym. Sci.* **2004**, *94* (2), 635–642.

Dobos, A. M., et al. Impact of Surface Properties of Blends Based on Quaternized Polysulfones on Modeling and Interpretation the Interactions with Blood Plasma. *Proc. ISI EHB Int. Conf. E-Health Bioeng.* **2016**, *7391423*, 1–4.

Dunne, Jr. W. M. Bacterial Adhesion: Seen Any Good Biofilms Lately? *Clin. Microbiol. Rev.* **2002**, *15* (2), 155–166.

Eren, T., et al. Antibacterial and Hemolytic Activities of Quaternary Pyridinium Functionalized Polynorbornenes. *Macromol. Chem. Phys.* **2008**, *209* (5), 516–524.

Filimon, A., et al. Electrospun Quaternized Polysulfone Fibers with Antimicrobial Activity. *Proc. ISI EHB Int. Conf. E-Health Bioeng.* **2016**, *7391562*, 1–4.

Filimon, A.; Ioan, S. Antimicrobial Activity of Polysulfone Structures. In *Functionalized Polysulfones: Synthesis, Characterization, and Applications*; Ioan, S., Ed., Taylor and Francis/CRC Press: Boca Raton, FL, 2015; pp 255–280.

Filimon, A., et al. Surface and Interface Properties of Functionalized Polysulfones: Cell-material Interaction and Antimicrobial Activity. *Polym. Eng. Sci.* **2015**, *55* (9), 2184–2194.

Filimon, A., et al. Surface Properties and Antibacterial Activity of Quaternized Polysulfones. *J. Appl. Polym. Sci.* **2009**, *112* (3), 1808–1816.

Fuchs, A. D.; Tiller, J. C. Contact-active Antimicrobial Coatings Derived from Aqueous Suspensions. *Angew. Chem. Int. Ed.* **2006**, *45* (40), 6759–6762.

Gao, Y.; Cranston, R. Recent Advances in Antimicrobial Treatments of Textile. *Text. Res. J.* **2008**, *78* (1), 60–72.

Goldberg, J. Biofilms and Antibiotic Resistance: A Genetic Linkage. *Trends Microbiol.* **2002**, *10* (6), 264.

Gorbunova, M. Novel Guanidinium and Phosphonium Polysulfones: Synthesis and Antimicrobial Activity. *J. Chem. Pharm. Res.* **2013**, *5* (1), 185–192.

Gozzelino, G.; Dellaquila, A. G.; Romero, D. Hygienic Coatings by UV Curing of Diacrylic Oligomers with Added Triclosan. *J. Coat. Technol. Res.* **2010**, *7* (2), 167–173.

Grunlan, J. C.; Choi, J. K.; Lin, A. Antimicrobial Behavior of Polyelectrolyte Multilayer Films Containing Cetrimide and Silver. *Biomacromolecules* **2005**, *6* (2), 1149–1153.

Han, J.; Castell-Perez, M. E.; Moreura, R. G. The Influence of Electron Beam Irradiation of Antimicrobial Coated LDPE/polyamide Films on Antimicrobial Activity and Film Properties. *LWT Food Sci. Technol.* **2007**, *40* (9), 1545–1554.

Hart, E., et al. Efficacy of Antimicrobial Polymer Coatings in an Animal Model of Bacterial Infection Associated with Foreign Body Implants. *J. Antimicrob. Chemother.* **2010**, *65* (5), 974–980.

Ioan, S.; Filimon, A. Biocompatibility and Antimicrobial Activity of Some Quaternized Polysulfones. In *Antimicrobial Agents*; Bobbarala V., Ed.; InTech: Rijeka, 2012; pp 249–274.

Isquith, A. J.; Abbott, E. A.; Walters, P. A. Surface-bonded Antimicrobial Activity of an Organosilicon Quaternary Ammonium Chloride. *Appl. Microbiol.* **1972**, *24* (6), 859–863.

Jacobs, T., et al. Plasma Surface Modification of Biomedical Polymers: Influence on Cell-Material Interaction. *Plasma Chem. Plasma Proc.* **2012**, *32* (5), 1039–1073.

Jaiswal, S.; Mchale, P.; Duffy, B. Preparation and Rapid Analysis of Antibacterial Silver, Copper and Zinc Doped Sol-gel Surfaces. *Colloids Surf. B. Biointerfaces* **2012,** *94* (1), 170–176.

Jampala, S. N., et al. Plasma-enhanced Synthesis of Bactericidal Quartenary Ammonium Thin Layers on Stainless Steel and Cellulose Surfaces. *Langmuir* **2008,** *24* (16), 8583–8591.

Klibanov, A. M. Permanently Microbicidal Materials Coatings. *J. Mater. Chem.* **2007,** *17* (24), 2479–2482.

Landini, P., et al. Molecular Mechanisms of Compounds Affecting Bacterial Biofilm Formation and Dispersal. *Appl. Microbiol. Biotechnol.* **2010,** *86* (3), 813–823.

Laporte, R. J. *Hydrophilic Polymer Coatings for Medical Devices-Structure, Properties, Development, Manufacture and Application*; CRC Press: Boca Raton, FL, 1997.

Lazar, V. *Microbial Adhesion*; Romanian Academy Ed.: Bucharest, 2003.

Lazar, V.; Chifiriuc, M. C. Medical Significances and New Therapeutical Strategies for Biofilm Associated Infections. *Rom. Arch. Microbiol. Immunol.* **2010,** *69* (3), 125–138.

Lee, S. B., et al. Permanent, Nonleaching Antibacterial Surfaces. 1. Synthesis by Atom Transfer Radical Polymerization. *Biomacromolecules* **2004,** *5* (3), 877–882.

Li, G.; Shen, J. A Study of Pyridinium-type Functional Polymers. IV. Behavioral Features of the Antibacterial Activity of Insoluble Pyridinium-type Polymers. *J. Appl. Polym. Sci.* **2000,** *78* (3), 676–684.

Li, Z., et al. Two-level Antibacterial Coating with Both Release-killing and Contact-killing Capabilities. *Langmuir* **2006,** *22* (2), 9820–9823.

Lichter, A. J., et al. Design of Antibacterial Surfaces and Interfaces: Polyelectrolyte Multilayers as a Multifunctional Platform. *Macromolecules* **2009,** *42* (22), 8573–8586.

Lin, J., et al. Bactericidal Properties of Flat Surfaces and Nanoparticles Derivatized with Alkylated Polyethylenimines. *Biotechnol. Prog.* **2002,** *18* (5), 1082–1086.

Lufrano, F., et al. Sulfonated Polysulfone as Promising Membranes for Polymer Electrolyte Fuel Cells. *J. Appl. Polym. Sci.* **2000,** *77* (6), 1250–1256.

Martin, T. P., et al. Initiated Chemical Vapor Deposition of Antimicrobial Polymer Coatings. *Biomaterials* **2007,** *28* (6), 909–915.

Meyer, B. Approaches to Prevention, Removal and Killing of Biofilms. *Int. Biodeter. Biodegr.* **2003,** *51* (4), 249–253.

Milovic, N. M., et al. Immobilized N-alkylated Polyethylenimine Avidly Kills Bacteria by Rupturing Cell Membranes with no Resistance Developed. *Biotechnol. Bioeng.* **2005,** *90* (6), 715–722.

Moustafa, M.; Fouda, G. Antibacterial Modification of Textiles Using Nanotechnology. In *A Search for Antibacterial Agents*; Bobbarala, V., Ed., InTech: Rijeka, 2012; pp 47–72.

Munch, D.; Sahl, H. G. Structural Variations of the Cell Wall Precursor Lipid II in Gram-positive Bacteria—Impact on Binding and Efficacy of Antimicrobial Peptides. *Biochim. Biophys. Acta.* **2015,** *1848* (11), 3062–3071.

Muñoz-Bonilla, A.; Fernández-García, M. Polymeric Materials with Antimicrobial Activity. *Prog. Polym. Sci.* **2012,** *37* (2), 281–339.

Nonaka, T., et al. Synthesis of Water-soluble Thermosensitive Polymers Having Phosphonium Groups from Methacryloyloxyethyl Trialkyl Phosphonium Chlorides-N-isopropylacrylamide Copolymers and Their Functions. *J. Appl. Polym. Sci.* **2002,** *87* (3), 386–393.

North, S. H., et al. Plasma-based Surface Modification of Polystyrene Microtiter Plates for Covalent Immobilization of Biomacromolecules. *ACS Appl. Mater. Interfaces* **2010,** *2* (10), 2884–2891.

Noshay, A.; Robeson, L. M. Sulfonated Polysulfone. *J. Appl. Poly. Sci.* **1997,** *20* (7), 1885–1903.

Novel, G. Les Biofilms. *Bull. Soc. Fr. Microbiol.* **1999,** *14* (2), 103–126.

Onofrei, M. D.; Filimon, A. Cellulose-based Hydrogels: Designing Concepts, Properties, and Perspectives for Biomedical and Environmental Applications. In *Polymer Science: Research Advances, Practical Applications and Educational Aspects, Polymer Science Book Series No 1*; Mendez-Vilas, A., Solano, A., Eds.; Formatex Research: Spain, 2016; pp 108–120.

Park, E. S., et al. Antimicrobial Activity of Phenol and Benzoic Acid Derivatives. *Int. Biodeterior. Biodegrad.* **2001,** *47* (4), 209–214.

Peng, J. S., et al. Inactivation and Removal of *Bacillus cereus* by Sanitizer and Detergent. *Int. J. Food Microbiol.* **2002,** *77* (1–2), 11–18.

Piccririllo, C., et al. Antimicrobial Activity of Methylene Blue Covalently Bound to a Modified Silicon Polymer Surface. *J. Mater. Chem.* **2009,** *19* (34), 6167–6171.

Podsiadlo, P., et al. Layer-by-layer Assembly of Nacre-like Nanostructured Composites with Antimicrobial Properties. *Langmuir* **2005,** *21* (25), 11915–11921.

Rajendra, R., et al. Use of Zinc Oxide Nano Particles for Production of Antimicrobial Textiles. *Int. J. Eng. Sci. Technol.* **2010,** *2* (1), 202–208.

Rathnayake, W. G. I. U., et al. Enhancement of the Antibacterial Activity of Natural Rubber Latex Foam by the Incorporation of Zinc Oxide Nanoparticles. *J. Appl. Polym. Sci.* **2014,** *131* (1), 39601–39609.

Ren, X., et al. Antimicrobial Coating of an N-halamine Biocidal Monomer on Cotton Fiber via Admicellar Polymerization. *Colloids Surf. A.* **2008,** *317* (1–3), 711–716.

Rizzotto, M. Metal Complexes as Antimicrobial Agents. In *A Search for Antibacterial Agents*; Bobbarala, V., Ed.; InTech: Rijeka, 2012; pp 73–88.

Shaikh, S., et al. Antibiotic Resistance and Extended Spectrum Beta-lactamases: Types, Epidemiology and Treatment. *Saudi J. Biol. Sci.* **2015,** *22* (1), 90–101.

Shi, Z.; Neoh, K. G.; Kang, E. T. Surface-grafted Viologen for Precipitation of Silver Nanoparticles and Their Combined Bactericidal Activities. *Langmuir* **2004,** *20* (16), 6847–6852.

Siedenbiedel, F.; Tiller, J. C. Antimicrobial Polymers in Solution and on Surfaces: Overview and Functional Principles. *Polymers* **2012,** *4* (1), 46–71.

Tashiro, T. Antibacterial and Bacterium Adsorbing Macromolecules. *Macromol. Mater. Eng.* **2001,** *286* (2), 63–87.

Thomassin, J. M., et al. Grafting of Poly[2-(Tert-butylamino)Ethyl Methacrylate] onto Polypropylene by Reactive Blending and Antibacterial Activity of the Copolymer. *Biomacromolecules* **2007,** *8* (4), 1171–1177.

Tiller, J. C., et al. Polymer Surfaces Derivatized with Poly(Vinyl-N-hexylpyridinium) Kill Airborne and Waterborne Bacteria. *Biotechnol. Bioeng.* **2002,** *79* (4), 465–471.

Tiller, J. C., et al. Designing Surfaces that Kill Bacteria on Contact. *Proc. Natl. Acad. Sci. USA* **2001,** *98* (11), 5981–5985.

Timofeeva, L.; Kleshcheva, N. Antimicrobial Polymers: Mechanism of Action, Factors of Activity, and Applications. *Appl. Microbiol. Biotechnol.* **2011,** *89* (3), 475–492.

Turos, E., et al. Antibiotic-conjugated Polyacrylate Nanoparticles: New Opportunities for Development of Anti-MRSA Agents. *Bioorg. Med. Chem. Lett.* **2007,** *17* (1), 53–56.

Vasilev, K.; Cook, J.; Griesser, H. J. Antibacterial Surfaces for Biomedical Devices. *Expert Rev. Med. Devices* **2009,** *6* (5), 553–567.

von Eiff, C., et al. Infections Associated with Medical Devices: Pathogenesis, Management and Prophylaxis. *Drugs* **2005,** *65* (2), 179–214.

Zhang, L., et al. Mechanistic Investigation into Antibacterial Behaviour of Suspensions of ZnO Nanoparticles against *E. coli. J. Nanopart. Res.* **2010,** *12* (5), 1625–1636.

Zilberman, M.; Elsner, J. J. Antibiotic-eluting Medical Devices for Various Applications. *J. Control Release* **2008,** *130* (3), 202–215.

BIOMIMETIC AND SMART BIOMATERIALS FOR ORTHOPEDIC APPLICATIONS: MORE THAN THE SUM OF THEIR COMPONENTS

DANIELA IVANOV[1*] and IULIAN VASILE ANTONIAC[2]

[1]*Department of Bioactive and Biocompatible Materials, "Petru Poni" Institute of Macromolecular Chemistry, 41 A Grigore Ghica Voda Alley, 700487 Iasi, Romania*

[2]*Materials Science and Engineering Faculty, Biomaterials Group, University Politehnica of Bucharest, Splaiul Independentei 313, Building J, Room JK109, Sector 6, 060032 Bucharest, Romania*

*Corresponding author. E-mail: dani@icmpp.ro

CONTENTS

ABSTRACT

Bone is a highly complex and a dynamic smart composite material, with remodeling and self-healing capability, permanently adapting in response to variable environmental physical factors (e.g., mechanical stress), chemical factors (e.g., specific ions), and biological factors (e.g., growth factors). The classic functional bone tissue engineering highlights the synergistic combination of a biocompatible scaffold that closely mimics the natural bone extracellular matrix, bone cells capable to generate the bone extracellular matrix, morphogenetic signals to direct the cells to differentiate into the specific phenotype, and vascularization to provide nutrient supply and clearance the wastes.

Current options to treat bone injuries and critical-sized defects present significant limitations that lead to the need for the development of innovative reconstruction synthetic grafts to support skeletal tissue repair. These biodegradable, bioresorbable, and bioactive biomaterials may also include smart components capable to change their properties depending on outer stimuli, based on the principles of biomimetic bioresponsive design, in terms of composition and structure. Different physical and chemical strategies have been developed correlated with suitable fabrication techniques, to design new biomimetic smart biomaterials for successful bone grafting.

12.1 INTRODUCTION

Bone injuries and defects mainly resulting from congenital deformities, degenerative diseases, complex traumatic injuries, or tumor resection represent a very significant clinical problem of all times. These problems are worsening especially with aging, obesity, and poor physical activity significantly compromising the quality of life and representing an increasingly significant socio-economic problem. Despite the remarkable native regenerative capacity of bone, the regeneration of critical-sized bone defects represents a major challenge in orthopedics as this capability is limited with the size of the trauma and/or the patient health condition.

Tissue engineering has the potential to address these clinical needs and develop new grafts and substitutes that structurally and morphologically mimic the native tissue, and moreover, perform similar biological functions (Correlo et al., 2011). Tissue engineering approaches suppose the appropriate combination of biomaterial scaffolds, biochemical, and physical stimuli to encourage new bone formation, and cells. Bone grafting is possible because bone tissue, unlike most other tissues, has the ability to regenerate completely and as bone recovers will gradually replace the graft material, resulting in a fully integrated region of new bone. According to their biological origin, bone grafts can be classified into autografts—fresh patient's own bone grafts, allografts—harvested from human donor or cadaver, xenografts—from an animal bone, and alloplast or nonbiological graft, artificial, or synthetic human-made substitutes. The autograft is ubiquitously referred to as the gold standard for tissue engineering, in spite of limitation and association with donor-site morbidity. Allografts and xenografts can also be used, yet they present the potential risks of immune recognition and rejection, besides disease transmission. Bone grafting represents the second most frequent tissue transplantation right after blood transfusion (Campana et al., 2014). A growing number of synthetic bone substitutes are nowadays commercially available for orthopedic applications.

12.1.1 MARKET FOR BONE GRAFTS AND SUBSTITUTES

The market for orthopedic implants is globally tremendously growing. The major factors driving the global bone grafts and substitutes toward development are the technological evolution besides the demand for minimally invasive surgery, and the rising number of orthopedic surgeries.

The global bone grafts and substitutes market was valued at over USD 2.3 billion in 2015 and is expected to reach over USD 3.6 billion, growing at an estimated compound annual growth rate (CAGR) of 5.2% during the forecast period of 2016–2024, according to a study by Grand View Research, Inc. *"Bone Grafts And Substitutes Market Analysis By Material (Natural - Autografts, Allografts; Synthetic - Ceramic, Composite, Polymer, Bone Morphogenetic Proteins (BMP)), By Application (Craniomaxillofacial, Dental, Foot & Ankle, Joint Reconstruction, Long Bone, Spinal Fusion) Forecasts To 2024."* Allografts segment dominated the global market in 2015 owing to the availability of an array of shapes, being the preferred natural bone graft across the world. The studies point out market shifts toward demineralized bone matrix grafts with better osteoconductivity and osteoinductivity compared to other allografts and synthetic grafts (ceramic, composite, and polymer, besides BMPs). The availability of synthetic substitutes at a cheaper cost is challenging the

growth of the natural bone grafts segment. Furthermore, synthetic substitutes present the advantage of no risk of disease transmission, hindering
the growth of the natural grafts market. As a consequence, synthetic
substitutes segment is anticipated to show significant growth during the
forecast period. The major participants of this market study include (but
not limited to) Baxter, Medtronic, Stryker Corporation, Orthofix Holdings,
Inc., AlloSource, DePuy Synthes, NuVasive, Inc., Wright Medical Group
N. V., Smith & Nephew, Inc., and Orthovita, Inc.

Medical composite biomaterials offer greater performance compared
to their corresponding individual components. It is worth mentioning that
any artificial material that comes in contact with skin (e.g., wearable artificial limbs) is not considered biomaterial since the skin acts as a protective
barrier. Some of the major benefits of medical composite materials include
high strength-to-weight ratio, excellent fatigue endurance, corrosion resistance, good impact resistance, and malleability. These biomaterials can be
easily manipulated in complex designs, owing to their different biomechanical and biocompatible properties and are used in the production of
orthopedic implants. The global medical composite material market can be
divided into three divisions, fiber composite, polymer–metal composite,
and polymer–ceramic composite. Fiber composites provide great external
impact resistance and are lighter, being used in manufacturing various
components such as implants and prosthetics. The Medical Composite
Material for Orthopedics Market *"Global Medical Composite Material for
Orthopedics Market 2016 Industry, Analysis, Research, Share, Growth,
Sales, Trends, Supply, Forecast to 2021"* provides also a basic overview of
the Medical Composite Material for Orthopedics industry, including defi-
nitions, classifications, applications, as well as manufacturing processes.

The orthopedic trauma fixation devices market is also the witness of
significant global growth, due to both increasing incidences of fractures
and lack of substitutes. This market is expected to grow at a CAGR of
about 7.2% during 2014–2020. Internal fixators have the largest area in
the orthopedic trauma fixation devices market and the shift toward bioabsorbable material is increasing the interest of orthopedic trauma fixation
devices companies, primarily in plate and screw systems as the most
widely used, besides intramedullary nails and screws. The leader players
in the global market for orthopedic trauma fixation are DePuy Synthes,
a Johnson & Johnson company, Stryker Corporation, Smith & Nephew,
Zimmer Holding Inc., and Orthofix Holding Inc., etc., according to the

recent Transparency Market Research report "*Orthopedic Trauma Fixation Devices Market – Global Forecast, Share, Size, Growth and Industry Analysis 2014–2020.*"

The field of bone grafts and substitutes requires a transdisciplinary approach between material sciences, cell and molecular biology, biochemistry, bioengineering, and clinical research. So, it comes clear to witness an increasing proportion of investments driven toward future research and development.

The FDA Orthopedic Device Advisory Panel has recommended the indication for recombinant human bone morphogenetic protein rhBMP-2, in conjunction with a collagen sponge, for the treatment of long bone fractures and rhBMP-7 (OP-1) as an autograft substitute for tibial nonunions. The administration methods include direct placement in the surgical site, with more promising results in the presence of the growth factors in combination with substrates, providing a material scaffold for bone formation.

The interest in biomaterials for orthopedic applications has been characterized besides a significant increase in the number of commercially available bone grafts and substitutes applications, by an ever-increasing number of publications, patents, and major international conferences and themed meetings.

12.1.2 NATIVE BONE TISSUE

Bone is hard osseous connective tissue with a complex structure with multiple functions, providing structure and support, producing blood cells, and storing minerals. The relative ratios of bone extracellular matrix components vary with site, gender, age, disease, and conditions. The properties of bone tissues are strongly dependent on both chemical structure and architectural organization of bone extracellular matrix and bone cells—osteoblasts (involved in the formation and mineralization of bone tissue) and osteocytes derived from osteoprogenitor cells, and osteoclasts (involved in the reabsorption of bone tissue).

Bone extracellular matrix is a biocomposite composed of an inorganic mineral phase (hydroxyapatite), an organic phase mainly consist of collagenous proteins (about 90% tropocollagen), noncollagenous proteins, chondroitin sulfate, keratin sulfate, lipids, and water. The noncollagenous proteins (osteocalcin, osteonectin, osteopontin, and bone sialoprotein, growth factors—cytokines and hormones) comprise only 10–15% of the organic matrix and appear to play a crucial role in the formation of bone structure (Fisher et al., 1985). Water, even the minor constituent, contributes to the overall toughness of the bone biocomposite, acting like a plasticizer (Granke et al., 2015).

Architectural structure of bone extracellular matrix has a sophisticated organization at seven hierarchical levels (Weiner et al., 1998), ranging from macroscopic to nanometric scale with specific interfaces critical for cellular normal activity (Fig. 12.1).

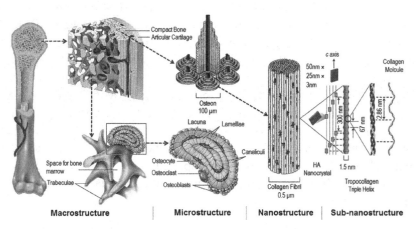

FIGURE 12.1 (See color insert.) The seven levels of hierarchy in bone structure. Reprinted from permission from Wang, X., et al. Topological Design and Additive Manufacturing of Porous Metals for Bone Scaffolds and Orthopaedic Implants: A Review. *Biomaterials* **2016**, *83*, 127–141. © 2016 Elsevier.

As the macroscale bone structure is represented by two osseous tissue types, compact or cortical bone (lamellar bone) and a spongy, cancellous or trabecular bone (woven bone) determined by interconnected pores and canals of different dimensions of extracellular matrix. These types differ at a microstructural level on the arrangement of protein fibrils and osteocytes, with cortical bone being composed of osteons or Haversian systems of concentric cylindrical lamellar elements, whereas trabecular bone with random organization. The nanoscale structure is given by the structural collagen (tropocollagen) fibrils reinforced with hydroxyapatite nanocrystals. Moreover, collagen itself contains multiple levels of hierarchy, being a right-handed superhelical structure, it is composed of three parallel left-handed helical polypeptide chains made of the repetitious amino acid sequence. The high elasticity of collagen fibrils confer bone and its considerable intrinsic tensile strength, while the hydroxyapatite crystals are responsible for hardness, rigidity, and the great compressive strength. Moreover, their effects are synergistic; the dispersion of a rigid but brittle material on elastic collagen matrix prevents the propagation of stress failure, also guiding bone cells migration. The crystals appear as nanosized platelets, with significant variations in size based on experimental results. The crystals are disposed either intra- or extra-fibrillar, with intra-fibrillar crystals disposed in the gap regions of the collagen fibril, and extra-fibrillar crystals surrounding the fibrils (Georgiadis et al., 2016; Reznikov et al., 2014; Florencio-Silva et al., 2015). The highest mineral loading is achieved by intra-fibrillar mineralization (Currey, 2003), while expressed in volumetric concentration, consists of about 33–43% apatite minerals, 32–44% organics, and 15–25% water because of the collagenous hydrogel matrix constitution.

Bone extracellular matrix is not only a highly complex but also dynamic tissue, being constantly formed and replaced in the bone remodeling process, along scarless self-healing ability. Moreover, it is a smart material as bone volume is balanced by the rates of bone formation and bone resorption, controlled by stress sensitive cells under the influence of bone-derived growth factors, permanently adapting in response to the applied mechanical loads (according to Wolff's law). It has been also hypothesized that this is a result of bone's piezoelectric properties, which cause bone to generate small electrical potentials under stress. This is why bone has been previously, and most appropriately, referred to as the ultimate smart material (Sommerfeldt et al., 2001).

12.2 BIOMIMETIC AND SMART BIOMATERIALS FOR ORTHOPEDIC APPLICATIONS

12.2.1 THE EVOLUTION OF ORTHOPEDIC BIOMATERIALS

The selection of a material for implants in orthopedics is governed by matching the material properties with specific requirements of this application. Bones are subjected to approximately 4 MPa stresses that are repetitive and fluctuate depending on the nature of the activities. Furthermore, body fluids represent a highly corrosive environment. Consequently, very stringent requirements are imposed on all types of potential candidate biomaterials to face attacks of a great variety of aggressive conditions.

The evolution of implantable materials is the best described according to their induced biological host interactions, as biomaterials chemically react with their environment. From biological reactions consideration, biomaterials can be biotoxic, bioincompatible, bioinert or biostable (no material is actually inert), biotolerant or biocompatible, and bioactive and bioresorbable. Biotoxic materials, such as metal alloys containing cadmium, vanadium, and lead, release substances in toxic concentrations that may cause immune reactions ranging from simple allergies to inflammation and to septic rejection due to chemical, galvanic, or other processes.

Bioinert (e.g., zirconia, alumina, titanium, and ultrahigh molecular weight polyethylene) and biotolerant (e.g., stainless steel, titanium, Co–Cr alloys, Co–Cr–Mo alloys, sintered hydroxyapatite, polymethylmethacrylate, polyether ether ketone, and carbon reinforced composites with polymers like polyethylene) materials do not release any toxic compounds. Moreover, these materials neither determine positive interaction with living tissue as may evoke a physiological response to form a fibrous capsule nor determine thickness of the layer of fibrous layer that can serve as a measure of bioinertness. NiTi shape memory alloys seemed to open a whole new range of applications, due to their special mechanical behavior, but their allergenic effect due to Ni has impeded their use. These materials represented the first generation of implanted material according to some authors (Hench, 1980).

Biomimetic, biodegradable, bioresorbable, and bioactive materials like synthetic and naturally-derived biodegradable polymers (e.g., collagen and hyaluronic acid), ceramics (synthetic or derived calcium phosphates such as hydroxyapatite, natural or synthetic calcium carbonate, and bioactive glasses), and recently developed biodegradable metals are considered the second generation of biomaterials. These materials degrade over time with formation of interface layers with the tissue at the atomic level, and allow a newly formed tissue invasion; consequently, the functions of bioresorbable materials are to participate in dynamic processes of formation and reabsorption. The bioactivity phenomenon is generated by both physical factors (surface roughness, porosity, and hydrophilic–hydrophobic surface) and chemical factors (molecular structures and chemical functionalization). These materials allow initiation of a healing process shortly after implantation by either physicochemical degradation or cellular activity, as well as by a combination of both processes, controlled by some factors, such as implant chemical composition, surface area to volume ratio, local acidity, fluid convection, temperature, etc.

Bioresponsive biomaterials designed to induce favorable cellular response are considered the third generation biomaterials. These biomaterials are particularly designed to better serve a specific function by including soluble factors (growth factors), actively or adaptively responding to dynamic external stimuli inducing smart behavior, adapt cell microenvironments to enhance particular characteristics (e.g., differentiate, proliferate, and migrate), or act as supportive matrices in regenerative medicine (e.g., embedding stem cells in bioactive scaffolds).

Different scientists classified the evolution of implantable materials in certain generations that should not be interpreted as chronologically but the conceptually, since each generation represents an evolution of the materials and techniques involved, facing the limitation of the resolution of investigation tools at the moment. The permanent development of novel biomaterials and technologies by introducing new concepts and finding solutions represents a creative and iterative process.

According to Anderson, history of biomaterials can be divided into three stages, bioMATERIALS within 1950–1975, BIOMATERIALS within 1975–2000, and since 2000 BIOmaterials, the capital letters emphasizing the major direction of the research efforts in biomaterials (Anderson, 2006). During the first stage, the long-term integrity of the biomaterials, as well as its nontoxic nature were considered of importance. After 1970s, biological interactions with biomaterials started to be more extensively investigated, also due to advances in biological mechanisms response. The revolution in the study of cell and molecular biology fields led to the investigation of interactions occurring at biological environment–material interfaces. Recently, with the advent of the areas of tissue engineering and regenerative medicine, the accent has been placed on biological interactions with biomaterials, including modifications in form and integrity of biomaterial with corresponding changes in the inflammatory and foreign body reactions.

Three generations of biomaterials were also assessed by some other authors (Hench, 1980). Hence, the first generation was considered targeting a suitable combination of physical properties with a minimal immune response and the foreign body reaction. The second generation of biomaterials was correlated with the development of bioactive materials with ability to interact with the biological environment by the tissue surface bonding, and bioabsorbable materials with progressive degradation rate, while new tissue regenerates and heals. The third generation aimed at development of new materials able to stimulate specific cellular responses at the molecular level directing cell proliferation, differentiation, and extracellular matrix production and organization and supporting self-healing of native tissue (Hench et al., 2002); thus third generation of biomaterials appeared together with tissue engineering concept development.

In some other authors understanding the evolution of biomaterials was more detailed and divided into four generations (Jandt, 2007). The first generation included bioinert biomaterials such as stainless steel, titanium alloys, or polyethylene. The second generation of biomaterials consisted of bioactive and bioresorbable biomaterials, such as bioceramics and polymers (collagen, hyaluronic acid, polylactide, polyglycolide, and polycaprolactone) and aimed body adjustments on the materials. The third generation was represented also by bioactive and bioresorbable biomaterials but of more complex structure, such as nanocomposites and inorganic–organic hybrid materials with similar compositions to those found in biological systems; these materials were conceived to adjust to the body milieu. The fourth generation of biomimetic materials biomaterials was correlated to tissue engineering scaffolding that interacts with the biological systems, such as biodegradation, regulation of biological processes, or real integration into the biological system.

12.2.2   BONE TISSUE RESPONSE INDUCES BY BIOMATERIALS

To achieve the goal of tissue reconstruction, the ideal scaffolds must meet several biomimetic requirements such as be manufactured from the materials with controlled biodegradability with the resorption rate to coincide as much as possible with the rate of bone formation; sufficient mechanical strength and stiffness to oppose loading forces and for the tissue remodeling; tailored surface properties such as surface energy, surface charge, and surface roughness critical in determining the cell–material interaction at the interface; high porosity with adequate interconnection and different pore dimensions to allow cell migration, vascularization, and diffusion of nutrients; easily fabricated into a variety of shapes and sizes; sterilization with no loss of properties; and long shelf life and at reasonable cost (Fig. 12.2).

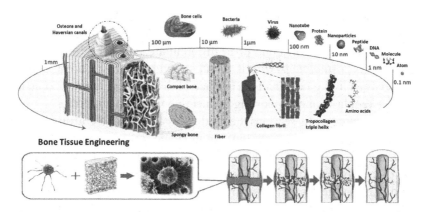

FIGURE 12.2 (See color insert.) Bone tissue engineering concepts of biomimetic microstructure and microcomposition. Reprinted with permission from Hao, Z., et al. The Scaffold Microenvironment for Stem Cell Based Bone Tissue Engineering. *Biomater. Sci.* **2017,** *5,* 1382–1392. © 2017 Royal Society of Chemistry.

Fixation of an implant represents a dynamic cellular response process that remodels the interface zone between the implant and living tissues at all dimensional levels, from the molecular up to the cell and tissue morphology level, at all-time scales, to ensure an optimal balance in terms of cell recognition, mass transport properties, and mechanical response to reproduce the morphological and functional features of natural tissues at the microscopic and nanoscopic level (Guarino et al., 2013). Moreover, nanotechnology was applied in bone materials to mimic even closer the nanostructured hierarchal self-assembly of native bone and support incorporation of nanomaterials to obtain nanocomposites to promote healing and functional recovery, surface modification at nanolevel for improving cell adhesion and functions, etc. (Sahoo et al., 2013; Gu et al., 2013).

The ideal bone tissue scaffolds should also stimulate the mechanisms of bone formation process and integration into host tissue of osteoconduction, osteoinduction, osteogenesis, and osteointegration. *Osteoconduction* represents scaffolds physical property to guide the infiltration of osteogenic precursor cells into the graft and neovasculature ingrowth; while the scaffold gradually degrades, this process supports the formation of integrated bone tissue. For example, hydroxyapatite is frequently used in orthopedic applications due to excellent biocompatibility and osteoconductivity. *Osteoinduction* refers to scaffolds ability to drive *osteogenesis,* stimulating the stem cells to differentiate into active osteoblasts which spread and generate new bone, typically associated with the presence of bone growth factors; osteogenic scaffolds already contain osteogenic stem cells able to form bone or to differentiate into bone-forming cells. Furthermore, *osteointegration* is direct structural and functional connection between living bone and the surface of the graft containing pores into which osteoblasts and supporting connective tissue can migrate.

To date, autografts serve as the gold standard for bone grafting because they are histocompatible and nonimmunogenic, osteoconductive, possess the essential components to achieve osteoinduction (i.e., BMPs and other growth factors), and osteogenic (i.e., osteoprogenitor cells). Decellularization of hard tissue presents a scaffolding alternative; it supposes removing the cells from a bone tissue while preserving the complex extracellular matrix (Lee et al., 2016).

The diamond concept for successful clinical bone graft integration states the mechanical stability beside the triangular standard tissue engineering concept of an osteoconductive scaffold upon which new bone can be formed, osteogenic cells to initiate bone repair and vascularization, and osteoinductive stimuli that differentiate the stem cells along the bone repair pathway (Giannoudis, 2007).

There is continuous ongoing research to develop next generation of advanced biomaterials like bioabsorbable materials, bioactive and smart biomaterials, shape memory biomaterials, etc.

12.2.3 BIOMIMETIC AND SMART BIOMATERIALS

The main driving force to develop biomimetic materials acting as smart implants is their tunable degradation feature in the physiological environment. The main advantage afforded by this class of material is the temporal clinical function over permanent implants and, once complete their function to biodegrade completely, while the new tissue is regenerated, avoiding the follow-up surgery to remove the implant after the tissue healed. The design criteria for the biomimetic implants require the materials to provide appropriate mechanical properties, suitable corrosion rate and excellent biocompatibility, and bioactivity.

There is a wide range of materials with biomimetic and smart features used for orthopedics applications, from biodegradable metals and their alloys, ceramics, and polymers, to hybrid composite systems. Such an example, biocompatible nitinol alloys, discovered in 1960s, exhibit both shape memory effect and superelasticity or pseudoelasticity, properties suitable for use in orthopedic implants (Bansiddhi et al., 2008). It is of great importance to know the performance of each of these materials used in clinical practice in the development of new biomaterials.

12.2.3.1 BIODEGRADABLE METALS

The early interest of material scientist to develop permanent highly corrosion resistant implantable metals was drastically shifted nowadays on temporary biodegradable metallic materials, one of the latest developments in biomaterials science for orthopedic applications. Hence, Mg-based, Fe-based, and Zn-based alloys have been proposed as biodegradable materials for load-bearing applications due to their superior combination of strength and ductility (Zheng, 2014).

Mg-based biodegradable materials are attractive for biodegradable orthopedic implants because of their good mechanical properties similar to the human bone, excellent biocompatibility, and stimulation the new bone formation (Witte, 2010; Zhang et al., 2016). Despite their remarkable properties, the extensive applications of Mg-based alloys (including the ones with Ca, Sr, Zn, Si, Sn, Mn, Y, Ag, etc.), are still restricted by fundamental shortcomings in physiological environment and commercial implants containing Mg and its alloys are still expected. The shortcomings are represented mainly by their high degradation rates and consequent loss in mechanical integrity, especially at pH levels between 7.4 and 7.6 (Kannan et al., 2008). On the other hand, the rapid reaction with water of Mg with formation of more hydrogen gas bubbles that can be dissolved in the physiological environment or diffuse from the implant surface, usually within the first week after surgery that could negatively affect Mg-based implants performance (Poinern et al. 2012) leading to implantation failure. The released hydrogen gas impede the good connectivity of osteocytes, interfering with bone healing process, and resulting in defects in callus formation, decreasing optimistic expectations for its use in osteosynthesis (Kraus et al., 2012). These materials due to the high rate of their biodegradation in physiological environment can be coated with less reactive materials, such as bioactive glasses or glass–ceramics (Rau et al., 2016). Mg-based alloys implants have been proven to stimulate the new bone formation when implanted as bone fixation devices, such as plates, screws, pins, nails, wires, and needles, have a huge potential market such as ZEK

100, LAE442, MgCa0.8, and MgYREZr (Reifenrath et al., 2013; Wolters et al., 2013; Erdmann et al., 2011). Recently, MAGNEZIX® screw produced by Syntellix AG represents the first commercially available MgYREZr alloy orthopedic product, was shown to be clinically equivalent to a standard titanium screw for the treatment of mild deformities (Waizy et al., 2013; Windhagen et al., 2013).

Fe and its alloys started to draw the interest as biodegradable material since more than a decade ago, such Fe–Mn, Fe–Pd, Fe–W, Fe–CNT, and Fe–C alloys due to their similar mechanical properties to stainless steel. However, compared to Mg-based materials, Fe and its alloy have generally, slower corrosion rates than clinical needs and persisting for a long time in human body even after completing their clinical needs. Another serious limitation is its ferromagnetism negatively impacts with certain imaging methods. In order to improve Fe-based material performances, research has focused on the chemical composition, microstructure, and surface of Fe modification, using different manufacturing process techniques, including casting, powder metallurgy, electroforming, and inkjet three dimensional (3D) printing. Future research should develop different strategies to accelerate the Fe-based biomaterials biodegradation. Very recent resorbable, porous iron–manganese–hydroxyapatite biocomposites with suitable degradation rates for orthopedic applications were prepared for the first time using salt-leaching technique. The newly developed porous $Fe_{30}Mn$ alloys and $Fe_{30}Mn$-10HA composites address inadequate resorption rates of Fe-based degradable materials previously proposed for orthopedic applications, inducing an enhance tissue ingrowth and vascularization within the implant (Heiden et al., 2015; Heiden et al., 2017).

Zn-based metals present a better corrosion rate compared with Fe and Mg alloys suitable for clinical applications. Besides the ideal corrosion rate, the significance of zinc in human nutrition has been intensively studied demonstrating as an essential element for basic biological function. It is also assessed that zinc stimulates bone formation, mineralization, and plays a role in the preservation of bone mass as a highly selective inhibitor of in vitro bone resorption by osteoclasts. However, as pure Zn is soft, brittle, and has low mechanical strength it needs to be combined in different alloys for orthopedic applications (Li et al., 2015). For instance, Zn is associated with the grain refinement and formation of secondary phases, influencing the mechanical, and corrosion properties of Mg–Zn and Mg–Zn–Mn alloys, allowing new bone formation (Chen et al., 2011).

12.2.3.2 BIODEGRADABLE AND BIOACTIVE CERAMICS

Biomedical applications of ceramic materials or bioceramics raised a strong interest initially as to increase biocompatibility of metallic implants. In short time it has grown into a distinct class of biomaterials, presently including relatively bioinert, bioactive or surface reactive, and bioresorbable bioceramics. Bioinert bioceramics represented the first generation of biomaterials designed to withstand mechanical stress without developing any specific cellular responses. Second generation shifted toward the opposite, to bioactive and bioresorbable bioceramics, able to elicit biological responses by stimulating new bone formation. These two generations of bioceramics represented the majority of the marketable bone substitute biomaterials, both of them being extensively commercialized in porous and dense forms in bulk, as well as powders, granules, and coatings. The distinction between the bioactive and bioresorbable bioceramics might be associated only with internal structure, for example, nonporous, dense, and highly crystalline hydroxyapatites behave as a bioactive bioinert material compared with highly porous form that can be resorbed within a year, including them in the concept of smart biomaterials for the clinical application (Yuan et al., 2010; Boyan, 2011).

The most important bioactive bioceramics advantages compared to all other implantable materials are inclusion in the metabolic processes, adaptation of either surface or the entire material to the physiological environment, integration of a bioactive implant with bone tissues at the molecular level or the complete replacement of resorbable bioceramics by healthy bone tissues. Surface reactivity contributes to their bone bonding ability and their stimulating effect on bone tissue formation (Bueno et al., 2009). The mechanical properties decrease significantly with increasing content of an amorphous phase, microporosity, and grain sizes, while a high crystallinity, a low porosity, and small grain sizes tend to give a higher stiffness, a higher compressive and tensile strength and a greater fracture toughness. The main shortcoming of bioceramics is their brittle nature that is attributed to high strength ionic bonds, impeding plastic deformation to happen prior to failure; in response, an initiated crack will continue to propagate, rapidly resulting in a catastrophic failure.

Native bone tissue is made of hydroxyapatite and therefore, both hydroxyapatite and related calcium phosphates (e.g., β-tricalcium phosphate) have been intensively utilized as the major component of scaffold materials for bone tissue engineering, especially for their osteoconductive properties, although have not shown osteoinductive ability. A significant finding was that bioactive glasses dissolution products (e.g., Bioglass®) upregulate the gene expression that control osteogenesis and the production of growth factors (Xynos et al., 2000), especially silicon has been found to play a key role in the bone mineralization and gene activation. In the last years, silica-based ordered mesoporous materials, composed of a silica network covered with silanol groups have been developed (Zhu et al., 2009; Vallet-Regí et al., 2011). These materials when placed in physiological fluid environment develop nanoapatite-like coatings with crystallinity similar to biological apatites, facilitating the bonding of implanted material to living bone.

Bone is nonuniform in composition and structure from cancellous to cortical bone cross-section. Different orthopedic biomedical applications require different configurations and shapes of implantable biomaterials. For instance, the structure suitable for implants of a high mechanical strength and with bone ingrowth is desirable a dense core–porous layer structure. Moreover, a layered structure with dense exterior and porous center mimics the structure of a calvarial bone graft. Bone biomimetic functionally gradient bioceramics consisting of calcium orthophosphates have been developed, with either compositional or structural gradient from the surface to the interior of the materials with the aim to combine mechanical strength and biocompatibility.

12.2.3.3   BIODEGRADABLE AND SMART POLYMERS AND POLYMERIC BIOMATERIALS

To better mimic native bone tissue, biodegradable material strategies have been developed, including the use of natural or synthetic biodegradable polymers with controlled structure, to obtain biocompatible, biodegradable, and multifunctional scaffolds with appropriate properties (Simionescu et al., 2016). Among the naturally derived materials, collagen type I as the major protein of bone extracellular matrix is the most often used due ability to support cell adhesion, while being rapidly biodegraded, allowing in vivo matrix remodeling. Synthetic polymers are also used as an alternative to minimize some of the possible risks associated with immune reaction and fast degradation rate. Natural scaffold materials are usually used either without alterations, in combination with synthetic polymers, or after chemical modifications. Stiffness and toughness, the two highly desirable properties of bone biomimetic biomaterials that can be reached by combining soft natural polymers, for example, collagen and stiff but brittle bioceramics, the properties of the resulted composite far surpass those of its individual components (Antoniac et al., 2016). To date, no resorbable orthopedic material fulfills all of the requisites of maintaining high strength needed for support during the full hard tissue recovering, with appropriate degrading period of time (generally accepted as 6 months to 2 years) (Pietrzak et al., 1996). Therefore, there is a need for new material designs that specifically address these issues to create a viable degradable and functional implant for hard tissues.

Stimuli-sensitive, intelligent, or smart polymers are polymers capable to mimic the behavior of living systems, by reversible physical or chemical large changes, in response to small environment modification—external stimuli, such as temperature, pH, pressure, light, magnetic or electric field, ionic factors, biological molecules, etc. (Aguilar et al., 2007; Liu et., al 2010). In human body, the polymer–polymer and the polymer–water interactions in small ranges of pH or temperature modifications are translated to a chain transition between extended and compacted coil states, including neutralization of charged groups by either a pH shift or the addition of an oppositely charged polymer, changes in the efficiency of the hydrogen bonding with an increase in temperature or ionic strength, and collapse of hydrogels and interpenetrating polymer networks. Research in smart polymers has undergone tremendous progress in the past few years; however, the effectiveness of these biomaterials as a valuable tool for tissue engineering clinical applications is still in its early stages. Major limitations such as rapid and well controllable response, reproducibility, and biocompatibility still need to be overcome (Custódio et al., 2014).

pH-sensitive polymers are polyelectrolytes (i.e., chitosan) that bear in their structure pendant weak acidic or basic groups that either accept or release protons in response to make changes in environmental pH. The presence of water from biological environment determine an increase in the hydrodynamic volume of the polymer, corresponding to a transition between tightly coiled and expanded state, influenced by any condition that modifies electrostatic repulsion (pH, ionic strength, and type of counterions). For example, biomimetic composites of chitosan with hydroxyapatite were found to be potential bone implant biomaterials with good osteoconductive, osteoinductive, and osteogenic properties (Venkatesan et al., 2010).

Temperature-responding polymers are characterized by a fine hydrophobic–hydrophilic balance in their structure around the critical temperature when the chains collapse or expand due to the hydrophobic and hydrophilic interactions between the polymeric chains and the aqueous media. There are two main types of thermoresponsive polymers, the first present a lower critical solution temperature (LCST) and the second present an upper critical solution temperature (UCST); LCST and UCST are, respectively, the critical temperature below and above which the polymer and solvent are completely miscible. Thermosensitive polymers with critical temperature close to the physiological value, that is, poly(N-isopropyl acrylamide) with LCST of around 32°C, were intensively studied for applications in the biomedical field (Ren et al., 2015). The change in surface properties of the thermosensitive polymers with hydrophobicity above the critical temperature to hydrophilic below critical temperature has been used in designing stimuli-responsive surfaces that favor cell adhesion. Thermoresponsive polymers are commonly used in tissue engineering as substrates that enable the cell attachment and detachment from a surface and proliferation, and as injectable gels for in situ of the scaffold. Both of these gels have the ability to swell in a solvent, with the difference that a physical gel will dissolve in the solvent in time, whereas cross-linked gels will not. Thermoresponsive polymers as physical gels present a sol–gel transition, while covalently linked networks exhibit a change in their degree of swelling in response to temperature. Polymeric structures sensitive to both temperature and pH have been conceived by simple combination of ionizable and hydrophobic (inverse thermosensitive) functional groups.

Shape memory polymers represent one of the most attractive areas in material science due to their processability compared with shape memory metals or ceramics. This type of smart polymers possesses the ability to recover their predefined shape in the presence of an external stimulus. The two prerequisites for shape memory polymer effect consist of stable network responsible for the original shape, and a reversible switching transition that fixes the temporary shape. The switching transition can be liquid crystal anisotropic–isotropic transition, reversible molecule cross-linking, crystallization-melting transition, and supramolecular association–disassociation. In addition, other stimuli can also trigger shape memory effect changing chain mobility, such as light, pH, moisture, pressure, electric or magnetic field, etc. (Pretsch, 2010). Significant researches success on various types of shape memory polymers have been achieved in many biomedical applications (Cabanlit et al., 2007; Yakacki et al., 2008; Serrano et al., 2012). Composites from shape memory polymers with multishape memory effect were synthesized and intelligent novel adaptive fracture fixation devices have been designed with different tightness. The unique properties of these materials allow the introduction of the shape memory polymer-based implants in a compressed form, by minimally invasive surgery procedures, followed by expansion once implanted in the desired site.

Smart hydrogels, 3D polymeric networks, constitute promising materials for scaffolds in tissue engineering as they can offer on-demand control the physical properties (Rahman et al., 2016). There are two main types of gels, physical gels and covalently linked (cross-linked) gels. The in situ formation of scaffold allows the delivery of encapsulated cells, growth factors, and nutrients to defects of any shape, using minimally invasive techniques. Once injected into the body, due to the temperature increase, the polymer forms a physical gel with 3D structure. Polymer interpenetrating networks, another group of cross-linked hydrogels, consist of two covalently linked polymer networks bounded together by physical entanglement as two intermixed networks that can only be separated by breaking bonds. These materials are of interest due to their ability to induce new properties when the networks interact.

Self-assembly is ubiquitous present in biological systems at both macroscopic and microscopic levels and represent the spontaneous association and organization of numerous individual entities into coherent and well-defined structures. Supramolecular biomimetic strategies have also been designed to generate dynamic biomaterials to promote the regeneration of hard tissues, such as bones. *Supramolecular chemistry* relies on self-assembling of specific, directional, highly tunable, reversible, dynamic noncovalent molecular structures based on physical interactions as hydrogen bonding, hydrophobic interactions, π–π interactions, van der Waals interactions, and metal chelation. Supramolecular biomaterials are imagined as inductive materials for prohealing functions or bioactive molecules delivery that stimulate endogenous tissue regeneration. Moreover, the self-healing properties of supramolecular materials have also been exploited in pro-osteogenic supramolecular hydrogels for application in regeneration of the osteochondral junction (Webber et al., 2015). For example, peptide-based self-assembly is a powerful strategy for designing templates for biomineralization, to replicate the regulatory role of the bone extracellular matrix and to facilitate osteogenic cell differentiation and bone deposition (Hosseinkhani et al., 2013).

Multifunctional scaffolds based on smart polymer materials, which can actively participate in the process to provide the biological signals that guide and direct cell function and promote new bone formation, have recently received particular attention in the field of bone tissue engineering. Such scaffolds based on novel functional and smart materials allow tuning of their properties and behavior and can simultaneously perform multiple crucial tasks (Kaliva et al., 2017).

12.3 BIOMIMETIC PHYSICAL AND BIOCHEMICAL STRATEGIES

12.3.1 PHYSICAL STRATEGIES IN BONE REGENERATION

The importance of the extracellular matrix as a modulator of cell activity has been greatly recognized, being of particular interest in the designing of smart biomimetic scaffolds. It follows that the physical properties of engineered scaffolds play an important role in regulating cellular infiltration, differential function, regeneration, and scaffold degeneration. Therefore, variable extracellular mediated signals such as porosity, surface topography, mechanical strain, and electrical fields, have been proposed as promising strategies for directed in vitro and in vivo bone regeneration. In addition, of critical importance is also the direct structural and functional integration between an implant and the native bone surface at implantation site (osteointegration) in sustaining prolonged functional success of orthopedic implants.

12.3.1.1 NANOMETRIC SCALE ARCHITECTURE

From the structural architecture approach, bone tissue represents a nanocomposite consisting of nanoblocks (type-I collagen nanofibers and hydroxyapatite nanocrystals) that endow bones with the self-remodeling and self-regenerative properties. It is known that in composites under repetitive loading, the flow of stress put mineral inclusions under tension and interfaces under shear, therefore, the mineral inclusions should resist high levels of tensile stress to prevent brittle fracture, nanometer size of the inclusions maximizing their fracture resistance. In this regard, nano-materials can be ideal candidates for the development of functional bone grafts since nanostructures can mimic the organization and mechanic performances of the extracellular bone matrix. Specifically, nanomaterials may continuously support bone biomineralization, mediate favorable cell behaviors, allow cellular infiltration, nutrient and waste transportation, bone ingrowth, and vascularization; also may change the way of growth factors delivery to provide the scaffolds with an ability to recruit endog-enous cells and accelerate the formation of new bone and blood vessels (Li et al., 2017).

Nanoparticles can affect bone regeneration in different ways. For example, nanoparticles can be incorporated into materials to obtain nanomaterials with adjustable mechanical strength for a preferential osteogenic differentiation, may possess the ability to improve osteogen-esis, are able to adjust the conformation of growth factors to increase their bioactivity for bone regeneration, etc. Moreover, the geometry of nanoparticles induces different osteogenic guidance effects. The nanotop-ographical surfaces may also exhibit an important influence in bone regen-eration (Kim et al., 2013). Therefore, the increase in surface roughness of implants can promote the interlocking between implants and native bones and direct osteogenic differentiation resulting in better osteointegration, may influence the differentiation of cells via cellular mechanotransduc-tion, providing a larger surface area. Furthermore, highly ordered nanopat-terns have been reported to promote osteogenic differentiation compared to disordered nanotopographies and variations of nanostructure depth that preferentially guide osteogenic differentiation. Interestingly, the influence of nanotopography on bone cells specific function seems to be most effec-tive when applied in combination with microscale features, translated to an in vivo response.

12.3.1.2 INTERCONNECTED PORE STRUCTURES

The structural design of the scaffold considers factors such as porosity, pore size and shape, the orientation of the interconnected channel, and a hierarchical control of structure to influence the desired bone tissue regeneration. The scaffold with a porous structure favors native tissue and vasculature ingrowth, the mass-transportation of nutrients and osteointe-gration with the host bone, and long-term stable fixation of bone implants (Mathieu et al., 2006).

The different porosity scales in the bone biomaterials should range from 1 to 1000 µm, in bone tissue engineering coexisting macro-, meso- and micro-pores. Specifically, pore diameters between 1and 20 µm support the cellular behavior and the type of attached cells, porosity ranges between 100 and 1000 µm is responsible for cellular growing, fluid flow, and mechanical resistance, while pore diameters larger than 1000 µm would determine the shape of the implant and its functionality (Rho et al., 1998).

Modulating the size and density of macropores may promote desir-able cellular processes at different stages of tissue development, but also could reduce the mechanical strength with compromising their structural integrity. This is why the total porosity of the biomaterial should be at least 40% and not over 80%. Furthermore, it has been emphasized the impor-tance of pores form, scaffolds with a spherical pore being demonstrated to exhibit higher elastic moduli compared to the ones with a cylindrical pore (Hollister, 2005). Moreover, many authors have drawn the attention to scaffolds with internal anisotropic microstructure (directional porosity) designed to mimic the bone microstructure (Lin et al., 2003; Mathieu et al., 2006) and it has been experimentally observed that bone tends to develop along the principal directions of porosity (Holy et al., 2003).

12.3.1.3 MECHANICAL AND ELECTRICAL STIMULATION FOR BONE REGENERATION

The cells of the bone tissue are able to sense and respond to biophysical stimuli from the in vivo dynamic environment to which they are exposed, that includes strain, stress, shear, pressure, fluid flow, streaming poten-tials, and acceleration (Salter et al., 1997).

The in vitro studies on seeded bone cells concluded that specific differentiation and phenotypic maintenance is enhanced by *biomimetic mechanical stimulation* (low-power ultrasound, fluid flow, centrifuga-tion, applied static load, vibration, or electromagnetic field) (Sittichock-echaiwut et al., 2009). Various types of bone tissue bioreactors have been designed, such as spinner flask, perfusion, and rotating wall bioreactors (Plunkett et al., 2011). Perfusion bioreactors provide bone construct viability using a constant, pulsating, or oscillating fluid flow through the scaffold for a homogeneous distribution of cell culture media and, in the same time, applies mechanical stimulation to cells within the scaffold. Mechanical loading has been extensively used in the tissue engineering of bone constructs, bioreactor systems capable of applying cyclic mechanical loads being commercially available (Kopf et al., 2012).

It is well known that bone matrix could be considered an amphoteric ion exchanger, with both positive and negative charges. Fukada and Yasuda first demonstrated piezoelectric properties in dry bone, particularly mechanical stress results in electric polarization while an applied electric field causes strain (Fukada and Yasuda, 1957). Moreover, the electrical potential induced by stress controls both bone cell activity and the orientation of bone deposition, suggesting that Wolff's law can be also explained in terms of electrical biocurrents generated within the bone tissue. Later, studies have shown that the electric fields induced in bone by cyclic loading and the piezoelectric effect are caused by the activation of mechanosensitive ion channels, inducing ion flux in bone cells, resulting in a polarization of the cell membrane. The bone cells can identify the characteristics of mechanical stimuli and respond electrophysiologically through the plasma membrane, resulting in the hyperpolarization associated with osteogenesis or depolarization associated with bone resorption. Furthermore, it is considered that mechanical information through activation of ion channels in osteocytes is also biochemically transmitted to neighboring osteocytes, osteoblasts, and osteoclasts bone cells (Yavropoulou et al., 2016). In orthopedic applications, electrical stimulation by pulsing electromagnetic fields, magnetic stimulation, and ultrasound stimulation have been demonstrated to enhance in vivo bone regeneration. Synthetic piezoelectric materials, such as ceramic piezoelectric materials (Feng et al., 1997) or different polymeric materials (Beloti et al., 2006) have shown promise applications as scaffolds in bone regeneration.

12.3.2   BIOMIMETIC CHEMICAL STRATEGIES

Among various phenomena associated with interaction of the most of the cell types with biomaterial, the first stage is adhesion, followed by growth, proliferation, adoption of phenotypic expression, synthesis of extracellular matrix, and eventually its mineralization (Owen et al., 2004). Smart interfaces are considered the interfaces capable of triggering favorable biochemical events, based on stimuli-responsive mechanism and self-organizing, by dynamically biological functions. A smart interface associated with bone tissue regeneration supposes the ability to elicit a favorable interaction between the cells and scaffold surface that simultaneously support specific cell binding and biomineralization. Moreover, direct bonding between the implant and the host bone (osteointegration) can only be achieved through reactive surface. This biomimetic approach could be achieved at the interface through one or more of intelligent functional designing strategies, like biomolecular immobilization, superior functionalization, stimuli-responsive spacer groups, etc.

12.3.2.1 BIOMINERALIZATION THROUGH BIOACTIVE MOLECULES

Biomimetic mineralization represents the process of mimicking biomineralization conditions by synthetic approaches (Palmer et al., 2008). This aim can be usually accomplished with the aid of organic templates like macromolecular strategies, cell walls, or membranes, based on specific or selective interaction between the organic moieties and the biomineral precursors.

Bioactivity of scaffold inert surfaces can be improved by chemical modification of the surface, by coating with a thin layer of a ceramic, integration of bioceramic nanoparticles in the scaffold, or by coating a polymer with appropriate functional groups. In bone regeneration, it is important to promote the production of bone-like apatite onto the biomaterial surface (Alves et al., 2010). Examples of induced biomineralization found in nature have inspired the production of biomimetic advanced materials and coatings for a wide range of biomedical and technological applications (Aizenberg, 2004). It has been shown that surface biomimetic mineralization may be triggered by either temperature (Shi et al., 2007) or pH (Dias et al., 2008). As the mineralization of bone is controlled by a number of regulated proteins, carboxylic acids and phosphorylated amino acids frequently appearing in such apatite-mineralizing responsible proteins, these acidic groups are used in many of the macromolecular strategies for surface functionalization. To reproducibly mimic the conditions of apatite formation in vitro mineralization studies on artificial substrates, simulated body fluid was developed as an organic-free mixture of reagent-grade salts buffered at pH 7.4; while this solution contains the ions necessary for mineralization, additional organic or inorganic nucleators are required.

Self-assembly systems offer the possibility to become basic models for mineralization with biomimetic features and tunable properties since they mimic the architecture of collagen fibrous matrices and also have potentially higher order parameters relative to polymers (Newcomb et al., 2012). Supramolecular systems can also be designed to combine multiplex biological signals with mineralization ability and may create a suitable niche for regeneration of mineralized tissues.

12.3.2.2 BIOMIMETIC THERAPEUTIC FUNCTIONALIZATION STRATEGIES

The dynamic process of bone remodeling involves replacement of existing bone with newly synthesized bone, based on orchestrated activity of the specific cell types, osteoblasts and osteoclasts, influenced by a cascade or signaling molecules—growth factors, mitogens (stimulate cell division), and morphogenesis (control generation of tissue form). The major challenge of tissue engineering is the need for more complex functionality through critical signaling molecules that direct cells activity and stimulate tissue regeneration. Various biomaterial scaffolds have been biomimetically modified to include growth factors BMPs, insulin-like growth factors (IGFs), and transforming growth factor-β (TGF-βs) related with bone healing. Biodegradable scaffolds allow for the controlled sequential temporal release strategies, for the delivery of incorporated growth factors, successful inducing differential cellular function, and regenerative phenotype in promoting bone regeneration. In addition, studies concluded that releasing of individual growth factors are less effective compared with the synergistic effect of combination of two or more growth factors. A facile method to incorporate biological factors into biodegradable scaffolds is physically entrapping by steric hindrance during manufacturing. Another method consists in encapsulation of growth factors into microsphere, nanoparticle or fibrous formulations, biological factor being released gradually, with the degradation of the vehicle. However, effective delivery of these biochemical stimuli is challenging because are quite expensive, may exhibit side effects, and limited control over dose administration as losing their bioactivity in time.

The most recent therapy of regenerative medicine consists of gene delivery to induce a regenerative phenotype (osteospecific and angiogenic) at the site of disease, typically in conjunction with induction of cellular infiltration and differentiation within a 3D engineered scaffold. Two methods of gene delivery have been generally explored in bone tissue engineering, in vitro direct transfection of cells followed by transplantation and, direct delivery of nucleic acids, either naked or via a vector, encapsulated into a scaffold, or through surface functionalization. However, local release is extremely important to avoid off-site transfection related with side effects in the surrounding tissues. Nonviral gene transfer vectors offer potential advantages over first designed viral vectors such as low toxicity and reduced immunogenicity. The efficiency of transfection of nonviral vectors depends on chemical structure, size, and composition of polyplexes. Efficient gene therapy could have a huge impact especially on noncurable diseases (Hu, 2014).

Bone is a highly vascularized tissue, therefore, angiogenesis or neovascularization play a key role in skeletal development and bone regeneration (Kanczler et al., 2008). However, human-made bone tissue engineered scaffolds are deprived of vasculature. One particular approach suggested that angiogenesis and bone morphogenesis promotion in vivo is controlled by gene activating ions release from bioactive glasses embedded in scaffolds. Recent studies have reported that the combination of angiogenic and osteogenic factors can stimulate bone healing and regeneration. Therefore, the design of bone scaffolds with controlled composition of bioactive glass, as well as with controlled local ion release, is considered a promising advanced strategy to enhance the repair mechanism of critical-sized bone defects.

12.3.2.3 BIOMIMETIC SELF-HEALING

The natural bone healing is a complicated process involving internal bleeding with forming of a fibrin clot, development of unorganized collagen fiber mesh, calcification of fibrous cartilage, and mineralization into fibrous bone and lamellar bone. Having been inspired by these findings, continuous efforts are being made to develop biomimetic smart biomaterials that have structurally incorporated the ability to self-repair mechanical damages. Generally, self-healing materials are designed to sense, stop, and even reverse damage, ideally without the intercession of external physical or chemical stimuli.

The evolution of self-healing polymer materials design was described in three stages, zeroth generation self-healing composites materials that retard but do not repair mechanical damages (particulate and fibers composite that increase the toughness, impact strength, and wear resistance by absorbing a greater fraction of the load, inhibiting pathways for crack propagation, and preventing void formation), first generation self-healing materials composites that irreversibly repair but do not restore damages (releasing a healing agent in response to the presence of damage) and, second generation self-healing materials that reversibly restore damaged matrix such as supramolecular polymer based systems (Brochu et al., 2011).

Depending on the healing mechanism, self-healing approach can be generally classified into two main groups, autonomous or extrinsic and intrinsic self-healing mechanisms, according to differentiated self-healing chemistries that the polymer and polymer composite undergo in response to the damage (Blaiszik et al., 2010, Billiet et al., 2013).

Autonomous or extrinsic self-healing mimic blood clotting and scar formation by embedding a healing agent included in capsule and vascular-like healing systems as an isolated phase that is released during crack formation, initiating one-time healing event through polymerization. The type of sequestration of the healing agent dictates the damage volume that can be healed, the repeatability of healing process, and the recovery rate.

Capsule-based self-healing materials contain the healing agent in discrete capsules (White et al., 2001). When the capsules are broken by damage, the self-healing mechanism is triggered through release and reaction of the healing agent, leading to only a singular local healing event. The healing efficiency that could be achieved by this system is quite limited as the capsules sizes are within the range of 100–200 µm; the ratio between the monomer and hardener has been shown to be crucial. Instead of polymer resin and catalyst healing agent-filled capsules, dual-microcapsule self-healing system containing monomer and hardener encapsulated separately and embedded into the polymer composite have been designed. As the crack forms, capsules are broken releasing the monomer which will polymerize when it comes in contact with hardener and eventually heal the cracks. A variety of techniques exist for encapsulation of healing agents based on the mechanism of wall formation, such as interfacial, in situ, coacervation, meltable dispersion, etc.

Vascular self-healing materials contain the healing agent in capillaries or hollow channels network, which may be interconnected one-, two-, or three-dimensionally until damage triggers self-healing process. After the vasculature is damaged and the release of healing agent occurs, the network may be refilled by an external source, allowing for multiple local healing events. As opposed to capsule-based systems, the healing agents for vascular materials are introduced after the network has been already integrated into the matrix. Therefore, the properties that determine the selection of healing agents include besides chemical reactivity, also surface wettability, and viscosity. One-dimensional vascular network self-healing polymer composite has been developed by embedding or creating one-dimensional hollow channels within a polymer composite matrix, filled with healing system, and hardener and polymerized during healing process. This self-healing system is superior to the capsule-based self-healing systems as besides a healing agent container, the vascular network act as reinforcement. Two- and three-dimensional network self-healing system featuring microvascular networks have recently attracted tremendous attention. Compared to the microcapsule self-healing and hollow fiber, this microvascular network, due to its interconnectivity, ensure a constant flow of healing agent within the microvascular structure, hence multiple healing events can be achieved. The practical challenge in these types of self-healing biomaterials is to identify nontoxic healing agent or catalyst system, efficient healing time, complex loading patterns, and functional lifetime of encapsulated agents.

Intrinsic self-healing represents self-healing through external intervention and is based on inherent ability of materials to increase its structure mobility, followed by restoring of chemical or physical properties, in principle for unlimited times. Intrinsic self-healing can be achieved through thermally reversible reactions, hydrogen bonding, ionomeric coupling, dispersed meltable thermoplastic phase, dynamic bonds, or molecular diffusion. Although no monomer or catalyst is required, different external stimuli to trigger the self-healing are required, depending on the types of self-healing mechanism. Unlike extrinsic self-healing mechanisms, this simpler concept could be more competitive in terms of end products manufacturing (Diesendruck et al., 2015).

Self-healing materials based on reversible reactions include components that can be reversibly transformed from the monomeric state to the cross-linked polymeric state through the addition of external energy—heat or photoirradiation—triggering enhanced mobility in the damaged region and bond reformation. The most common self-healing systems are based on the Diels–Alder and retro-Diels–Alder reactions. Self-healing in thermoset materials are obtained by incorporating a meltable thermoplastic additive, self-healing occurring by the thermoplastic material melting, and subsequent re-dispersion into the crack site, the healing being achieved at the molecular level. Ionomeric self-healing materials are based on copolymers with ionic segments that can form clusters by reversible cross-linking, under activation by external stimuli such as temperature or UV irradiation; because the formation of the clusters is reversible, multiple local healing events are possible. Dynamic bond self-healing supposes any type of bonds that are able to undergo repetitively breaking and reformation, under an equilibrium state. Dynamic bond self-healing can be segregated into two divisions, supramolecular and dynamic covalent self-healing; the healing of supramolecular polymer systems is normally under equilibrium, while dynamic covalent systems necessitate supplementary intervention (heat, salt, etc.). A more complex strategy includes the use of noncovalent interactions and transient bonds such as π–π stacking, hydrogen bonding, and coordination of ligands to metal ions to generate networks in order to support multiple healing events.

12.4 FABRICATION TECHNIQUES OF BIOMIMETIC BONE SCAFFOLDS

For biomimetic bone regeneration, the mechanical properties of the scaffold biomaterial are critical to successful repair of load-bearing bone defects. The characteristics such as stiffness, strength, toughness, strain, and hardening at the interfaces, are representative guidelines in the scaffold design. Another indispensable aspect is the pore structure, interconnectivity, and pore size of the scaffolds necessary to recreate the macroscale and microscale to nanoscale properties of native bone tissues. Some of these mechanical and structural properties can be controlled by manufacturing.

Numerous fabrication methods have been developed in time and utilized to produce 3D scaffolds for bone tissue engineering applications. These methods can be divided into two principal categories: conventional fabrication techniques and additive manufacturing technologies—3D printing, each of these providing different features and characteristics such as internal architecture as well as mechanical properties (Thavornyutikarn et al., 2014).

12.4.1 CONVENTIONAL FABRICATION TECHNIQUES

Conventional fabrication techniques (Morsi et al., 2008) can be classified in porogen assisted fabrication and textile techniques—woven or nonwoven fibers.

12.4.1.1 POROGEN ASSISTED TECHNIQUES

Porogen assisted techniques utilize solid biomaterials dissolved in appropriate solvents or incorporated with porogens, followed by processing; porogens are further removed and porous structure is generated within the scaffold. Such techniques include solvent casting, particulate leaching, thermal induced phase separation, gas foaming, and ice-templating.

Solvent casting utilizes polymer–ceramic particles mixture in an organic solvent, the mixture is casting into a predefined 3D mold; the solvent is subsequently evaporated, generating the porous scaffold. *Solvent casting particulate leaching* involves the casting of polymer solution and porogen particles (e.g., sieved salt or inorganic granules) mixtures and after solvent removing the dried scaffolds are fractionated in a suitable solvent to leach out the porogen, while porous structure is obtained. The internal pore size and porosity of the final scaffold are controlled by the size of porogen particles and the ratio of porogen to polymer.

Thermally induced phase separation uses a solution of polymer, eventually mixed with ceramic particles, in a volatile organic solvent of a low melting point. The polymer solution is first rapidly cooled to induce phase separation; subsequently, a porous scaffold is obtained after the sublimation of the solvent. The use of thermally induced phase separation followed by freeze-drying can produce scaffolds of high porosity; however, the small pore size produced by this technique limits its utility in bone tissue engineering.

Gas foaming processing employs a polymeric biomaterial saturated with a foaming agent as porogen such as carbon dioxide, nitrogen, or water at high pressures. This technique was developed to avoid the organic solvents that might induce an inflammatory response after scaffold implantation. The polymer is first placed in a chamber that is further saturated with high-pressure foaming agent; pores formation occurs as the pressure is rapidly dropped, as a result of the thermodynamic instability. However, it has been reported that only 10–30% of the pores are interconnected and a nonporous skin layer at the surface of the final product is formed. The combination of gas foaming with particulate leaching technique has been developed to achieve highly interconnected structures and to improve overall porosity.

Ice-templating techniques, including *freeze-drying* and *freeze casting* use water or an appropriate organic solvent as porogens avoiding utilization of porogen particles. The ceramic slurry, polymer solution, or colloid system in a suitable solvent is poured into molds and frozen for a definite time; the frozen structure (cryogel) is further lyophilized to generate porous scaffolds of highly interconnected pores and high porosities. The size of the pores can be adjusted by the size of the ice crystals that depends on the freezing temperature, freezing rate, and concentration. Recently, ice template scaffolds with sophisticated biomimetic porous structure have been generated by directional freezing and ice-segregation-induced self-assembly technique (Gutiérrez et al., 2008). Specifically, ceramic suspensions in water are directionally frozen to promote the formation of lamellar ice crystals that expel the ceramic particles as they grow; after ice crystals sublimation, a layered homogeneous ceramic scaffold is generated. The scaffold can then be further filled with a second soft phase so as to create a hard–soft layered composite. Controlling the composition of the suspension and the freezing kinetics various architecture of the biomaterial can be adjusted at several length scales. The combination of different templates allowed for the achievement of bimodal and even trimodal pore structured materials. Ice-templating techniques allow the obtaining of biomaterials with hierarchical of different levels of spatial organization, ultimately providing functionality to the final scaffold. The use of biomolecules (e.g., proteins and liposomes) as templates may impart functionalities and increased levels of organization to the final structures. The combination between hierarchy and functionality would provide the biomimetic and smart features to these biomaterials, opening the possibility to a variety of orthopedic applications.

12.4.1.2   TEXTILE TECHNIQUES—WOVEN OR NONWOVEN FIBERS

Woven or nonwoven fibers textile techniques, such as fiber bonding or generated by *electrospinning*, represent processing techniques that involve application of a voltage to electrically charge a polymeric solution (or melted polymers) loaded into a syringe. When the electrostatic repulsion forces from the surface charges combined with the Coulomb force exerted by the external electric field overcome the surface tension of the solution, a uniaxial stretch jet erupts through a nozzle. The solvent simultaneously evaporates and the generated fibers are deposited on the grounded collector into nonwoven materials that can further be casted, molded, coated, or cross-linked (Morsi et al., 2008). The parameters influencing the final material characteristics are typically classified into three categories, polymer material parameters (molecular weight, branching, blends of two or more polymers, and ceramics or metals polymer mixtures), polymer solution parameters (conductivity, surface tension, viscosity, and polymer concentration), and device parameters (voltage, flow rate, distance from nozzle tip to collector, collecting plate material, static or dynamic type of collector, type of the nozzle, and ambient conditions during electrospinning). Hence, by modifying these parameters, a large dimensional variety of fibers can be obtained, from nanometric to micrometric fibers according to specific application. Electrospinning technique allows the fabrication of various fiber patterns with a high porosity with interconnected pores and different morphology. The varieties of electrospinning techniques include electrospinning for uniaxially aligned nanofibers, centrifugal electrospinning, coaxial electrospinning, melt electrospinning, etc. This technique has been widely used to fabricate scaffolds for tissue regeneration applications; the main disadvantage is the use of organic solvents, which could be cytotoxic if not completely removed.

Although the conventional fabrication techniques offer the possibility to produce scaffolds of various tissue engineering applications, most of the resulted materials are lacking continuous interconnectivity and controlled pore size, pore geometry, and spatial distribution. The need for an organic solvent or a porogen represents one of the main limitations, most of them being cytotoxic and their traces in the scaffold may cause in vivo severe inflammatory responses. Moreover, some of conventional techniques present poor reproducibility.

12.4.2 ADDITIVE MANUFACTURING TECHNOLOGIES—3D PRINTING

Additive manufacturing technologies—3D printing represent recent advancements in fabricating next generation of biomimetic scaffolds for tissue engineering (Guvendiren et al., 2016). Additive manufacturing technologies offer a versatile pathway for the fabrication of complex anisotropic architectures like native osseous tissues, with dynamic functionalities like remodeling and self-healing, that are not accessible by conventional processing techniques. Biomimetic potential of additive manufacturing technologies consists in their intrinsic ability to control the local microstructure in a layer-by-layer approach, from the bottom-up perspective, with increasing levels of complexity (Martin et al., 2015).

3D printing, also known as rapid prototyping, is an additive manufacturing technique or computer-aided manufacturing (CAM) that enables fabrication of high structural complex materials following a computer-aided-design (CAD) acquired based on the images of patient-specific tissue captured with medical imaging techniques (such as computed tomography and magnetic resonance imaging). The most commonly used 3D printing technologies for tissue engineering applications can be classified in three main groups, light-assisted 3D printing, inkjet 3D printing, and extrusion-based methods (direct ink writing and fused deposition modeling), and particle fused deposition modeling (selective laser sintering and particle binding).

12.4.2.1 LIGHT-ASSISTED 3D PRINTING

Light-assisted 3D printing, also known as stereolithography, the original additive manufacturing method, involves patterning with a UV or laser radiation beam over a bath containing the photopolymerizable viscous polymer solution. Radiation beam induces a single polymerized layer that submerges into the polymer solution, allowing photopolymer solution to flow over, so the next layer is polymerized on top of the previous. The incorporation of filler inside the liquid photopolymerizable solution makes possible to also additive manufacture biocomposites; furthermore, if anisotropic magnetic particles are added, preferential orientation control can be acquired if process is driven into a magnetic field (Studart, 2016). Light-assisted 3D printing offers of the highest resolutions of all the printing methods, with feature resolutions in the single micrometer range.

However, this method has been limited in its biomedical applications by the lack of suitable biocompatible and biodegradable materials, harsh nature of UV radiation, extensive post-processing, liquid resin entrapping within the end product, etc.

Indirect 3D printing technique was developed to produce 3D scaffolds especially based on sensitive natural polymers, such as collagen or gelatin (Yeong et al. 2006). Compared to direct 3D printing where scaffold is produced directly, indirect 3D printing utilizes a temporary negative mold precisely designed, in which the solution of the desired polymer is casted under pressure or vacuum, the materials are hardened and, finally the mold is selectively removed to generate the desired scaffold. Moreover, indirect 3D fabrication can be combined with a foaming process to produce highly porous scaffolds with both macro- and micro-morphological features.

Recent developments in cross-linkable polymers and higher resolution machines, pave the way to tissue engineering applications, such as *microstereolithography* and *two-photon polymerization (2PP)* stereolithography. In *microstereolithography*, the laser beam is focused more precisely, improving the resolution of the process. Nowadays 2PP stereolithography is the most advanced technique of ultrafast fabrication of 3D constructs, with submicron resolution, by using a femtosecond near-IR wavelength radiation to induce polymerization. The polymerization is limited to the focal point voxel of the femtosecond laser that allows for a precise control over polymerization spot in a 3D environment. 2PP polymerization is based on the excitation through multiphoton nearly simultaneous absorption with ultrashort laser pulses. The resolution of this technique, from micrometers down to hundreds of nanometers, is dependent on the laser spot size, laser wavelength and energy, pulse width, pulse duration, pulse frequency, and pulse peak intensity all affect the size of the polymerized. Multiphoton polymerization represents a promising technology platform for the development of standard biomimetic microenvironments for developing 3D cell cultures (Ovsianikov et al., 2012).

12.4.2.2   INKJET 3D PRINTING AND EXTRUSION-BASED METHODS

Inkjet 3D printing and bioprinting enables disposition of individual droplets of very small volumes from a nozzle onto a printing surface while generating porous structures post solidification; to accelerate the printing process multinozzle inkjet print heads, containing several hundred individual nozzles, have been developed. Continuous inkjet printing and drop-on-demand inkjet printing commonly used for tissue engineering applications have been developed, based on the mechanism of droplet generation. Printable ink formulations are constrained by the stages of inkjet printing, drop generation, drop – substrate interaction, and drop solidification. The inks suitable for tissue engineering applications include polymer photocurable solutions, colloidal suspensions, polymer melts, as well as ceramic suspensions; biological materials, including cells, can be also incorporated into the ink formulation—bioinks. The most important technological ink properties are the viscosity and the surface tension. Inkjet printing has been used for tissue engineering *bioprinting* (scaffolds, bioadhesives, and living cells) (Saunders et al., 2014). Some other 3D cell bioprinting techniques such as microextrusion and laser-assisted printing have been developed, depending on the principles of releasing the cells from the printing head.

Direct ink writing, similarly, to the fused deposition modeling is an extrusion-based technique that delivers the material as a continuous viscoelastic filament through a syringe nozzle. Material hardens after deposition in a printing tray, via a physical or chemical mechanism, resulting objects with a wide range of geometries and sizes. The biomaterials used as ink can be also biomimetic composites reinforced with cells, stiff fibers, and whiskers. Multimaterial magnetically assisted 3D printing (MM-3D printing) new technology, represent an extrusion-based method that allows obtaining of biomimetic gradient multicomponent materials. Multiple materials can be printed by loading the inks containing different monomer and particles in distinct syringes. If the mixing units are placed before the extrusion nozzle, continuous gradients in composition are also possible. The magnetic manipulation of anisotropic particles has also been recently implemented in a direct ink-writing platform to enable the fabrication of intricate 3D composite architectures that combine particle orientation control. Microstructural control is possible through magnetic directed assembly of particulate filler (Studart, 2016).

Extrusion-based 3D printing methods, such as *direct ink writing* and *fused deposition modeling*, are among of the most widespread methods to fabricate scaffolds for tissue engineering applications (Turner et al., 2014). The extrusion additive manufacturing utilizes a viscous liquid or melts ink that is forced through a nozzle. The computer-generated model is sent to the printer and the selected material loaded into the device flow through the printer nozzle; the printer head moves up and down, side to side, and forward and back, successive depositing layers of chosen biomaterials are transformed into a solid form. New layers adhere to previous ones, creating the final product. The process has been successfully used to fabricate polymer, ceramic, metal, and composite scaffolds for bone tissue engineering. The fabrication of 3D bone tissue constructs including vasculature, suitable for repairing large bone defects, utilizing an extrusion-based direct-writing bioprinting strategy has been reported very recently (Byambaa et al., 2017).

12.4.2.3   PARTICLE FUSION PRINTING METHODS

Particle fusion printing methods, such as *selective laser sintering* and *particle binding*, have the ability to print polymers, ceramics, metals, and composites in unique and complex geometries (Shirazi et al., 2015) but are not suitable for bioprinting. *Selective laser sintering* uses a directed laser beam, according to the computer-modeled object, to increase the temperature of the particulate polymer or metal particles above their melting temperature, causing the particles fuse together, and the process is repeated. The suitable ink materials should be processable into a fine powder form with good particle flow dynamics, must have a convenient melting temperature, and bind together when heated. However, the process is slow, expensive, and requires a large amount of material. Particle binding or indirect selective laser sintering is a technique similar to selective laser sintering but uses a liquid binding solution to fuse particles together followed by a high-temperature sintering step instead of melting particles together with a laser beam. The resolution for particle fusion printing methods is significantly affected by the particle size, the particle size distribution, material binding properties, and laser or binder width. Particle fusion printing methods have been used to create devices for orthopedics hard-tissue engineering applications (Duan et al., 2010).

12.5   CONCLUDING REMARKS

Biomimetics concept was first introduced by Otto Schmitt in 1950s (Schmitt, 1969) who pointed out the important progression from biomimicry which involves superficial imitation of the complex biological systems to biomimesis and ultimately to bioinspiration, by translation from exploratory research and in basic science to technological implementations (Ivanov Vullev, 2011).

The impact of biomimetic biomaterials in regenerative medicine future will greatly depend on the ability to orchestrate multiple signals in a spatio-temporally controlled manner on specific cells in a smart materials and multifunctional scaffolds. However, there are still multiple critical aspects and challenges to be overcome before this constructs could be transferred to the clinic. In bone tissue regeneration, the main challenge is the development of smart multifunctional scaffolds with multiple cues and mimicking the functions of the natural bone tissues to promote cell activity and vascularization.

Tissue engineering initially approached a top-down for porous scaffold fabrication to allow the seeded cells to migrate into the scaffold, afterward creating their own matrix. This technique faced several challenges due to its limitations. Therefore, bottom-up biomimetic approaches have been developed, scaffold construct being realized simultaneously with cell encapsulation. Next strategy in tissue regeneration, so-called 4D printing, will take into account time as the fourth dimension. This emerging approach will allow producing biomimetic hierarchical hybrid and dynamic structures, from the nano- to macro-scale, containing adaptive self-assembling, self-shaping, or self-healing components, by specific manufacturing techniques. The development of biomimetic smart scaffolds able of modulated responses to the dynamic physiological and mechanical environments and displaying self-healing ability remains the important challenge in bone tissue regeneration.

KEYWORDS

- bone graft
- bone scaffold
- biomimetic
- biocomposite
- smart biomaterial
- orthopedic application

REFERENCES

Aguilar, M. R., et al. Smart Polymers and Their Applications as Biomaterials. In E-book: *Topics in Tissue Engineering*; Ashammakhi, N., Reis, R., Chiellini, E., Eds.; 2007; Vol. 3, pp 1–27.

Aizenberg, J. Crystallization in Patterns: A Bio-inspired Approach. *Adv. Mater.* **2004**, *16*, 1295–1302.

Alves, N. M., et al. Designing Biomaterials Based on Biomineralization of Bone. *J. Mater. Chem.* **2010**, *20*, 2911–2921.

Anderson, J. M. The Future of Biomedical Materials. *J. Mater. Sci. Mater. Med.* **2006**, *17*, 1025–1028.

Antoniac, I. V., et al. Collagen–bioceramic Smart Composites. In *Handbook of Bioceramics and Biocomposites*; Antoniac, I. V., Ed.; Springer International Publishing: Switzerland, 2016; pp 301–324.

Bansiddhi, A., et al. Porous NiTi for Bone Implants: A Review. *Acta Biomater.* **2008**, *4* (4), 773–782.

Beloti, M. M., et al. In Vitro Biocompatibility of a Novel Membrane of the Composite Poly(Vinylidenetrifluoroethylene)/Barium Titanate. *J. Biomed. Mater. Res. A.* **2006**, *79A*, 282–288.

Billiet, S., et al. Chemistry of Crosslinking Processes for Self-healing Polymers. *Macromol. Rapid Commun.* **2013**, *34*, 290–309.

Blaiszik, B. J., et al. Self-healing Polymers and Composites. *Annu. Rev. Mater. Res.* **2010**, *40*, 179–211.

Brochu, A. B. W.; Craig, S. L.; Reichert, W. M. Self-healing Biomaterials. *J. Biomed. Mater. Res. A* **2011**, *96* (2), 492–506.

Boyan, B. D.; Schwartz, Z. Regenerative Medicine: Are Calcium Phosphate Ceramics "Smart" Biomaterials? *Nat. Rev. Rheumatol.* **2011**, *7*, 8–9.

Bueno, E. M.; Glowacki, J. Cell-free and Cell-based Approaches for Bone Regeneration. *Nat. Rev. Rheumatol.* **2009**, *5*, 685–697.

Byambaa, B., et al. Bioprinted Osteogenic and Vasculogenic Patterns for Engineering 3D Bone Tissue. *Adv. Healthc. Mater.* **2017**, *6*, 1700015.

Cabanlit, M., et al. Polyurethane Shape-memory Polymers Demonstrate Functional Biocompatibility In Vitro. *Macromol. Biosci.* **2007**, *7*, 48–55.

Campana, V., et al. Bone Substitutes in Orthopaedic Surgery: From Basic Science to Clinical Practice. *J. Mater. Sci. Mater. Med.* **2014**, *25* (10), 2445–2461.

Chen, D., et al. Biocompatibility of Magnesium-zinc Alloy in Biodegradable Orthopedic Implants. *Int. J. Mol. Med.* **2011**, *28* (3), 343–348.

Correlo, V. M., et al. Natural Origin Materials for Bone Tissue Engineering and Properties, Processing, and Performance. In *Principles of Regenerative Medicine*; Atala, A., Lanza, R., Thomson, J. A., Nerem, R. M., Eds.; Elsevier Inc.: London, 2011; pp 557–586.

Currey, J. D. Role of Collagen and Other Organics in the Mechanical Properties of Bone. *Osteoporosis Int.* **2003**, *14*, S29.

Custódio, C. A., et al. Smart Instructive Polymer Substrates for Tissue Engineering. In *Smart Polymers and Their Applications*; Aguilar De Armas, M. R., Román, J. S., Eds.; Woodhead Publishing Ltd.: Cambridge, UK, 2014; pp 301–326.

Dias, C. I.; Mano, J. F.; Alves, N. M. pH-responsive Biomineralization onto Chitosan Grafted Biodegradable Substrates. *J. Mater. Chem.* **2008**, *18*, 2493–2499.

Diesendruck, C. E., et al. Biomimetic Self-healing. *Angew. Chem. Int. Ed.* **2015**, *54*, 10428–10447.

Duan, B., et al. Three-dimensional Nanocomposite Scaffolds Fabricated Via Selective Laser Sintering for Bone Tissue Engineering. *Acta Biomater.* **2010**, *6* (12), 4495–4505.

Erdmann, N., et al. Biomechanical Testing and Degradation Analysis of MgCa0.8 Alloy Screws: A Comparative In Vivo Study in Rabbits. *Acta Biomater.* **2011**, *7*, 1421–1428.

Feng, J.; Yuan, H.; Zhang, X. Promotion of Osteogenesis by a Piezoelectric Biological Ceramic. *Biomaterials* **1997**, *18*, 1531–1534.

Fisher, L. W.; Termine, J. D. Noncollagenous Proteins Influencing the Local Mechanism of Calcification. *Clin. Orthop. Relat. Res.* **1985**, *200*, 362–385.

Florencio-Silva, R., et al. Biology of Bone Tissue: Structure, Function, and Factors that Influence Bone Cells. *Biomed. Res. Int.* **2015**, *2015*, 421746.

Fukada, E.; Yasuda, I. On the Piezoelectric Effect of Bone. *J. Phys. Soc. Japan* **1957**, *12* (10), 1158–1162.

Georgiadis, M.; Müller, R.; Schneider, P. Techniques to Assess Bone Ultrastructure Organization: Orientation and Arrangement of Mineralized Collagen Fibrils. *J. R. Soc. Interface* **2016**, *13* (119), pii: 20160088.

Giannoudis, P. V.; Einhorn, T. A.; Marsh, D. Fracture Healing: The Diamond Concept. *Injury* **2007**, *38* (4), S3–6.

Granke, M.; Does, M. D.; Nyman, J. S. The Role of Water Compartments in the Material Properties of Cortical Bone. *Calcif. Tissue. Int.* **2015**, *97*, 292–307.

Gu, W., et al. Nanotechnology in the Targeted Drug Delivery for Bone Diseases and Bone Regeneration. *Int. J. Nanomed.* **2013**, *8*, 2305–2317.

Guarino, V., et al. Scaffold Design for Bone Tissue Engineering: From Micrometric to Nanometric Level. In *Biologically Responsive Biomaterials for Tissue Engineering*; Antoniac, I., Ed.; Springer Series in Biomaterials Science and Engineering, Springer: New York, 2013; Vol. 1, pp 1–16.

Gutiérrez, M. C.; Ferrer, M. L.; del Monte, F. Ice-templated Materials: Sophisticated Structures Exhibiting Enhanced Functionalities Obtained after Unidirectional Freezing and Ice-segregation-induced Self-assembly. *Chem. Mater.* **2008**, *20* (3), 634–648.

Guvendiren, M., et al. Designing Biomaterials for 3D Printing. *ACS Biomater. Sci. Eng.* **2016**, *2* (10), 1679–1693.

Hao, Z., et al. The Scaffold Microenvironment for Stem Cell Based Bone Tissue Engineering. *Biomater. Sci.* **2017**, *5*, 1382–1392.

Heiden, M., et al. Evolution of Novel Bioresorbable Iron–Manganese Implant Surfaces and Their Degradation Behaviors In Vitro. *J. Biomed. Mater. Res. Part A.* **2015**, *103A*, 185–193.

Heiden, M.; Nauman, E.; Stanciu, L. Bioresorbable Fe-Mn and Fe-Mn-HA Materials for Orthopedic Implantation: Enhancing Degradation through Porosity Control. *Adv. Healthc. Mater.* **2017**, *6*, 1700120.

Hench, L. L.; Biomaterials. *Science* **1980**, *208*, 826–831.

Hench, L. L.; Polak J. M. Third-generation Biomedical Materials. *Science* **2002**, *295* (5557), 1014–1027.

Hollister, S. J. Porous Scaffold Design for Tissue Engineering. *Nat. Mater.* **2005**, *4*, 518–524.

Holy, C. E., et al. Use of a Biomimetic Strategy to Engineer Bone. *J. Biomed. Mater. Res. A.* **2003**, *65*, 447–453.

Hosseinkhani, H.; Hong, P. D.; Yu, D. S. Self-assembled Proteins and Peptides for Regenerative Medicine. *Chem. Rev.* **2013**, *113* (7), 4837–4861.

Hu, Y. C. Gene Therapy for Bone Tissue Engineering. In *Gene Therapy for Cartilage and Bone Tissue Engineering*; Springer-Verlag Berlin Heidelberg; Springer Briefs in Bioengineering Series: Berlin, Germany, 2014; pp 33–53.

Ivanov Vullev, V. From Biomimesis to Bioinspiration: What's the Benefit for Solar Energy Conversion Applications? *J. Phys. Chem. Lett.* **2011**, *2* (5), 503–508.

Jandt, K. D. Evolutions, Revolutions and Trends in Biomaterials Science—A Perspective. *Adv. Eng. Mat.* **2007**, *9* (12), 1035–1050.

Kaliva, M.; Chatzinikolaidou, M.; Vamvakaki, M. Applications of Smart Multifunctional Tissue Engineering Scaffolds. In *Smart Materials for Tissue Engineering: Applications*; Wang, Q., Ed.; Royal Society of Chemistry: Croydon, UK, 2017; pp 1–38.

Kanczler, J. M.; Oreffo, R. O. Osteogenesis and Angiogenesis: The Potential for Engineering Bone. *Eur. Cell. Mater.* **2008**, *15*, 100–114.

Kannan, M. B.; Raman, R. K. S. In Vitro Degradation and Mechanical Integrity of Calcium-containing Magnesium Alloys in Modified-simulated Body Fluid. *Biomaterials* **2008**, *29*, 2306–2314.

Kim, H. N., et al. Nanotopography-guided Tissue Engineering and Regenerative Medicine. *Adv. Drug Deliv. Rev.* **2013**, *65*, 536–558.

Kopf, J., et al. BMP2 and Mechanical Loading Cooperatively Regulate Immediate Early Signalling Events in the BMP Pathway. *BMC Biol.* **2012**, *10* (37), 1–12.

Kraus, T., et al. Magnesium Alloys for Temporary Implants in Osteosynthesis: In Vivo Studies of Their Degradation and Interaction with Bone. *Acta Biomater.* **2012**, *8* (3), 1230–1238.

Lee, D. J., et al. Decellularized Bone Matrix Grafts for Calvaria Regeneration. *J. Tissue Eng.* **2016**, *7*, 1–11.

Li, H. F., et al. Development of Biodegradable Zn-1X Binary Alloys with Nutrient Alloying Elements Mg, Ca and Sr. *Sci. Rep.* **2015**, *5*, 10719.

Li, Y.; Liu, C. Nanomaterial-based Bone Regeneration. *Nanoscale* **2017**, *9* (15), 4862–4874.

Lin, A. S., et al. Microarchitectural and Mechanical Characterization of Oriented Porous Polymer Scaffolds. *Biomaterials* **2003**, *24*, 481–489.

Liu, F.; Urban, M. W. Recent Advances and Challenges in Designing Stimuli-responsive Polymers. *Prog. Polym. Sci.* **2010**, *35* (1–2), 3–23.

Mathieu, L. M., et al. Architecture and Properties of Anisotropic Polymer Composite Scaffolds for Bone Tissue Engineering. *Biomaterials* **2006**, *27* (6), 905–916.

Martin, J. J.; Fiore, B. E.; Erb, R. M. Designing Bioinspired Composite Reinforcement Architectures Via 3D Magnetic Printing. *Nat. Commun.* **2015**, *6*, 8641.

Morsi, Y. S.; Wong, C. S.; Patel, S. S. Conventional Manufacturing Processes for Three-dimensional Scaffolds. In *Virtual Prototyping of Biomanufacturing in Medical Application*; Bidanda, B., Bartolo, P. J., Eds.; Springer: New York, 2008; pp 129–148.

Newcomb, C. J., et al. The Role of Nanoscale Architecture in Supramolecular Templating of Biomimetic Hydroxyapatite Mineralization. *Small* **2012**, *8* (14), 2195–2202.

Ovsianikov, A., et al. Engineering 3D Cell-culture Matrices: Multiphoton Processing Technologies for Biological and Tissue Engineering Applications. *Expert Rev. Med. Devices* **2012,** *9* (6), 613–633.

Owen, G. R., et al. Focal Adhesion Quantification—A New Assay of Material Biocompatibility? Review. *Eur. Cell Mater.* **2004,** *9*, 85–96.

Palmer, L. C., et al. Biomimetic Systems for Hydroxyapatite Mineralization Inspired by Bone and Enamel. *Chem. Rev.* **2008,** *108* (11), 4754–4783.

Pietrzak, W. S.; Sarver, D.; Verstynen, M. Bioresorbable Implants-practical Considerations. *Bone.* **1996,** *19* (1), 109S–119S.

Plunkett, N.; O'Brien, F. J. Bioreactors in Tissue Engineering. *Technol. Health Care* **2011,** *19*, 55–69.

Poinern, G. E. J.; Brundavanam, S.; Fawcett, D. Biomedical Magnesium Alloys: A Review of Material Properties, Surface Modifications and Potential as a Biodegradable Orthopaedic Implant. *Am. J. Biomed. Eng.* **2012,** *2*, 218–240.

Pretsch, T. Review on the Functional Determinants and Durability of Shape Memory Polymers. *Polymers* **2010,** *2*, 120–158.

Rahman, M. M., et al. Stimuli-responsive Hydrogels for Tissue Engineering. In *Smart Materials for Tissue Engineering: Fundamental Principles*; *Smart Materials Series*; Schneider, H. J., Shahinpoor, M., Eds.; Royal Society of Chemistry: Cambridge, UK, 2016; pp 62–99.

Rau, J. V., et al. Glass-ceramic Coated Mg-Ca Alloys for Biomedical Implant Applications. *Mater. Sci. Eng. C.* **2016,** *64*, 362–369.

Reifenrath, J., et al. Degrading Magnesium Screws ZEK100: Biomechanical Testing, Degradation Analysis and Soft-tissue Biocompatibility in a Rabbit Model. *Biomed. Mater.* **2013,** *8* (4), 045012.

Ren, Z., et al. Effective Bone Regeneration Using Thermosensitive Poly(N-isopropylacrylamide) Grafted Gelatin as Injectable Carrier for Bone Mesenchymal Stem Cells. *ACS Appl. Mater. Interface* **2015,** *7* (34), 19006–19015.

Reznikov, N.; Shahar, R.; Weiner, S. Three Dimensional Structure of Human Lamellar Bone: The Presence of Two Different Materials and New Insights into the Hierarchical Organization. *Bone.* **2014,** *59*, 93–104.

Rho, J. Y.; Kuhn-Spearing, L.; Zioupos, P. Mechanical Properties and the Hierarchical Structure of Bone. *Med. Eng. Phys.* **1998,** *20* (2), 92–102.

Sahoo, N. G., et al. Nanocomposites for Bone Tissue Regeneration. *Nanomedicine (Lond)* **2013,** *8* (4), 639–653.

Salter, D. M.; Robb, J. E.; Wright, M. O. Electrophysiological Responses of Human Bone Cells to Mechanical Stimulation: Evidence for Specific Integrin Function in Mechanotransduction. *J. Bone Miner. Res.* **1997,** *12*, 1133–1141.

Saunders, R. E.; Derby, B. Inkjet Printing Biomaterials for Tissue Engineering: Bioprinting. *Int. Mater. Rev.* **2014,** *59* (8), 430–448.

Schmitt, O. H. In *Some Interesting and Useful Biomimetic Transforms*, Proceeding, Third International Biophysics Congress, Boston, MA, Aug. 29–Sept 3, 1969; p 297.

Serrano, M.; Ameer, G. Recent Insights into the Biomedical Applications of Shape-memory Polymers. *Macromol. Biosci.* **2012,** *12*, 1156–1171.

Shi, J.; Alves, N. M.; Mano, J. F. Thermally Responsive Biomineralization on Biodegradable Substrates. *Adv. Funct. Mater.* **2007,** *17* (16), 3312–3318.

Shirazi, S. F. S., et al. A Review on Powder-based Additive Manufacturing for Tissue Engineering: Selective Laser Sintering and Inkjet 3D Printing. *Sci. Technol. Adv. Mater.* **2015,** *16* (3), 033502.

Simionescu, B. C.; Ivanov, D. Natural and Synthetic Polymers for Designing Composite Materials. In *Handbook of Bioceramics and Biocomposites*; Antoniac, I. V., Ed.; Springer International Publishing: Switzerland, 2016; pp 233–286.

Sittichockechaiwut, A., et al. Use of Rapidly Mineralising Osteoblasts and Short Periods of Mechanical Loading to Accelerate Matrix Maturation in 3D Scaffolds. *Bone* **2009,** *44,* 822–829.

Sommerfeldt, D. W.; Rubin, C. T. Biology of Bone and How it Orchestrates the form and Function of the Skeleton. *Eur. Spine J.* **2001,** *10* (2), S86–S95.

Studart, A. R. Additive Manufacturing of Biologically-inspired Materials. *Chem. Soc. Rev.* **2016,** *45* (2), 359–376.

Thavornyutikarn, B., et al. Bone Tissue Engineering Scaffolding: Computer-aided Scaffolding Techniques. *Prog. Biomater.* **2014,** *3,* 61–102.

Turner, B. N.; Strong, R.; Gold, S. A. A Review of Melt Extrusion Additive Manufacturing Processes: I. Process Design and Modeling. *Rapid Prototyping J.* **2014,** *20* (3), 192–204.

Vallet-Regí, M.; Ruiz-Hernández, E. Bioceramics: From Bone Regeneration to Cancer. *Nanomed. Adv. Mater.* **2011,** *23* (44), 5177–5218.

Venkatesan, J.; Kim, S. K. Chitosan Composites for Bone Tissue Engineering—An Overview. *Mar. Drugs.* **2010,** *8,* 2252–2266.

Waizy, H., et al. Biodegradable Magnesium Implants for Orthopedic Applications. *J. Mater. Sci.* **2013,** *48,* 39–50.

Wang, X., et al. Topological Design and Additive Manufacturing of Porous Metals for Bone Scaffolds and Orthopaedic Implants: A Review. *Biomaterials* **2016,** *83,* 127–141.

Webber, M. J., et al. Supramolecular Biomaterials. *Nat. Mater.* **2015,** *15,* 13–26.

Weiner, S.; Wagner, H. D. The Material Bone: Structure-mechanical Function Relations. *Annu. Rev. Mater. Sci.* **1998,** *28,* 271–298.

White, S. R., et al. Autonomic Healing of Polymer Composites. *Nature* **2001,** *409,* 794–797.

Windhagen, H., et al. Biodegradable Magnesium-based Screw Clinically Equivalent to Titanium Screw in Hallux Valgus Surgery: Short Term Results of the First Prospective, Randomized, Controlled Clinical Pilot Study. *Biomed. Eng. Online* **2013,** *12,* 62–72.

Witte, F. The History of Biodegradable Magnesium Implants: A Review. *Acta Biomat.* **2010,** *6* (5), 1680–1692.

Wolters, L., et al. Applicability of Degradable Magnesium LAE442 alloy Plate–Screw Systems in a Rabbit Model. *Biomed. Tech.* **2013,** *58* (1), 4015–4016.

Xynos, I. D., et al. Ionic Products of Bioactive Glass Dissolution Increase Proliferation of Human Osteoblasts and Induce Insulin-like Growth Factor II mRNA Expression and Protein Synthesis. *Biochem. Biophys. Res. Commun.* **2000,** *276,* 461–465.

Yakacki, C., et al. Strong, Tailored, Biocompatible Shape-memory Polymer Networks. *Adv. Funct. Mater.* **2008,** *18,* 2428–2435.

Yavropoulou, M. P.; Yovos, J. G. The Molecular Basis of Bone Mechanotransduction. *J. Musculoskelet. Neuronal Interact.* **2016,** *16* (3), 221–236.

Yeong, W. Y., et al. Indirect Fabrication of Collagen Scaffold Based on Inkjet Printing Technique. *Rapid Prototyping J.* **2006,** *12* (4) 229–237.

Yuan, H., et al. Osteoinductive Ceramics as a Synthetic Alternative to Autologous Bone Grafting. *Proc. Natl. Acad. Sci. USA* **2010,** *107,* 13614–13619.

Zhang, Y., et al. Implant-derived Magnesium Induces Local Neuronal Production of CGRP to Improve Bone-fracture Healing in Rats. *Nat. Med.* **2016,** *22* (10), 1160–1169.

Zheng, Y. F.; Gu, X. N.; Witte F. Biodegradable Metals. *Mater. Sci. Eng. R.* **2014,** *77,* 1–34.

Zhu, Y.; Kaskel, S. Comparison of the In Vitro Bioactivity and Drug Release Property of Mesoporous Bioactive Glasses (MBGs) and Bioactive Glasses (BGs) Scaffolds. *Micropor. Mesopor. Mat.* **2009,** *118* (1–3), 176–182.

CHAPTER 13

QUANTUM CHEMISTRY SIMULATIONS: A COMPUTATIONAL TOOL TO DESIGN AND PREDICT PROPERTIES OF POLYMER-BASED SMART SYSTEMS

DUMITRU POPOVICI*

Laboratory of Polycondensation and Thermostable Polymers, "Petru Poni" Institute of Macromolecular Chemistry, 41 A Grigore Ghica Voda Alley, 700487 Iasi, Romania

*E-mail: dumitru.popovici@icmpp.ro

CONTENTS

ABSTRACT

Reaction mechanism of the partially alicyclic polyimides base on bicyclo-2,2,2-oct-7-ene-2,3,5,6-tetracarboxylic dianhydride—BOCA was investigated on an aniline model compound using quantum simulation at density functional theory B3LYP level. For each stage of reaction, the energetic profile was determined. The obtained results are in good correlation with the experimental observation. Synthesis of the model compound and isolation of the intermediate products was confirmed by spectroscopic methods (H^1-RMN and FT-IR). The good correlation between the theoretical calculations and the experimental results confirm that using the computational models of molecular design the reaction parameters can be adjusted to obtaining new compounds, thus reducing costs and engineering time.

13.1 INTRODUCTION

Each distinct stage in the development of humankind has been associated with advances in materials technology that have helped shape today's world. Within the last 20 years, after long periods of experimentation in finding solutions, the research scientists and engineers have recognized the need for a more systematic approach in order to develop new materials. This approach was one that to combined interdisciplinary research, new advances in computational modeling and simulation, and critical laboratory experiments to rapidly reduce the time from concept to end product (Eschrig, 2004; Gates and Hinkley, 2003).

Computer simulation is a discipline oriented toward to description of materials in the physical or chemical world using mathematical models for the estimating of macroscopic properties. This was the new paradigm by which all future materials research would be conducted and have embraced the concepts that have come to be known simply as *"computational materials"* (Janssens et al., 2007). Computational material systems with tailored capabilities are a topic trend in plethora of research. Computational materials approach increases the confidence that new materials will possess the desired properties when scaled up from the laboratory level so that lead-time for the introduction of new technologies is reduced and also, lowers the likelihood of conservative or compromised designs, which might have resulted from reliance on less-than-perfect materials.

In the above context, polymer-based smart materials help the research scientists and engineers to incorporate smartness functionality into their design through programming cycles. The smartness functionality offers the adjustment between inherent properties of these materials with their industrial applications through modeling techniques. Analytical and numerical models become an essential part of the analysis in order to minimize the experimental trial-and-errors efforts for finding the optimum range of design. Modification of this methodology is obtainable through understanding the relationship between the basic constituents of the polymers, chains and crystalline segments, and the effects of their interactions in the response of the whole structure. Therefore, a model that can predict the evolution of a targeted property in the form of microscale evolution of chains in multiple length scales can be very useful as a tool to design and analyze polymeric actuation mechanism (Kröger, 2005).

Recognizing the importance and potential benefits of the use of smart materials, as well as modeling their behavior, polyimides have been identified this as one of the special polymeric class. From this reason, the challenges posed by the complex behavior of such materials in creating the models are highlighted and illustrated with a case study.

Polyimides are one of the most important classes of thermostable polymers that can be used as smart systems in many applications, such as bio-MEMS devices biomaterials or sensors. This type of polymers are very versatile materials, accepting a wide variety in their chemical structures with allow the tailoring of the materials properties to match the requirements of many different applications (Wilson and Atkinson, 2007). In addition to high thermal stability, the polyimides possess other very interesting properties such as chemical stability, radiation resistance, very good mechanical, dielectrics, and optics properties, as well as very good adhesion on various supports, which recommends them as high performance polymers in many areas, such as aerospace, electronics and microelectronics, alternative energy sources, liquid or gas separation membranes, etc. (Abadie and Rusanov, 2007; Mittal, 2009; Sroog, 1996). Moreover, polyimides exhibit a low coefficient of thermal expansion (CTE)—close to that of metals (Numata et al., 1987; Pogge, 1996), therefore, it is used in microelectronics as dielectric layers.

These special qualities, however, have a number of inconveniences such as

- most of the aromatic polyimides are insoluble and infusible in their totally imidized form due to the rigid structure of the macromolecular chain, which make their processing to be very difficult (St. Claire, 1990);

- due to inter- and intramolecular charge-transfer-complex interactions (CTC), most of the aromatic polyimides are colored from pale yellow to dark brown, their use in applications where color absence is mandatory being difficult (Ando et al., 1997; Wakita et al., 2009);
- although it exhibits a relatively low dielectric constant ($\varepsilon' = 3.2$–2.9), in many applications this value is insufficient, the accepted values being $\varepsilon' < 2.6$, which is characteristic of materials with a low dielectric constant, or even $\varepsilon' \leq 2.2$, characteristic of materials with an ultralow dielectric constant (Chen et al., 2004).

One of the methods used to eliminate or diminution the abovementioned disadvantages is to modify the structure of macromolecular chain by introducing fluorinated groups (Bruma et al., 1995; de Abajo and de la Campa, 1999), bulky substituents (Yang et al., 2004), using asymmetric monomers (Chung and Kim, 2000; Hulubei et al., 2007) or flexible bridges (Yang et al., 2003). Lately, polyimides having linear or cyclic aliphatic segments into the backbone of macromolecular chain have been increasingly studied (Hasegawa et al., 2013; Koning et al., 1998; Varganici et al., 2015).

Compared to aromatic polyimides, the partially aliphatic polyimides have a number of advantages, such as increased solubility in various organic polar solvents, low dielectric constant ($\varepsilon' < 2.8$), less intense color or even color absence, lower glass transition temperature, which easily permit the processability of them (Barzic et al., 2012; Barzic et al., 2013; Popovici et al., 2012a, 2012b). Such properties recommend these materials for use in high-performance applications, such as flexible solar cells, color filters, dielectric materials, liquid crystal alignment layers, etc.

It has also been observed that polyimides having a partially cycloaliphatic structure exhibit remarkable biological properties. An intrinsic antibacterial activity (Li et al., 2008) as well as very good blood compatibility (Nagaoka et al., 2001; Popovici et al., 2012a, 2012b) of some of these materials has been observed. Moreover, these properties are preserved over time, the studied materials preserving antibacterial activity after a controlled aging process that simulated sun exposure for one year (Popovici et al., 2015).

A particularly interesting behavior has been observed for a series of partially aliphatic polyimides obtained from a cycloaliphatic dianhydride, namely bicyclo-2,2,2-oct-7-ene-2,3,5,6-tetracarboxylic dianhydride—BOCA (Fig. 13.1). In respect to our studies, these materials exhibit biological activity for both pristine material and surface-modified samples by plasma treatment and immersion in $AgNO_3$ solution or by treatment with diamine alcohol solutions (in work).

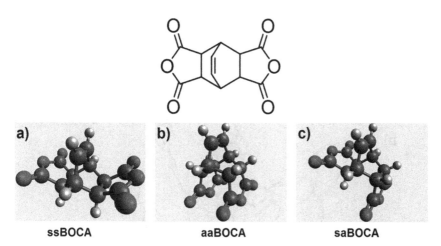

FIGURE 13.1 (See color insert.) Spatial conformation of bicyclo-2,2,2-oct-7-ene-2,3,5,6-tetracarboxylic dianhydride—BOCA: (a) *sin*-conformation; (b) *anti*-conformation; (c) *sin-anti* conformation.

This dianhydride can adopt three spatial conformations relative to the double bond, as can be noticed from Figure 13.1. Crystallographic studies have been highlighted the fact that, in a solid state, BOCA adopts a W-type conformation (Fig. 13.1a), both anhydride cycles being on the same side with the double bond (Hu, 2008). However, such spatial conformation does not confirm the experimental results obtained for BOCA-based polyimides.

It was noticed that XPS analyzes for the films made from BOCA and 4-(4-{[4-(4-aminophenoxy)phenyl]sulfonyl}phenoxy)aniline (pBAPS) polyimide (Fig. 13.2) the amount of sulfur at the surface of the film is less than the theoretical determined based on the chemical structure of the structural unit (Barzic et al., 2013). It has been concluded that for this polymer the sulfone groups are located inside of the macromolecular ball (Fig. 13.2b—hydrogen removed for clarity), the BOCA anhydride adopting an *anti* – *anti* conformation, the imidic cycles adopting an opposed position to the double bond. Moreover, in another study, it has been observed that the double bond possesses an increased reactivity, so it can easily participate in chemical reactions with various cross-linking agents (Hulubei et al., 2014). The increased reactivity of the double bond can only be explained if an *anti – anti* type conformation for the imidic cycles is accepted.

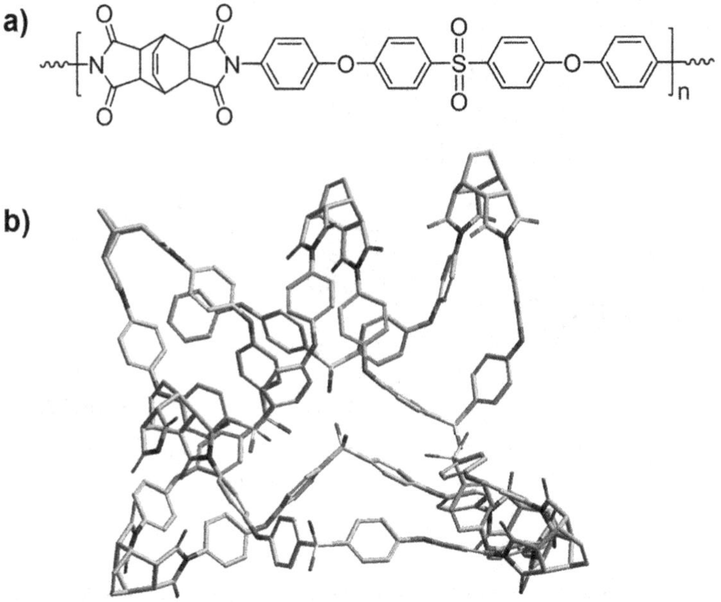

FIGURE 13.2 **(See color insert.)** Chemical composition of BOCA—pBAPS structural unit (a) and 3D conformation of 7SU chain (b).

As a result of these observations, it can be concluded that during the polycondensation reaction of BOCA with various diamines, the anhydride undergoes a conformational change that ultimately results in a spatial rearrangement of the macromolecular chain and the appearance of the presented properties.

13.2 REACTION MECHANISM FROM QUANTUM SIMULATION POINT OF VIEW

The way to provide an accurate check and balance of properties is to establish the analysis methods and validation criteria of simulations at both atomic and bulk scales. The transition from classical mechanics to molecular mechanics and then to quantum studies is the last step in the evolution of material analysis methods, semiempirical models or those based on wave function theory, being able to approximate with great accuracy macrophysical properties. For this reason, main theoretical concepts of modeling of a smart system with the use of quantum simulation methods were systematized.

13.2.1   CASE STUDY: PARTIALLY ALIPHATIC DIIMIDE MODEL COMPOUND

In order to be able to explain all of these observations made before, a mathematical model based on quantum calculations of each stage of the reaction mechanism of BOCA with aniline was used (Fig. 13.3). All calculations were made using the DFT B3LYP (Density Functional Theory, Becke's three-parameter hybrid exchange function with Lee–Yang–Parr gradient-corrected correlation functional) method at cc-pVDZ (Dunning correlation-consistent polarized valence only double-zeta basis sets) as the theory level, implemented in the GAMESS-US version 2013 (Gordon and Schmidt, 2005; Schmidt et al., 1993). The construction of the molecules was done using the Avogadro version 1.2 software and the visualization of the final results was done using Avogadro and wxMacMolPlt version 7.7 (Bode and Gordon, 1998; Hanwell et al., 2012). We chose the cc-pVDZ function set instead of the classical 6–31(d) because, compared to last one, the chosen one is based on correlation consisted of polarized of valence only base sets.

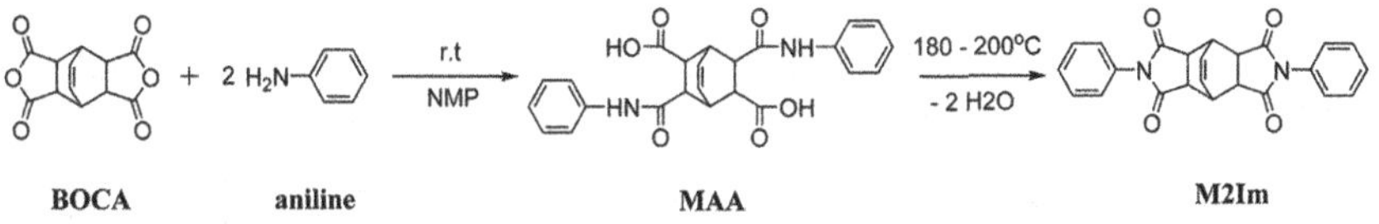

BOCA   **aniline**   **MAA**   **M2Im**

FIGURE 13.3   Generic synthesis of model compound.

As already mentioned, BOCA dianhydride can adopt three different spatial conformations reported in the double bond (see Fig. 13.1), but only spatial conformation ssBOCA has been experimentally confirmed so we used this confirmation for our calculations.

In the beginning, we optimized the confirmation of the starting reactants, aniline and the three spatial conformations of the dianhydride, and determining the distribution of the charge on the active reaction sites, respectively, for the aniline nitrogen atom and the four carbon atoms in the anhydride cycles. It has been observed that the partial charge at the four carbon atoms for the experimentally confirmed form is virtually identical, suggesting that the reaction can occur at any one of these. According to the reaction mechanism, the first step is the addition of the unpaired electrons from the aniline nitrogen to the carbon atom of the carbonyl group. The first computational attempt was made for the attack of the aniline by the opposite side of the double bond, but it was unable to obtain the transition state followed by the occurrence of one or more negative frequencies in the hessian calculation. After several unsuccessful attempts, a new approach was attempted, namely, the aniline attack at the anhydride carbon from the double bond. In this case, the transition state was obtained, the length of the C–N link calculated for the transition state being 2.29 Å (Fig. 13.4).

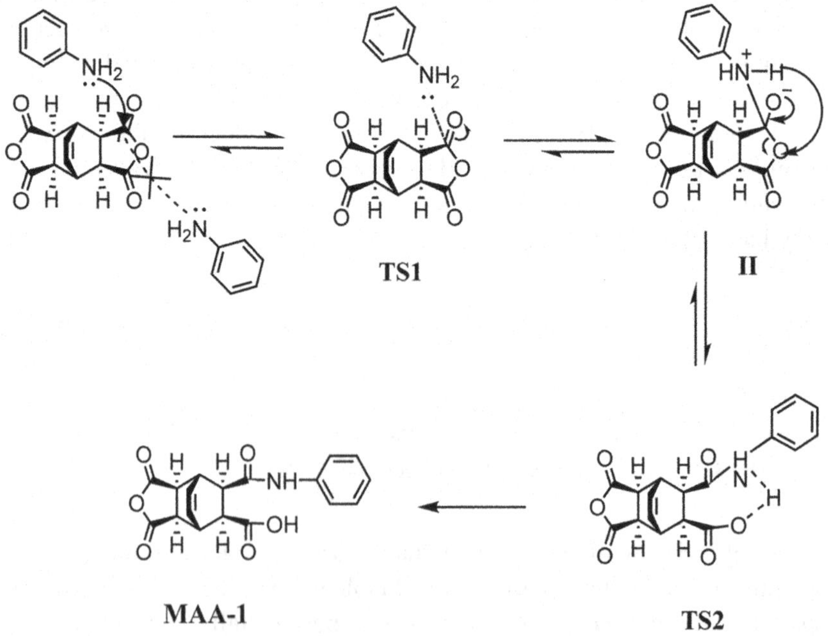

FIGURE 13.4   Reaction mechanism of addition of aniline to BOCA.

According to Parson's rules (Ivanov et al., 2011), if the difference between the HOMO–LUMO energy levels of the two reactants is less than 6 eV, it can be assumed that the reaction takes place under the control of atomic orbitals. In the presented case, the energy levels of the HOMO orbitals of aniline and LUMO orbitals of BOCA, obtained using the DFT method B3LYP/cc-pVDZ has been ($E_{ssBOCA}^{LUMO} = -1.474$ eV and $E_{aniline}^{HOMO} = -6.117$ eV), the energy difference being $\Delta E = 4.643$ eV (Table 13.1).

After addition of the aniline to the reaction site, the molecule is partially stabilized to a metastable state II. As a result of the thermal agitation, a proton from the amine group may be sufficiently close to the oxygen atom of the etheric bridge to form a new O—H bond. Concurrently with this proton transfer, the double bond from the carbonyl group is reformed and the C—O bond from the etheric bridge is broken. The energy of the TS2 transition state corresponding to these rearrangements inside the molecule is lower than the energy of the TS1 transition state, which leads to the conclusion that the reaction of broken the anhydride cycle to form the products is more favorable in the expense of restoring the starting reactants. Finally, the molecule is stabilized by formation of the amidic acid, the total energy of products being lower than the aggregate energy of the reactants. This observation is consistent with the experimental data, which shows that the formation of the amidic acid is an exothermic reaction. The schematic representation of the energy profile corresponding to the formation of the amidic acid for the opening of the first anhydride ring and the formed intermediate species is shown in Figure 13.5.

Following the same calculation algorithm, the energies of the transition states for the reaction of aniline with the second anhydride cycle could be determined (Figs. 13.6 and 13.7). By calculating the partial load distribution for the two carbon atoms in the remaining anhydride cycle, it was observed that there are no significant differences between them and therefore, the second aniline molecules can attack any of the two active sites (Fig. 13.6). However, the transition state energy for the *trans* configuration is slightly lower, indicating that the addition of the aniline to the C_6 carbon atom can be more favored. It cannot be concluded that the addition of aniline takes place only at the C_6 atom, the energy levels corresponding to the transition states of the *cis* and *trans* conformations being very close.

TABLE 13.1 Partially Charge of Reaction Center and Orbital Energy for Reactants and Intermediary.

Structure	Energy, kcal/mol	Partial charge						
		C1	C2	C5	C6	N	N5	N6
ssBOCA	–224.97	0.64	0.64	0.64	0.64	–	–	–
aaBOCA	–217.94	0.62	0.62	0.62	0.62	–	–	–
saBOCA	–222.47	0.65	0.62	0.63	0.63	–	–	–
Aniline	15.25	–	–	–	–	–0.66	–	–
MAA-1	–205.93	–	–	0.64	0.64	–	–	–
MAA-*cis*	–212.68	–	–	0.66	0.71	–	–0.55	–0.56
MAA-*trans*	–200.03	–	–	0.68	–	–	–0.59	–0.56

	LUMO energy of carbon active site, eV				HOMO energy of nitrogen, eV		
	C1	C2	C5	C6	N	N5	N6
ssBOCA	–1.474						
aaBOCA	–2.046						
saBOCA	–1.103		–1.127				
Aniline					–6.117		
MAA-1	–0.354		0.1272				
MAA-*cis*				0.737		–6.748	–7.129
MAA-*trans*			–0.238	–0.240		–6.298	–7.190

Source: Adapted from Popovici and Vasilescu (2015).

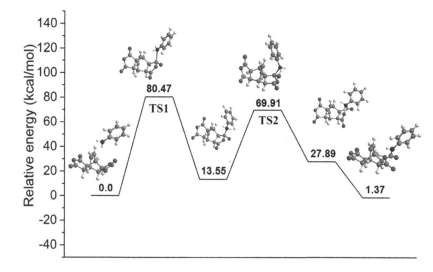

FIGURE 13.5 **(See color insert.)** Energetic profile of addition of one aniline molecule to BOCA.

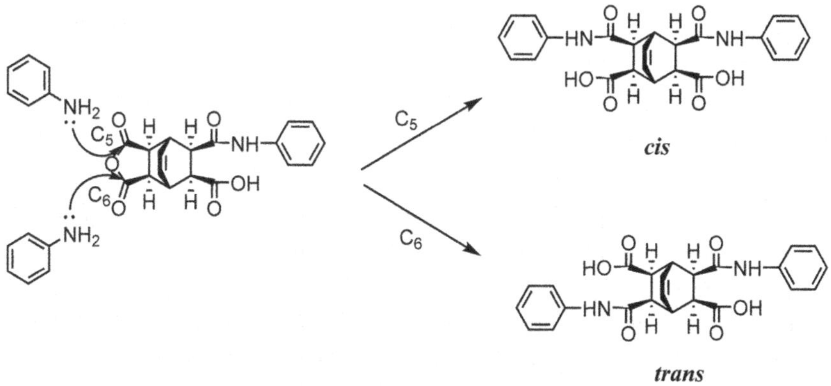

FIGURE 13.6 Reaction paths for second addition of aniline.

As stated earlier, a number of properties for the polymeric materials based on BOCA dianhydride can be explained only if it is accepted that the aliphatic bicycle conformation moieties forming the polyimide chain is in the form of the aaBOCA (see Fig. 13.1b). Therefore, during the heat treatment, a spatial rearrangement of this segment from *sin*-conformation to *anti*-conformation can occur. This transposition reaction can be also favored by the π–π stacking interactions between aromatic rings which are linked to imidic cycle as can be observed in Figure 13.2.

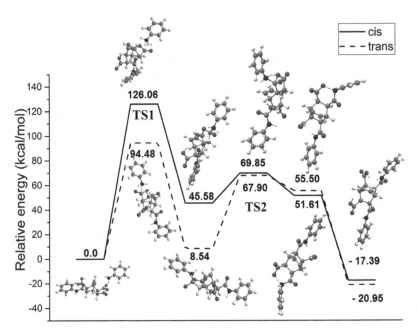

FIGURE 13.7 (See color insert.) Energetic profile for second addition of aniline molecule to BOCA.

The transposition mechanism has not yet been elucidated, but it has been considered an ideal case in which, under the action of thermal agitation resulting from the heating of the system, a transposition between carboxyl group and the corresponding hydrogen atom from the *sin*-position into the *anti*-position occur (Fig. 13.8). It has been observed that the calculated energy of the transition state is not very high, $\Delta E_{TS1}^{trans} = 110.09$ kcal/mol. This energy can be provided on the one hand by the high temperature and, on the other hand, by the intramolecular collisions that take place during the system agitation. The transposition reaction can be also favored by the fact that the calculated energy of the *anti*-conformation is lower than the corresponding *sin*-shape. The calculations of the transition state for this transposition reaction was done using semiempirical approximation implemented in MOPAC2016 program, the calculation method being PM6 as level of theory (Kromann et al., 2016; Stewart, 2016).

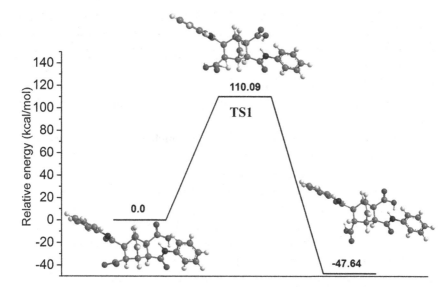

FIGURE 13.8   (See color insert.) Energetic profile of transposition reaction of one site.

In the second step of forming the imidic cycle, the cyclodehydration reaction is carried out, the reaction mechanism being shown in Figure 13.9. To determine the energetic profile of this reaction, it was considered the ideal case in which the diamidic acid obtained by reaction of aniline with the BOCA dianhydride is in *anti*-conformation, respectively, all the amidic and carboxyl functional groups being opposite to the double bond.

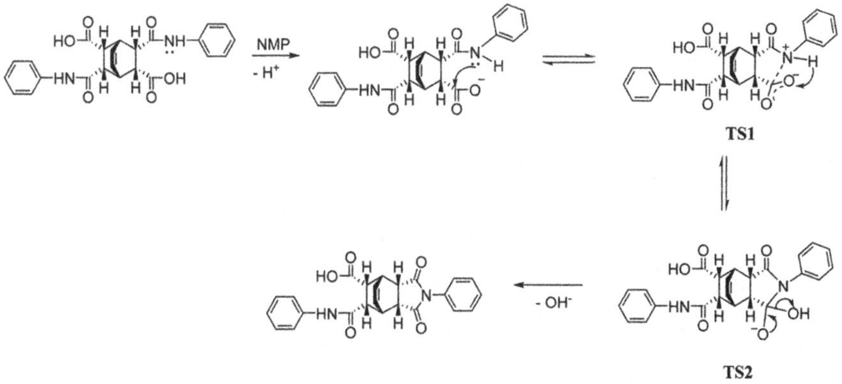

FIGURE 13.9   General dehydration mechanism of amidic acid to imide.

At laboratory scale, the cyclodehydration reaction may take place either by the action of chemical agents, such as acetic anhydride or under the influence of temperature. The most common method of cyclodehydration of amidic acid for the synthesis of polyimides is the thermal method, the normal working temperature being 180–200°C when N-methyl-2-pyrrolidone (NMP) is used as solvent or 150–160°C when using Dimethylacetamide (DMAc).

The theoretical calculations of the cyclodehydration reaction of an amidic acid derived from phthalic anhydride and L-leucine were widely discussed by Wu et al. (2003), based on experimental data obtained by Onofrio et al. (2001). They stipulated that the cyclodehydration reaction is catalyzed by the pH of solution.

Because we use NMP as a regular solvent, which has slightly alkaline behavior, we use this type of approaching for theoretical calculations. Under the solvent action, some of the carboxylic groups dissociate with formation of carboxylate ion, which further participate in the dehydration reaction.

If for the *cis* form of the amidic acid model the only possible path of dehydration is with the formation of the five-membered imide cycles (Fig. 13.10a); in the case of the *trans* form the cyclodehydration reaction can follow two paths (Fig. 13.10b).

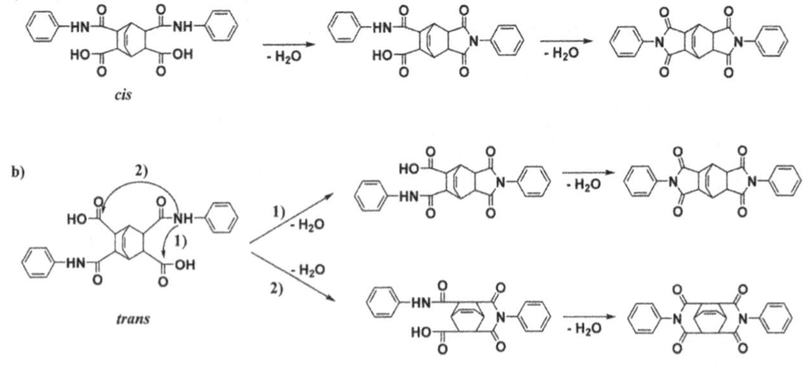

FIGURE 13.10 Reaction path of dehydration reactions: (a) dehydration of *cis* form; (b) dehydration of *trans* form.

The first step consists in the ionization of the carboxyl group under the weak basic character of the solvent, the ionic species so formed having a lower energy than the starting amidic acid, suggesting that in solution this one can be found in dissociated form. Under the reaction conditions (high temperature, stirring), the nitrogen atom from the amidic group approaches to the positively charged carbon atom from the carboxylate group (δ_{C5} = 0.71) forming a first TS1-a transition state. It is noticeable that the required energy for the transition state formation is high ΔE = 125.70 kcal/mol which justifies the extreme conditions necessary for the cyclodehydration reaction. The energy gap between HOMO orbital of the nitrogen atom and the LUMO orbital of the C_5 atom from carboxylic group has been found to be 6.833eV which means this reaction need a surplus of energy to cover the energy barrier. Under these conditions, a length of 2.4 Å was obtained for the C_5—N bond, a value of 0.11Å greater than that obtained for the addition of the aniline molecule to the anhydride ring. Due to the proximity of the proton from the secondary amine group to the pair of oxygen atoms in the carboxyl ion, it is transferred to one of them, forming a TS2-a transition state. The high temperature and the presence of the proton in solution attached to the solvent molecules favor the removal of a water molecule and the formation of the imide stable ring (Fig. 13.11).

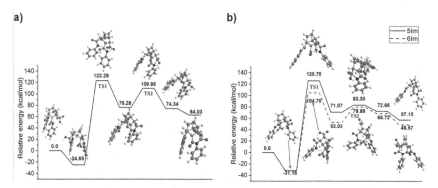

FIGURE 13.11   (See color insert.) Energetic profile of the first step of dehydration reaction: (a) dehydration of *cis* form and (b) dehydration of *trans* form.

It is possible also, in case of *trans* configuration the nitrogen atom to form a bond with the C_6 carbon atom. In this case, it has been found that the energy of the transition state TS1-b is lower $\Delta E_{TS1\text{-}b}{}^{C6}$ = 104.70 Kcal/mol and the length of the C_6—N bond are greater C_6—N=2.47 Å. The mechanism of elimination of the water molecule is similar to that shown in the formation of the five-membered imidic ring. The removal of the second water molecule is carried out by the same mechanism, the energy profiles for each case is shown in Figures 13.12 and 13.13.

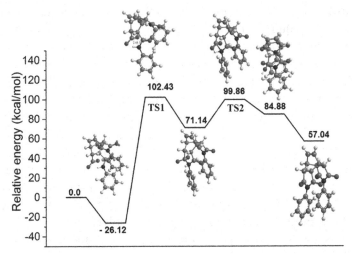

FIGURE 13.12 (See color insert.) Energetic profile of second step of dehydration reaction for *cis* conformation.

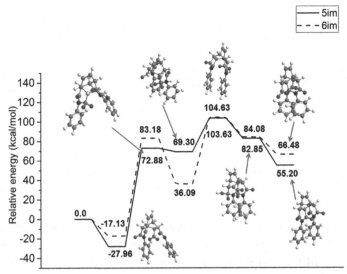

FIGURE 13.13 (See color insert.) Energetic profile of second step of dehydration reaction for *trans* conformation.

13.3 SYNTHESIS OF MODEL COMPOUND, THE WAY TO VALIDATE THE COMPUTER SIMULATIONS

It is necessary that for the validation of computational methods to use experimental data to establishing the performance range of method and to validate predicted behavior. To achieve this, validation requires advances measurements, as well as advanced theory and models, coupled with integrated, interdisciplinary research.

In order to validate the theoretical calculations, the model compound obtained from the reaction of BOCA anhydride with aniline was synthesized, using the two-step synthesis method with separation of the amidic acid and thermal cyclodehydration reaction (Popovici and Vasilescu, 2015).

13.3.1 CHEMICAL SYNTHESIS AND INSTRUMENTAL CHARACTERIZATION: FROM THEORY TO PRACTICE

In a two-neck round bottom flask, the required amount of aniline was dissolved in the freshly distilled NMP. After complete homogenization, the stoichiometric amount of dianhydride was added in small portions. The amounts of the two reactants were calculated so that a final 15% solution was obtained. The homogeneous reaction mixture was stirred for 36 h at room temperature under inert atmosphere (N_2). It is necessary to work in an inert atmosphere because traces of water from the solvent or from the air can hydrolyze the anhydride group with the restoration of dicarboxylic acid, thus deactivating the anhydride and stopping the growth of macromolecular chain—in the case of the synthesis of imidic polymers.

After obtaining the amidic acid, referred as MAA, a quantity from the obtained solution was precipitated in water, the remaining solution is heated at 180–200°C for 16 h to complete the cyclodehydration reaction and obtaining the final diimide product. The water resulting from the cyclodehydration reaction was removed from the system with a moderate flow of nitrogen. The protocol for obtaining the model compound was the same as that normally used for obtaining the imide base polymers, precisely to compare the obtained results of the polyimides with those obtained for the model compound. Finally, diimide M2Im was precipitated in a mixture of water and methanol and dried under reduced pressure at 200°C for 20 h (see Fig. 13.3). In addition to removing water and solvent traces from the final product, heating at 200°C allowed the completion of cyclo-dehydrating reaction of any unreacted acid-amidic groups. The completion of the cyclodehydration reaction in the drying step is particularly useful in the case of obtaining imidic polymers, the probability of having not fully closed imidic cycles being higher due in particular to the length of the macromolecular chain and its rigidity.

Drying of the amidic acid intermediate product was also made at low pressure for 10 days, the temperature, in this case, being maximum of 50°C. Raising the temperature above this value can initiate the cyclodehydration reaction for some amidic-acid pair, which should be avoided (Bessonov et al., 1987).

After precipitation of amidic acid, separation of two precipitated phases was observed, namely a well-defined granular phase, which has been deposited over time at the base of the Berzelius beaker, and a fluffy phase which remained as a flocculation suspension in the aqueous phase. The separation of those two phases was maintained after the system was stirring. Some samples were taken from the two phases and washed several times with water to remove any traces from the complementary phase. The two phases were noted as MAAs (the granular phase separated by deposition at the base of the beaker) and MAAf (the remaining phase), these were analyzed separately using FTIR (Bruker Vertex70 on KBr palettes) and ¹H-RMN (Bruker Avance DRX 400 in DMSO-d6) spectroscopic analysis (Fig. 13.14).

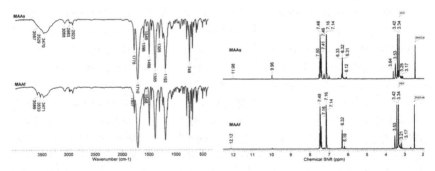

FIGURE 13.14   FTIR and ¹H-NMR spectra for synthesized MAAs and MAAf amidic acid.

From the analysis of FTIR spectra for the two phases of amidic acid, a significant difference was observed in domain 4000–3000 cm⁻¹, respectively, in the area where the vibration peaks of the NH groups from the amide sequence (3450–3250 cm^{-1}), O—H from the carboxyl sequence (3700–3500 cm^{-1}) and the hydrogen bonds that may form intra or intermolecular (3500–3100 cm^{-1}) occurring (Silverstein et al., 1991). Thus, in the case of MAAs, a broadband without clear peaks in the range of 3700–3150 cm^{-1} has been observed. This broadband is mainly due to hydrogen bonds that occur intramolecularly between the protons from the NH and COOH groups with the neighboring carbonyl groups, the vibration of these bonds covering the characteristic vibration peaks of O—H and N—H bonds. These peaks can, however, be distinguished only as a shoulder at 3529 cm^{-1} for N—H and 3587 cm^{-1} for O—H.

In the case of the MAAf phase, the hydrogen bonding band is not so strong, so the peaks characteristic of the NH groups (3589 cm^{-1}) and OH (3533 cm^{-1}) appear well defined. Moreover, two shoulders appear at 1807 and 1680 cm^{-1} characteristic vibration frequencies of the five-membered anhydride ring.

As a result of these observations, it can be concluded that the precipitate from the MAAs phase corresponds to an amidic acid in which both the anhydride cycles have been reacted, while the precipitate from the MAAf phase corresponds to an amidic acid in which only one anhydride cycle have been reacted. These observations were also confirmed by magnetic resonance spectroscopy (Fig. 13.14b, in which the ratio between proton integrals at 6.35 ppm (aliphatic double protons) and the total proton integrals in the aromatic zone (7.0–9.0 ppm) was 2:6.14 for the MAAf phase and 2:9.65 for the MAAs phase. Deviation from the ideal ratio of 2:5 for phase MAAf or 2:10 for the MAAs phase can be explained by complementary phase impurification.

Therefore, in the first step of obtaining the imides or polyimides derived from the BOCA dianhydride, if the reaction time is not long enough, unreacted anhydride groups may remain, which may lead to poly-acid-amides with low molecular weights, and the quality of the polyimide may be inappropriate for biomedical applications or smart systems where this type of polymers are used.

The spectroscopic analysis of the final M2Im diimide using the same methods revealed that the imidization reaction was complete (Fig. 13.15). From the analysis of the FTIR spectrum, it can be observed that the broadband band characteristic of the hydrogen bonds and the NH and OH groups in domain 3700–3250 cm^{-1} disappeared, leading to the conclusion that all acid-amidic sequences reacted with the removal of a water molecule. Moreover, due to the disappearance of the two shoulders characteristic of the anhydride cycles, it can be concluded that raising the temperature favored the reaction of these cycles with the aniline residue from the reaction system. This conclusion is also supported by the ¹H-NMR spectrum in which no unreacted acid or amine residues were observed, the ratio between the aliphatic double bond proton integrals and the total integrals for the aromatic zone being 2:10.17, very close to the ideal ratio of 2:10.

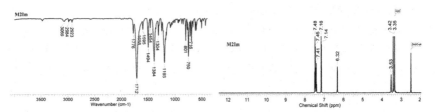

FIGURE 13.15   FTIR and ¹H-NMR spectra of synthesized M2Im diimide.

The formation of the six-atoms imide ring was revealed by FTIR spectroscopy. The six-membered imidic cycle exhibits a characteristic vibration at 1658 cm⁻¹, which is, however, partially masked by the strong vibration of the carbonyl group at 1712 cm⁻¹ (Barzic et al., 2014). Although in the same region the vibration of the carbonyl group C=O from the carboxylic acids appears, but the absence of the peak characteristic of the OH group in domain 3400–3200 cm⁻¹ and the lack of the broad peak at 12.5 ppm from the ¹H-NMR spectrum demonstrate that all the carboxyl groups have been reacted with the formation of imidic cycles.

Following these observations, it can be concluded that the production of high molecular weight polyamidic acids base on BOCA dianhydride is not only conditioned by a longer reaction time but also by a slight increase of temperature which helps the conversion of most of the anhydride cycles into amidic-acid moieties. Moreover, due to the higher temperature need for cyclodehydration reaction, some of the unreacted anhydride moieties can be transformed into imide ring if an amine group can be found nearby.

With the progress in computational capacities and materials sciences, building math-models for simulations the behavior of smart materials is being seen as a fruitful direction in achieving predictive capability so much needed to help sort out many of the issues mentioned above.

13.4 CONCLUSIONS

Computational materials research lead to propose development of new structured materials for various applications by bringing physical and microstructural information in the field of chemical design. Concurrently, a rational approach of the structural designer generates an integrated analysis tool that incorporates the fundamental behavior of smart material to minimize the trial-and-error numerical curve fitting processes. The approach is based on advances in measurement sciences and information technology to develop multiscale simulation methods that are validated by critical experiments. Currently, key structure–property relationships, include the constitutive relationships and effective-continuum representations of polymers and composite materials, are addressed by atomistic and continuum methods. From this stage, the knowledge about model translates to the knowledge about the smart materials by different methods related to microscopic evolution principles to the macroscopic response.

In conclusion, in our particular case, it was observed that the first stage of the reaction mechanism is that which determines the subsequent evolution of the system. Therefore, by knowing the detailed reaction mechanism and all possible reactions that can occur in such a system the reaction conditions can be estimated and optimized to maximize yield and to obtain a certain product with well-defined characteristics. Moreover, if it would opt for chemical cyclodehydration method, the reaction conditions should be gentler with the possibility of obtaining the BOCA-based imide polymers with radically different properties. It is possible that when using milder conditions the bicyclic aliphatic moieties do not undergo conformational changes and the polyimides thus obtained can no longer participate in cross-linking reactions to the double bond.

Finally, it can be concluded that the methods of theoretical simulation of reaction mechanisms can be successfully used to assess the reaction conditions so that, ultimately, materials with the desired properties and characteristics can be obtained. Simulation of reaction mechanisms proves to be a viable and inexpensive method of adjusting the reaction conditions, which can significantly reduce costs and the engineering time of new materials with improved properties. The future challenge includes seamless transfer of data between the nano-to-mesoscale models and experimentally validating simulations of atomistic behavior.

KEYWORDS

- **computational materials**
- **quantum simulation**
- **partially aliphatic diimide**
- **transition state**
- **reaction mechanism**
- **model compound**
- **energetic profile**

REFERENCES

Abadie, M. J. M.; Rusanov, A. L. *Practical Guide to Polyimides*; Smithers Rapra Technology: Shawbury, UK, 2007.

Ando, S.; Matsuura, T.; Sasaki, S. Coloration of Aromatic Polyimides and Electronic Properties of Their Source Materials. *Polym. J.* **1997,** *29*, 69–76.

Barzic, A. I., et al. Chain Flexibility Versus Molecular Entanglement Response to Rubbing Deformation in Designing Poly(oxadiazole-naphthylimide)s as Liquid Crystal Orientation Layers. *J. Mater. Sci.* **2014,** *49* (8), 3080–3098.

Barzic, A. I.; Stoica, I.; Hulubei, C. Semi-alicyclic Polyimides: Insights into Optical Properties and Morphology Patterning Approaches for Advanced Technologies. In *High Performance Polymers—Polyimides Based—From Chemistry to Applications*. Abadie, M. J. M., Ed.; InTech: Rijeka, Croatia, 2012; pp 167–198.

Barzic, A. I., et al. An Insight on the Effect of Rubbing Textile Fiber on Morphology of Some Semi-Alicyclic Polyimides for Liquid Crystal Orientation. *Polym. Bull.* **2013,** *70* (5), 1553–1574.

Bessonov, M. I., et al. *Polyimides–Thermally Stable Polymers*; Consultants Bureau: New York, 1987.

Bode, B. M.; Gordon, M. S. Macmolplt: A Graphical User Interface for Gamess. *J. Mol. Graphics Mod.* **1998,** *16* (3), 133–138.

Bruma, M.; Schulz, B.; Mercer, F. W. Polyamide Copolymers Containing Hexafluoroisopropylene Groups. *J. Macromol. Sci. A.* **1995,** *32* (2), 259.

Chen, Y. W., et al. Ultra-Low-κ Materials Based on Nanoporous Fluorinated Polyimide with Well-Defined Pores via the Raft-moderated Graft Polymerization Process. *J. Mat. Chem.* **2004,** *14* (9), 1406.

Chung, I. S.; Kim, S. Y. Soluble Polyimides from Unsymmetrical Diamine with Trifluoromethyl Pendent Group. *Macromolecules* **2000,** *33* (9), 3190.

de Abajo, J.; de la Campa, J. Processable Aromatic Polyimides. In *Progress in Polyimide Chemistry*; Kricheldorf, I. H., Ed.; Springer: Berlin/Heidelberg, 1999; Vol. 140, pp 23.

Eschrig, H. The Essentials of Density Functional Theory and the Full-potential Local-orbital Approach. In *Computational Materials Science: From Basic Principles to Material Properties*; Hergert, W., Däne, M., Ernst, A., Eds.; Springer: Berlin/Heidelberg, 2004; pp 7–21.

Gates, T. S.; Hinkley, J. A. *Computational Materials: Modeling and Simulation of Nanostructured Materials and Systems*; NASA Langley Research Center: Hampton, 2003; pp 1–22.

Gordon, M. S.; Schmidt, M. W. Advances in Electronic Structure Theory: Gamess a Decade Late. In *Theory and Applications of Computational Chemistry*; Frenking, G., Kim, K. S., Scuseria, G. E., Eds.; Elsevier: Amsterdam, 2005; pp 1167–1189.

Hanwell, M. D., et al. Avogadro: An Advanced Semantic Chemical Editor, Visualization, and Analysis Platform. *J. Cheminformatics* **2012**, *4* (1), 17.

Hasegawa, M., et al. Solution-Processable Colorless Polyimides Derived from Hydrogenated Pyromellitic Dianhydride with Controlled Steric Structure. *J. Polym. Sci. A.* **2013**, *51* (3), 575–592.

Hu, T.; Bicyclo[2.2.2]oct-7-ene-2,3,5,6-tetracarboxylic Dianhydride. *Acta. Crystallogr. Sect. E Struct. Rep.* **2008**, *64* (6), o1021.

Hulubei, C.; Hamciuc, E.; Bruma, M. New Polyimides Based on Epiclon. *Rev. Roum. Chim.* **2007**, *52* (11), 1063–1069.

Hulubei, C., et al. New Polyimide-based Porous Crosslinked Beads by Suspension Polymerization: Physical and Chemical Factors Affecting Their Morphology. *J. Polym. Res.* **2014**, *21* (9), 1–16.

Ivanov, D.; Maftei, D.; Constantinescu, M. A. Maleamic Acids Cyclodehydration with Anhydrides. DFT Study in Gas Phase and Solution. *Rev. Roum. Chem.* **2011**, *56* (2), 89–95.

Janssens, K. G. F., et al. *Computational Materials Engineering: An Introduction to Microstructure Evolution*; Academic Press: Burlington, 2007.

Koning, C., et al. Influence of Polymerization Conditions on Melt Crystallization of Partially Aliphatic Polyimides. *Polymer* **1998**, *39* (16), 3697.

Kröger, M. Shortest Multiple Disconnected Path for the Analysis of Entanglements in Two- and Three-Dimensional Polymeric Systems. *Comput. Phys. Commun.* **2005**, *168* (3), 209–232.

Kromann, J. C., et al. Prediction of pKa values using the PM6 semiempirical method. *Peer J.* **2016**, *4*, e2335.

Li, L., et al. Polyimide Films with Antibacterial Surfaces from Surface-initiated Atom-transfer Radical Polymerization. *Polym. Int.* **2008**, *57* (11), 1275–1280.

Mittal, K. L. *Polyimides and Other High-Temperature Polymers: Synthesis, Characterization and Applications*; VSP—An imprint of BRILL: Leiden, 2009; Vol. 5.

Nagaoka, S., et al. Evaluation of Blood Compatibility of Fluorinated Polyimide by Immunolabeling Assay. *J. Artif. Organs* **2001**, *4* (2), 107–112.

Numata, S.; Fujisaki, K.; Kinjo, N. Re-examination of the Relationship Between Packing Coefficient and Thermal Expansion Coefficient for Aromatic Polyimides. *Polymer* **1987**, *28* (13), 2282.

Onofrio, A. B., et al. Reactions of N-(O-carboxybenzoyl)-L-leucine: Intramolecular Catalysis of Amide Hydrolysis and Imide Formation by Two Carboxy Groups. *J. Chem. Soc. Perkin Trans. 2.* **2001**, (9), 1863–1868.

Pogge, H. B. *Electronic Materials Chemistry*; Marcel Dekker: New York, 1996.

Popovici, D., et al. Plasma Modification of Surface Wettability and Morphology for Optimization of the Interactions Involved in Blood Constituents Spreading on Some Novel Copolyimide Films. *Plasma Chem. Plasma Proc.* **2012a,** *32* (4), 781–799.

Popovici, D. et al. Polyimides Containing Cycloaliphatic Segments for Low Dielectric Material. *High Perform. Polym.* **2012b,** *24* (3), 194–199.

Popovici, D., et al. *Studies of Surface Modifications for Some Partial Cycloaliphatic Polyimides Films after Polychromatic Irradiation for Biomedical Applications*; Fourth Symposium Frontiers in Polymer Science: Riva del Garda, Italia, 2015.

Popovici, D.; Vasilescu, D. S. *Computational Study of Cyclodehydration of Bicyclo[2.2.2] Oct-7-Ene-2,3,5,6-Tetracarboxylic Dianhydride Base Amidic Acid,* 15th International SGEM GeoConference, Albena, Bulgaria; STEF92 Technology Ltd.: Sofia, 2015.

Schmidt, M. W., et al. General Atomic and Molecular Electronic Structure System. *J. Comput. Chem.* **1993,** *14* (11), 1347–1363.

Silverstein, R. M.; Bassler, G. C.; Morrill, T. C. *Spectrometric Identification of Organic Compounds*; Wiley: Hoboken, NJ, 1991.

Sroog, C. E. *Application of High Temperature Polymers*; CRC Press: Boca Raton, 1996.

St. Claire, T. L. Structure–Properties Relationship in Linear Aromatic Polyimides. In *Polyimides*; Wilson, D., Stenzenberger, H. D., Hergenrother, P. M., Eds.; Blackie: Glasgow, 1990; pp 58–78.

Stewart, J. J. P.; Stewart Computational Chemistry, *Mopac2016*: Colorado Springs, 2016.

Varganici, C. D., et al. On the Thermal Stability of Some Aromatic–Aliphatic Polyimides. *J. Anal. Appl. Pyrol.* **2015,** *113*, 390–401.

Wakita, J., et al. Molecular Design, Synthesis, and Properties of Highly Fluorescent Polyimides. *J. Phys. Chem. B.* **2009,** *113* (46), 15212.

Wilson, W. C.; Atkinson, G. M. *Review of Polyimides Used in the Manufacturing of Micro Systems*; NASA Langley Research Center: Hampton, 2007; pp 1–16.

Wu, Z.; Ban, F.; Boyd, R. J. Modeling the Reaction Mechanisms of the Imide Formation in an n-(o-carboxybenzoyl)-l-amino Acid. *J. Am. Chem. Soc.* **2003,** *125* (12), 3642–3648.

Yang, C. P.; Hsiao, S. H.; Wu, K. L. Organosoluble and Light-colored Fluorinated Polyimides Derived from 2,3-bis(4-amino-2-trifluoromethylphenoxy)naphthalene and Aromatic Dianhydrides. *Polymer* **2003,** *44* (23), 7067.

Yang, C. P.; Su, Y. Y.; Hsiao, F. Z. Synthesis and Properties of Organosoluble Polyimides Based on 1,1-bis[4-(4-amino-2-trifluoromethylphenoxy)phenyl]cyclohexane. *Polymer* **2004,** *45* (22), 7529–7538.

INDEX

W

Wet chemical treatments
  polymeric surfaces modifications
    chemical treatments for, 299

X

Xanthan, 76
  AFM images of, 80
  applications of, 95–97
  chemical modification
    grafting process, 93–95
  chemical structure and conformation of,
    77–80
  conformational transitions, structural
    modification

presence of others polymers, induced
  by, 85–87
salt addition and pH change, induced
  by, 83–84
temperature, induced by, 84–85
degradation, 87
  chemical, 91–93
  radiation, induced by, 89–90
  ultrasound, 88–89
outline of production process, 82
strategies for changing properties, 80
  chemical functionalization of, 93–95
  conformational transitions, 83–87
  degradation of, 87–93
  structural modifications, 81–83
Xanthomonascampestris, 76

Printed in the United States
by Baker & Taylor Publisher Services